STRUCTURED ELECTRONIC DESIGN

# Structured Electronic Design

## Negative-feedback amplifiers

by


**C.J.M. Verhoeven**
*Delft University of Technology, Delft, The Netherlands*

**A. van Staveren**
*Delft University of Technology, Delft, The Netherlands*

**G.L.E. Monna**
*SystemtematIC Design B.V., Delft, The Netherlands*

**M.H.L. Kouwenhoven**
*National Semiconductor Corporation, Delft, The Netherlands*

and

**E. Yildiz**
*Delft University of Technology, Delft, The Netherlands*



KLUWER ACADEMIC PUBLISHERS
BOSTON / DORDRECHT / LONDON

A C.I.P. Catalogue record for this book is available from the Library of Congress.

ISBN 978-4020-7590-2

---

Published by Kluwer Academic Publishers,
P.O. Box 17, 3300 AA Dordrecht, The Netherlands.

Sold and distributed in North, Central and South America
by Kluwer Academic Publishers,
101 Philip Drive, Norwell, MA 02061, U.S.A.

In all other countries, sold and distributed
by Kluwer Academic Publishers,
P.O. Box 322, 3300 AH Dordrecht, The Netherlands.

*Printed on acid-free paper*

# Contents

# Preface

Analog design is all about insight. More designs lead to better insight, which is built on experience. There is no shortcut to more experience than to carry out more designs. And yet, analog design automation has been promised to build up insight faster and more efficiently. For this purpose, analog design must first be structured. This is what this book is all about.

Terms such as structured design, algorithmic design, systematic design, etc. all refer to a better description of the design procedures used. They are key to the documentation of an analog circuit. They are key to the redesign of analog circuits. As a result structured design has become a necessity to efficient analog design. Structured design has become a necessity to provide optimal circuits, subject to a set of specifications.

This book explains how to structure the design process such that optimal circuits result. It is divided in a number of chapters which detail the design procedures. Each chapter contains a number of examples. It is aimed at amplifiers first, covering all aspects such feedback, frequency compensation, noise and distortion. Finally considerable attention is paid to biasing circuits. The structured design approach is mainly illustrated by means of amplifier designs. No sampling circuits nor converters are considered. On the other hand both CMOS and Bipolar transistors are used in parallel.

The first chapters introduce the fundamentals of structured design, showing how this can lead to the synthesis of analog integrated circuits. Concepts such as nullors, norators, conveyers, etc. are introduced and applied to circuits of higher complexity. Then feedback is discussed. The four types of feedback are detailed. It is shown how they naturally evolve from the basic circuit blocks of the first chapters.

The two main limitations are noise at low signal levels and distortion at high signal levels. Noise receives considerable attention in this book. All aspects are discussed in great detail, such as origin, models and ways to accurately describe their influence both in the circuits themselves as towards the inputs.

Amplifiers invariably use feedback to improve the accuracy and the bandwidth. Chapters on frequency compensation are thus inevitable. Terms such as pole splitting, pole-zero compensation, phantom zero, etc. are treated with great care and precision. This by itself is sufficient a reason to check this book out.

Finally a large chapter is devoted to biasing. Most books on analog circuits omit these circuits as they seem to be of less importance. However in analog design, biasing circuits are the backbone to set the transistor parameters right. This chapter is thus very welcome. It details many circuit solutions and design examples.

It can be concluded that this book is invaluable to whoever is serious in optimizing the design cycle. It shows how analog design can be structured, leading to faster design and less errors. As a result this book is key towards gaining insight in analog circuit design and performance.

WILLY SANSEN
KU-Leuven, Belgium
June 2003

*To Ernst Nordholt*

# 1

# INTRODUCTION TO STRUCTURED ELECTRONIC DESIGN

## 1.1 Searching the "circuit space"

The design of electronic circuits actually consists of a search through a large space of combinations of components with all different kinds of properties. It would take very much time to find a suitable circuit for an application by just performing an exhaustive search and testing every combination of components that can be made. Therefore it is necessary to use some strategy, to find a solution more quickly. The "oldest" strategy that can be used is evolution. Starting from a (randomly chosen) circuit, repeatedly new generations of circuits are generated of which a number of parameters is changed. The performance of the new circuits is then evaluated and compared to the performance of the parent generation. If the performance has improved, the new generation is maintained, if the performance has not improved, it is abandoned (becomes extinct).

From the great example, nature, we know that evolution takes a lot of time, so it is obvious that designing electronic circuits in this way takes a lot of time too. Still there are computer programs for designing electronic circuits that are based on the evolution principle that produce usable circuits. In some cases, evolution is "speeded up" a bit by looking at the sensitivity of the performance parameters with respect to circuit parameters. Then it can be found which parameter has the most influence on the performance at that moment and in which direction it should change. In this case, evolution becomes a bit less blind for the results of its actions.

Though the evolution strategy eventually will yield *usable* results and is therefore reliable in this sense, there are several major drawbacks from this strategy.

- Evolution never guarantees that an *optimal* solution to a problem will be found and does not give any clue about the distance of a given solution to the optimum.
- The relations between component parameters and circuit performance are never made completely explicit. Only from generation to generation derivatives may be taken, but the conclusions are only valid at that time. Purely coincidentally, circuit performance may become sensitive to an irrelevant parameter because of the fact that the circuit is actually unsuitable for the application.
- It will be hard to tell what should be changed to a circuit when, for example, the specifications are changed. Since there are no clues how the present circuit was found and why it is able to fulfill its present job, there are no clues for improvements either. Especially, when apparently no further improvements can be made, there is never certainty whether the new specifications are indeed beyond reasonable limits or whether the inadequate performance of the present circuit just happens to be the result of an unfortunate turn somewhere in the evolution process. In the latter case, a new drastic mutation could give a fresh start to the evolution process which might yield a suitable circuit, in the former case mutations would only result in useless time consuming calculations. No "proof" can be given that given specifications are fundamentally impossible to meet.

*Thinking* about design, *awareness* of relations between parameters and performance, and *adequate modelling* of circuit behavior, speeds up the design incredibly. Creating order in the chaos by separation of the design problem into smaller, if possible *orthogonal* (independent) problems, makes the design problem more clear and more easy to deal with.

This book is about a design strategy that tries to meet with these conditions. A number of assumptions will be made that enable us to formulate an "algorithmic" design strategy which quickly yields results that are at least optimal within the restrictions on the search space that were assumed. This may seem a very unfortunate consequence at first sight, but in practice the designs found with this strategy are usually among the best circuits known for their application. So, the penalty for obtaining a short design time seems to be very modest. Also it has appeared that the design method is very suitable for automated design, since most design rules are very clear and without any heuristics.

The method of "structured electronic design" is very suitable to find a solution to a design problem very quickly. However, it depends on a number of assumptions and a limited set of rules. There may be electronic circuits that are widely used, accepted and even valuated that would not be found with the design theory described in this book. Evolution of electronic circuits is based on "survival of the fittest", so instead of an immediate rejection of circuits like this, these survivors should be evaluated with great interest. It very often happens that it can be shown that the circuit is not optimal and that a better performing circuit is found with the design strategy, but in some cases these circuits reveal gaps in the design theory. In the end, this results in an improvement of the design theory. Actually the complete design theory originates from close inspection of the results of evolution and the interpretation of its "decisions". Therefore, the practical application of structured electronic design should never result in dogmatic rejection of evolutionary results (or results of any other design strategy), since this would also stop the evolution of the design theory itself.

The structured electronic design strategy is not focused on one particular type of circuit. Though it started with the structured design of amplifiers [1], [2], [3], [6], [7], it is applicable to any other (electronic) design problem, like bandgap references [8], [9] , oscillators [10], [11], [12], [13], [14], [15], filters [16], [17], [18], [19] and so on. However, the design rules for amplifiers are the most simple, so it is very convenient to show the theory being applied for amplifier design. Also the class of amplifiers to be design has been restricted for the sake of simplicity. Only the design of single-loop negative-feedback amplifiers will be treated. The design of multi-loop amplifiers is described in [1] and does, apart from the increased complexity, not differ from the strategy for single-loop negative-feedback amplifiers.

## 1.2 Circuit performance

The technical merits of an electronic circuit are valuated from the way it performs its function. Of course there are more factors involved than just the technical ones, like the costs to produce the circuit and so on. However, in this book only the pure technical merits are evaluated. In this way it is found what can be ultimately achieved, when all freedom is given. When due to other constrains some options are not allowed, it can easily be found how much the decrease of performance will be. It depends on the circumstances whether a manufacturer is willing to pay the price for extra performance or not, but it is at least known how much improvement can be expected and what price has to be paid to get it. In some cases, it can save the manufacturer from paying a price for finding an improvement that is fundamentally not feasible. There are, for example, circuits build in bipolar technology that would not have an improved performance when they would be build in BiCMOS technology, since the extra options given by the more expensive BiCMOS process play no role in the optimization of that particular circuit. Structured electronic design might produce "good reasons" for BiCMOS.

### 1.2.1 Fundamental specifications

There are many ways to specify the performance of electronic circuits, but there are only three fundamental aspects of the performance:

- Noise ($N$)
- Signal Power ($S$)
- Bandwidth ($B_w$)

related to each other via Shannons formula [20] :

$$\mathbf{C} = B_w \log_2 \left( \frac{S+N}{N} \right) \quad (1.1)$$

in which, **C** is the signal-handling capacity of the circuit. It is a measure for the information the circuit can process per second (for digital circuits it is expressed in bits per second). The ultimate goal of a designer is to maximize the signal-handling capacity within the constraints given by the environment. It can be seen from equation (1.1) that the bandwidth is linear in the expression and the signal-to-noise ratio is in the logarithm, so an increase of the bandwidth yields more improvement than a comparable increase of the signal-to-noise ratio. Circuits that "are good at bandwidth" together with signals that are properly coded—so they have little dynamic range and much bandwidth—are therefore likely to be favored when a large signal-handling capacity is required. Digital circuits and signals perfectly match to these requirements and they are indeed by far the "dominant species" in the signal processing world.

However, in areas where bandwidth is a problem, like it is under low-power conditions or at very high frequencies, or when there is no freedom to code the information properly, the signal-handling capacity has to be optimized via the signal-to-noise ratio. In this area high-performance analog circuits have to do the job. This book focuses on the design of these analog circuits and in particular on the design of negative-feedback amplifiers. Still the design strategy for these circuits differs in principle only from the design of digital circuits as far as the rules are concerned. The assumptions to enable the application of the strategy are basically the same.

The three fundamental aspects described above are*sufficient* to specify the performance of a circuit, but they are not the only specifications that are given in practice. Supply voltage, supply current, power consumption, chip area and so on, are, of course, also of great importance. These specifications can, however, be seen as a parameter (often the cost) for the three fundamental criteria. It is usually not too complicated to show, for example, the relation between power consumption and noise behavior. From this it can be found what are the power costs of optimal noise behavior or what the noise behavior will be at a given

power consumption.
In practice it is important to know both these relations and the optima if there were ultimate freedom. Then it can be evaluated if it is desired or justifiable to, for example, change a supply voltage or increase a supply current in order to improve the performance of a circuit. It can become clear what is the influence of the supply voltage on the performance of a circuit and what voltage becomes "critical", and marks the limit beyond which degradation becomes dramatic. This approach gives much more insight in the behavior of circuits than for example the blink design of 1V circuits because it is a fashion today.

## 1.3 Fast synthesis

To speed up the design process, several assumptions have to be made. They will be made on:

- how the fundamental specifications are interrelated;
- the validity of the predictions on circuit behavior found with simplified models;
- the relation between decisions taken at the different stages of the design process.

### 1.3.1 Orthogonality

As we have seen before, there are three fundamental aspects of the performance of a circuit that have to be specified. During the synthesis of a circuit, it is tried to *orthogonalize* the design with respect to these three aspects. Then the multi-dimensional design problem, which is hard to solve, can be split into three separate one-dimensional design problems that are more easy to solve.

Suppose there is a design problem with two different criteria, $A$ and $B$. All possible solutions are within the collection shown in figure 1.1. Suppose that all solutions that are on the same horizontal line have identical properties as far as property $A$ is concerned. Then, when there is an optimum for property $A$, this optimum is not a point in the collection, but a line. It does not matter which solution is chosen on this line, shown by the drawn line in figure 1.1a, as far as this property is concerned. When for property $B$ all solutions on a vertical line are identical, a vertical line of optimal circuits exists, as shown in figure 1.1b. It does not matter which solution on this line is chosen, property $B$ of the solution is always optimal. When the intersect point of both lines is chosen, a solution that is optimal for $A$ *and* $B$ is obtained. The orthogonal organization of the collection makes it very simple to find this optimum, since it is allowed to search just for the optimum of $A$ ignoring the behavior for $B$ completely. When this optimum is found, an independent search for the optimum of property $B$ can

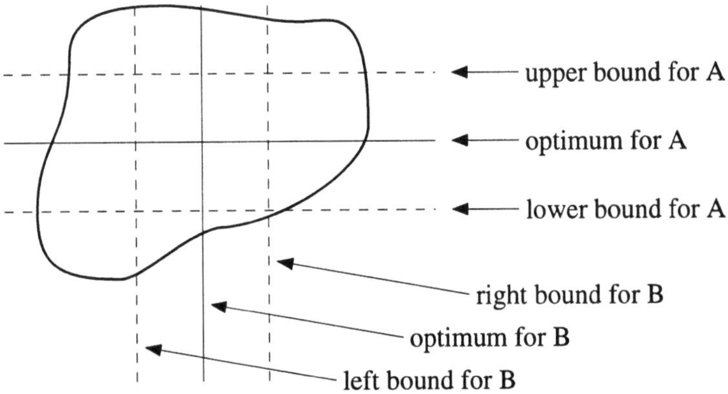


*Figure 1.1.* Optimal solutions for property $A$ and for property $B$.

be started. It is guaranteed that this search for the optimum of property$B$ does not change the performance for property $A$ in any way.

In practice, of course, the criteria are not truly orthogonal, but during the design *it will be assumed they are orthogonal.* Even *special measures* will be taken to make the assumption true. This strategy has as a result that some solutions, that have fundamental non-orthogonality, cannot be found. However, there is no efficient way to find these "absolute best" solutions. Only the slow multi-dimensional search strategies (like evolution)*might* find them, presumably after a long time. A combination of both strategies is the best option in those cases. The "orthogonal" solution should then be used as a starting value for the non-orthogonal search. In most cases the actual optimum is so close to the orthogonal solution that it can be found with ease by the slow strategies, but the (usually minor) improvement of the performance may not be worth the trouble anyway.

In the design strategy described in this book, the design for noise, signal power and bandwidth will be done orthogonally and several measures will be taken to make this possible. For the three design aspects, the following assumptions on orthogonality hold:

- When **noise** is evaluated, signal-power aspects, like distortion, are not considered. Therefore, the linear small-signal models of the components can be used. Frequency behavior is taken into account when the noise performance is evaluated, *but* the bandwidth demands on the complete circuit are not considered.

- When **signal power** is evaluated, neither noise nor frequency behavior are considered. Static large-signal models will be used. Noise is assumed to be small enough to obtain a negligible correlation with the non-linear behavior of a circuit and (unfortunately) theory for dynamic non-linearity is not mature enough for synthesis yet[1], so it will just be checked for later.

- When **bandwidth** is evaluated, signal-power (distortion) and noise behavior are not considered, so again small-signal models will be used.

### 1.3.2 Model simplification

Since there is a vast number of possible solutions, it is of great importance to have efficient selection criteria. It should be prevented that extensive calculations are necessary to detect the non-feasibility of a solution. As soon as possible, a non-feasible solution should be recognized and rejected. Therefore, several criteria will be formulated that are necessary (but not sufficient), to obtain a certain performance. When these criteria are not met, it is certain the final solution will not meet its specifications. If the criteria are met, there is *a chance* the specifications will be met. The risk of such a solution being non feasible in the end is reduced.

To obtain these fast evaluation criteria, simple models will be defined, that yield a superior behavior to the actual behavior. The simple models will ease the evaluation to a great extent and it is sure that the actual solution shall never perform better than the ideal solution. If the ideal solution does not meet the specifications, the actual solutions will not too.

Of course, it would be nonsense to define models at a level of ideality that is way beyond the actual performance of a solution. Then a proven feasibility of the ideal solution would have little predictive value for the behavior of the actual solution. Therefore, the models will be arranged such that the actual performance can come very close to the idealized behavior. Consequently, at some places in the design of a circuit, special measures have to be taken to *make the assumption on the validity of the simple models true.*

### 1.3.3 Hierarchy

During the design, different levels of complexity can be distinguished. Starting from simple models the design is more and more refined, which produces increasing accuracy but also increasing complexity for the calculations. At each level, decisions for the design are taken. To obtain an efficient design strategy, every decision that has been taken, has to remain correct and thus unaltered for the rest of the design. Also orthogonality is of great importance to enable hierarchy in the design process. It makes it possible to take a decision concerning one of the fundamental criteria without taking the others into account.

The use of hierarchy in the design itself is also a good means to obtain reduction of the complexity of the design process. For example, in a negative-

[1]This is probably the largest gap in the design theory at present.

feedback amplifier two main blocks can be distinguished: the feedback network and the active part. It is possible to design one of the two, assuming ideal behavior for the other. In this way, the design of one negative-feedback amplifier splits into the (independent) more simple design of two smaller sub-circuits. Also in these sub-circuits further divisions are possible. In this way, the large complete design problem is made more clear. For every aspect of the design, a responsible part can be pointed out, so it becomes very easy to distinguish the bottlenecks and to evaluate the measures to be taken for further optimization of a design. Of course, the orthogonality of the three main aspects noise, signal-power and bandwidth is of great help to bring hierarchy into the design.

### 1.3.4 Summary

For the design of electronic circuits there are three fundamental aspects to describe the performance completely:

- noise;
- signal power
- bandwidth.

The signal-handling capacity of a circuit is completely determined by these three aspects.

For the design theory described in this book three assumptions are made that enable the strategy to be efficient:

- Orthogonality
  The circuits will be arranged such that the behavior for the three fundamental aspects can be designed *orthogonally.*
- Simplicity
  Simple models will be defined to obtain quick predictions for the feasibility of a design. Non-feasible solutions can be detected in an early stage. Special arrangements will be made such that the predictions are*close* to the actual results.
- Hierarchy
  The hierarchy in the design makes it possible to reduce the complexity of the design problem, because the it can be split efficiently into smaller independent design problems. The design will be arranged such that each decision, taken on a certain hierarchical level, remains valid through the rest of the design.

## 1.4 Synthesis of electronic circuits

Any electronic system can be described by a differential equation. The task of an electronic designer is to "translate" the differential equation into silicon.

So, in principle the design of an electronic system starts with the specification of the differential equation. A differential equation is composed of various basic operators. For example the equation:

$$x * \frac{dx}{dt} + x + \frac{1}{x} = \int \sin(x)dx - a \tag{1.2}$$

comprises the operations:

- Generation of a well-defined constant: $a$
- Multiplication: $*$
- Addition: $+$
- Subtraction: $-$
- Non-linear operation: $\sin(x)$
- Division: $\frac{1}{x}$
- Integration: $\int dx$
- Differentiation: $\frac{d}{dx}$
- Equating: $=$
- ...

When for each mathematical operation an electronic implementation exists, every electronic system can be constructed by combining the correct basic functions.

### 1.4.1 Implementation limitations

In mathematics the range of (most) operators is unlimited. For example, addition is valid for any operand:

$$a + b = c \qquad a, b, c \in C \tag{1.3}$$

In electronics, usually conditions are imposed on the operands. Their value may, for instance, be limited to the supply voltages:

$$V_a + V_b = V_c \tag{1.4}$$

if:

$$V_a < V_{cc} \tag{1.5}$$
$$V_b < V_{cc} \tag{1.6}$$
$$V_c < V_{cc} \tag{1.7}$$

else:

$$V_c = V_{cc}, \tag{1.8}$$

independent of the inputs. Apparently, there is an *upper limit* to the operands. The output signal becomes distorted when the inputs become too large. The signal power ($S$) is limited and the limits are set by the amount of distortion that is accepted. Beyond these limits the circuit does not operate as an adder.

There is also a lower limit. In the mathematical description the accuracy of the addition is infinite. Consider the following addition:

$$1 + 1 = 2 \tag{1.9}$$

The result of the addition of two operands, that are *exactly* equal to one, result in *exactly* 2; there are no tolerances. In electronics however, operands cannot be given an exact value. Only a *range* can be specified in which the value of the operand is with a certain probability.

$$1' \in \, < 0.9, 1.1 > |_{90\%} \tag{1.10}$$

Also the result of the addition can be observed with a certain accuracy. So, the electronic equivalent of equation (1.9) could be:

$$0.98 + 1.09 \approx 2.03 \tag{1.11}$$

It is of no use to try to put information of higher accuracy through this circuit, all the corresponding details will be lost. The unpredictability[2] of the signals is called noise ($N$).

Apparently, every electronic circuit has a limited input range and a limited output range. It is limited in the upper side by distortion, which limits the maximally allowed signal power ($S$) and on the lower side by noise ($N$), which masks any further details of the information. The combination of these two limits is called the *dynamic range* ($DR$) of the circuits and signals, according to:

$$DR = \frac{S + N}{N} \tag{1.12}$$

The dynamic range is a very important quality measure for electronic circuits. It is one of the factors in Shannons formula, equation (1.1), which determines the signal-handling capacity. It is more important to look at the dynamic range than to look at the distortion and the noise separately. Occasionally an improvement of the noise behavior leads to an increase in the distortion, sometimes resulting in a net *decrease* of the dynamic range. There are only very rare cases in which

[2]Not only stochastic noise, like thermal noise of resistors results in unpredictability, but also, for example, errors in matching of devices.

this is really desired.
Only when the design of the noise behavior and the distortion behavior are made orthogonal, it is safe to do separate optimizations.

There is also a third practical limit to the performance of the electronic implementations of mathematical operators. This is*speed*. Speed is not defined for the operator " + " in equation (1.9). In the electronic implementation of " + " however, the electronic circuit needs some time to determine the result of the addition. So, the operands may not change faster than the implementation of " + " can add. Faster variations are not observed by the adding circuit and therefore do not contribute to the output result. The operator has a limited *bandwidth* ($B_w$), which should, of course, always be equal to, or larger than the bandwidth of the information.

The speed limitation together with the limited dynamic range determines the signal-handling capacity as given by equation (1.1). This expression suggests that bandwidth can be exchanged for dynamic range and vice versa without a loss of signal-handling capacity. So, optimization of the signal-handling capacity also comprises the optimal distribution of the information between bandwidth and dynamic range. However, this is not the topic of this book. This distribution depends on the way the information is coded in the signals and very often there is no freedom to choose the appropriate coding for each separate circuit. The way of coding is a "system level" decision and therefore usually one of the fixed constraints for a circuit being designed for that system. In this book, bandwidth and dynamic range are treated as orthogonal criteria, between which exchange is not considered.

### 1.4.2 Specifications

The first step of a design is to*find the complete set of specifications* for the electronic system. This implies that, at first, the designer has to specify the desired behavior of the system without the "professional blindness" that very often accompanies experience. Very often electronic designers start their design on a too low hierarchical level. They start from a well know system solution to their problem and design an improved version of it. An example of this is the radio receiver. Generally, a basic receiver is considered to be organized as depicted in figure 1.2. It consists of a pre-selection filter, a mixer, an oscillator,

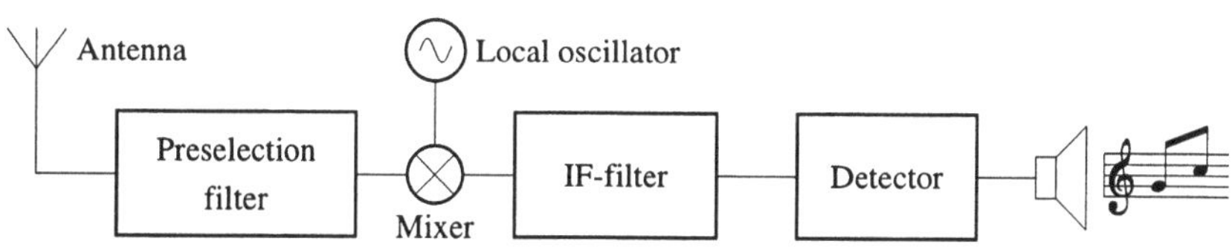


*Figure 1.2.* A heterodyne radio receiver.

an intermediate-frequency filter and a detector.

However, this is not the direct implementation of the receiver system. The primary task of a radio receiver is to select one spectral component out of a large frequency band of radio signals and convert the information in this signal to the audible base band. So, only a filter for the selection and a detector for the conversion are essential. The radio receiver depicted in figure 1.3 represents therefore the most direct implementation of the desired electronic system. In

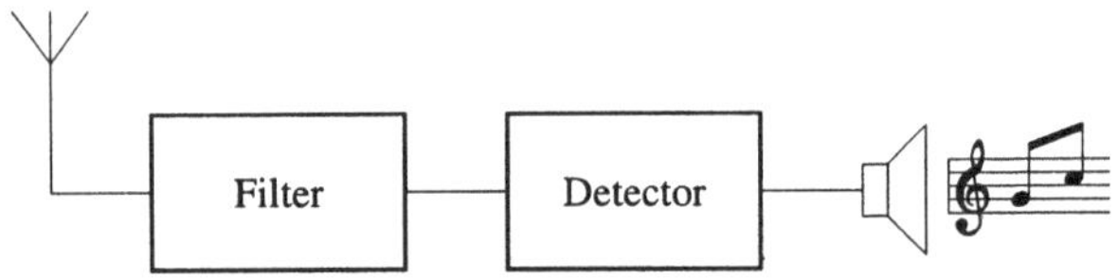


*Figure 1.3.* A direct receiver.

practice it is very often difficult to realize the filter for the direct receiver. Some specifications are in severe conflict with each other. The quality factor, $Q$, of the filter has to be high enough to select just one radio station at a time *and* it has to be tunable. For many years it has been technically not possible to combine both properties with sufficient quality in one filter. So, an optimal fixed filter is designed and the tuning is arranged with a mixer-oscillator system. Presently, it has become possible to implement active filters, that do posses both the desired $Q$ and the tunability, that can be used in the long-wave band. They offer the possibility to implement a direct receiver for the long wave range and it would be a shame if this option was forgotten due to "experience".

Another example is found in an amplifier with a frequency dependent transfer, like bandpass amplifiers. According to the strict definition, amplifiers just increase the power of the signal at their input on which the information is present. They just multiply the input signal with a *constant.* Frequency dependency of the transfer is not according to this definition, and indeed some degrees of freedom are lost when bandpass amplifiers are designed without any system consideration. Frequency selectivity and amplification are two different basic functions. A bandpass amplifier is actually a combination of these two functions. When both functions have to be performed, there are several different options for the implementation. When an amplifier and a filter are designed separately, there is more freedom to optimize each, and further, there are a number of ways to cascade these functions. Combinations are, for example, first filtering and then amplification, first amplification and then filtering or even first a part of the filtering, then amplification and the rest of the filtering after the amplification and so on. The combination of the filter and amplifier into one electronic circuit is just one of these options and it is obvious that, due to the correlation that then exists between the two functions, optimization becomes more complicated. Each option has its own advantages and disadvantages and their importance depends strongly on the specific design environment. A justified decision must be made for one of the options.

## 1.5 Small-signal models, biasing and distortion

### 1.5.1 Models

The electrical behavior of electronic components can be described with many different models. But since at the system level a linear behavior is expected from an amplifier, the first choice is to use just linear elements to implement it[3]. To have amplification, linear controlled sources could be used. These linear controlled sources do not exist in practice but using them in the first steps of the design makes life very easy. Later in the design, more practical (and complex) elements are used, that show this linear behavior in a sufficiently large signal range. So, further in the design, the assumption that linear controlled sources can be used in the first design steps is correct, is (must be) made true.

The linear controlled sources that are used in electronic design are depicted in figure 1.4 and figure 1.5. They are the so-called "small-signal models" of the bipolar transistor and the field effect transistor, respectively. *This merely means that when these models are used in the first design steps, it is likely that for a limited signal range the behavior of these sources can also be realized using a bipolar transistor or a field effect transistor.* Merely this, because more elements than just a transistor are needed to really implement something that may behave like the small-signal model. Inspecting figure 1.4 and figure 1.5, it can be seen that both models contain a source. A transistor does not contain a source. The sources in the models can generate power, a transistor can not. But a *biased* transistor can. Being a *non-linear* element, a transistor can convert power from a DC-source (a bias source, a battery) to any other frequency. So, the combination of a transistor and DC-source can generate the necessary power at the right frequency and thus perform the role of the controlled source that is present in the small-signal model. It is important to understand what the small-signal model really stands for, since this will provide the proper insight in what biasing is and what is its relation to distortion.

Apart from the small-signal model, also detailed so-called "large-signal models" for the detailed evaluation of a circuit are used. These models are typically

[3]Linearity has already been the main paradigm for electronic design for many decades. Although the world around us is not linear at all, the linear models help us very much to synthesize circuits with a predictable behavior. Deviations from the linear behavior are called distortion and are considered to be undesirable. The fact that these deviations can still be rather predictable is usually ignored in the sense that they are not used to improve circuit behavior. Circuits that truly exploit the dynamic non-linear capabilities of the electronic components are rarely designed. Probably the design theory for circuits like that is not mature enough.

used in computer programs like SPICE and are mainly used for *analysis* of designed circuits. The small-signal model is, because of its simplicity, also conveniently used for *synthesis*.

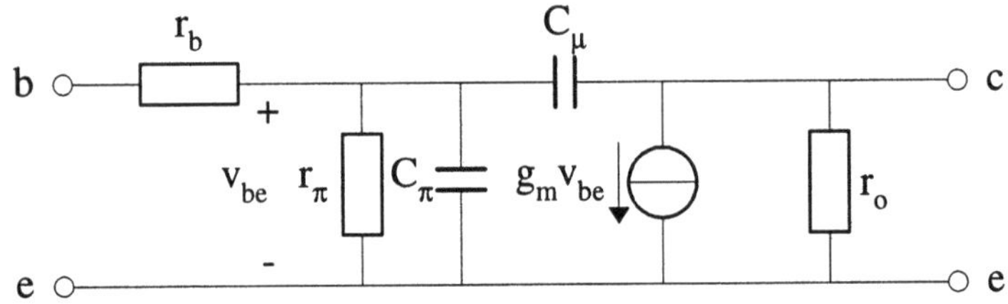


*Figure 1.4.* The small-signal model for the bipolar transistor.

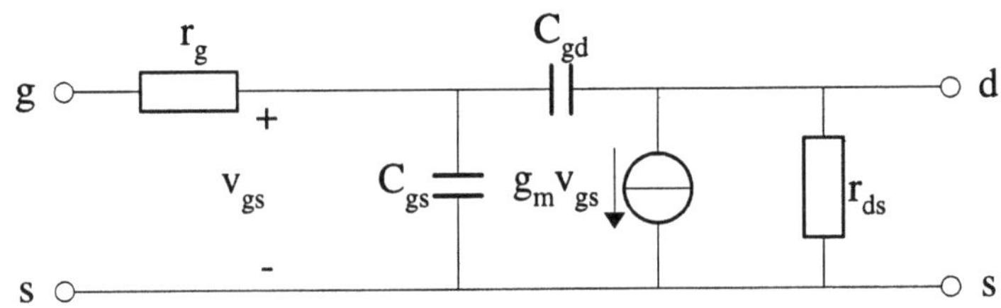


*Figure 1.5.* The small-signal model for the field effect transistor.

The small-signal models shown in figure 1.4 and figure 1.5 can be used efficiently to design the frequency behavior and the noise behavior (when adding the modelled noise sources) of a circuit. For designing the distortion behavior, these models cannot be used, since distortion is a measure for the deviation of the actual behavior from the small-signal models.

Note that the only difference between the two small-signal models is the presence of $r_\pi$ for the bipolar transistor. The gate bulk resistance $r_g$ finds its origin in parasitic resistances of interconnect to the gate. Very often poly-silicon is used for interconnect to the gates, which can easily result in a series resistance that is comparable to the base-resistance of a bipolar transistor.

### 1.5.2 Biasing

The parameters of the small-signal models are the derivatives of the transfers in the large-signal model in a chosen point of operation. For small signal-variations, the parameters can be considered constant. Any variation of the parameters will be called distortion. To have (for small signals) a transfer described by this derivative, the input signal for the device is translated to the operating point by adding the proper bias signal to it. At the output, the signal is translated back to the origin by subtracting a bias signal from it. In figure 1.6 this is shown in more detail. The operating point on the non-linear curve is chosen to be at $(X_Q, Y_Q)$, at which the derivative is determined. This is the desired small-signal transfer $f_{ss}(x_s)$ with:

$$y_s = f_{ss}(x_s) = \tilde{f}'(X_Q)x_s = A_t x_s \qquad (1.13)$$

in which $\tilde{f}'(X_Q)$ is the derivative of $\tilde{f}(x)$ in the operating point. $A_t$ is a constant. To have this transfer, first the input signal is translated to the operating point by adding $X_Q$ to it. After passing the device, the signal is translated back to the origin by subtracting $Y_Q$ from it. In this way a linear (small-signal) transfer, with the value $A_t$, for the information is achieved.

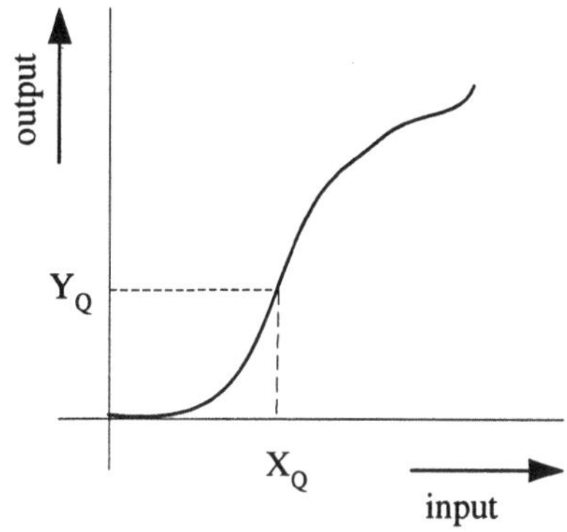


*Figure 1.6.* A non-linear transfer function.

Note that for a linear system with a transfer $H$ the following relation has to be true:

$$H(\alpha x_1 + \beta x_2) = \alpha H(x_1) + \beta H(x_2) \qquad (1.14)$$

So:

$$H(x_1 + x_2) = H(x_1) + H(x_2) \qquad (1.15)$$

$$H(\alpha x) = \alpha H(x) \qquad (1.16)$$

From this, it follows that the translation described above is necessary to have a linear transfer. If one of them is left out, this results in an offset which on its turn leads to non-linearity.

Every linear transfer function is centered around the origin. So, when a device with a non-linear characteristic $\tilde{f}(x)$ is used to implement a small-signal model with a transfer $A_t$, the operating point has to become the new origin. This is achieved by translating the signal, both at the input and the output, by adding bias signals. In figure 1.7 the principles of operation are shown. A new transfer $\tilde{g}(x)$ results that is centered around the origin ($\tilde{g}(0) = 0$). First, a signal $X_Q$ is added to the input signal $x_s$. Then, when the signal $x_s$ is small enough, it is

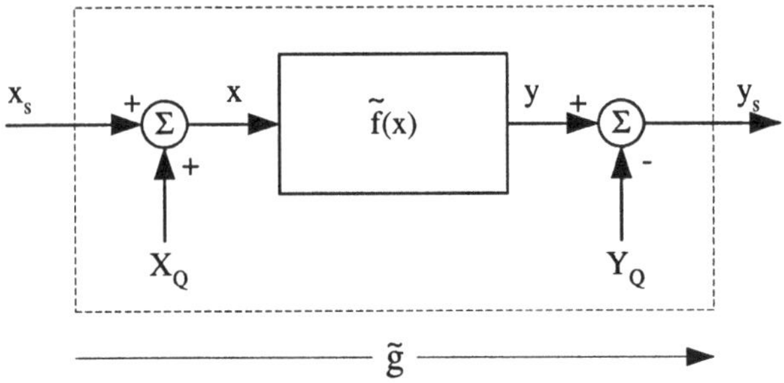


*Figure 1.7.* The principle of translation with bias signals.

amplified with a "constant" factor $A_t$. Then, a signal $Y_Q$ is subtracted and the output signal $y_s$ remains. The addition and subtraction of the bias signals $X_Q$ and $Y_Q$ have produced the required centering of the transfer around the origin. In figure 1.8 a practical example is shown. The two voltages $V_{iQ}$ and $V_{oQ}$

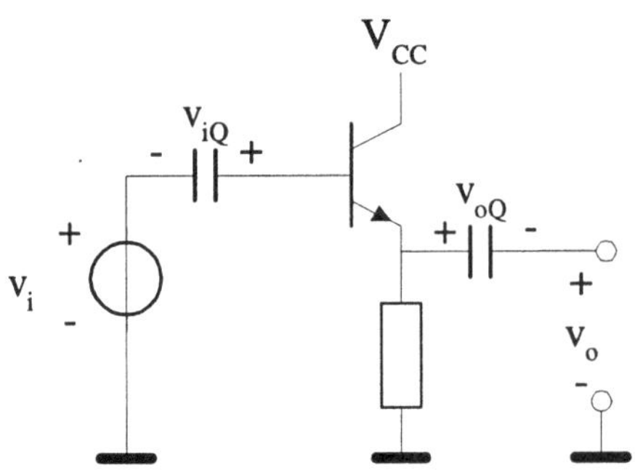


*Figure 1.8.* A practical implementation of bias voltages.

across the coupling capacitors form the two shifting informationless quantities $X_Q$ and $Y_Q$ that center the transfer of the complete circuit around the origin.

### 1.5.3 Distortion

Actually, $A_t$ is the linear term of the Taylor expansion [5] of $\tilde{f}(x)$ in the operating point $(X_Q, Y_Q)$:

$$\begin{aligned} y_s + Y_Q &= \tilde{f}(X_Q) \\ &+ \frac{\tilde{f}'(X_Q)}{1!}(x - X_Q) \\ &+ \frac{\tilde{f}''(X_Q)}{2!}(x - X_Q)^2 \\ &+ \frac{\tilde{f}'''(X_Q)}{3!}(x - X_Q)^3 \\ &+ \dots \end{aligned} \tag{1.17}$$

All other terms, including the DC ones, are distortion terms. The constant term is of a different nature compared to the higher-order terms. The constant term is signal independent, whereas the others are signal dependent. So, the influence of the constant term can be completely removed via the translations described in section 1.5.2. The influence of others cannot be completely removed because of the unpredictability of the input signal $x_s$. What remains after the translation is the non-linear system which is shown in figure 1.7 with a transfer $\tilde{g}(x)$ centered around the origin. When the information carrying signals are small enough, the linear term is the only significant term for amplification. The higher-order terms represent the distortion of the signal. The information that they contain is not used. The complete Taylor expansion for the system shown in figure 1.7 is:

$$y = A_t x_s + \frac{\tilde{g}''(0)}{2!}x_s^2 + \frac{\tilde{g}'''(0)}{3!}x_s^3 + \dots \tag{1.18}$$

with:

$$A_t = \frac{\tilde{g}'(0)}{1!} \tag{1.19}$$

There are several ways to describe the non-linearity of this circuit. For each different application, different ways of describing the non-linearity are used. In chapter 5 some of these measures will be discussed. For now, it is sufficient to know that when small-signal models are used during the design:

- the parameters of the small-signal models are the first-order derivatives of the actual non-linear transfer function in the operating point. Since the derivatives commonly change when the operating point is changed, bias quantities tend to appear as parameters in the noise performance and frequency behavior of a circuit. Though they are not applied as actual signals yet, *their value is already prescribed via the required small-signal behavior.*

- informationless biasing signals have to be added later to the information carrying signals in the practical circuit, in order to obtain the centering of the transfer around the operating point. *So, there will be a separate biasing step at the end of the design process.* The details will be described in chapter 8.

### 1.5.4 Checking device parameters

The parameters for the small-signal models are based on the parameters for the (non-linear) large-signal models. Therefore, it is good practice to check for the presence of all large-signal model-parameters that may be relevant in the operating point (and you have to know at least their order of magnitude to see if they are relevant). Sometimes, a model is delivered with a parameter-set that is not complete. The lacking parameters will be given a default value by the simulator. These default values usually represent an idealized behavior. So, then the small-signal model probably yields too optimistic results.

With only a very small number of basic simulations, it is already possible to perform a simple check, in which the parameterization of the DC-behavior and the frequency behavior can be tested. It hardly costs any time to do these simulations but they provide much insight in the behavior of the available devices. *It is therefore a good habit to always start a design with these simple checks.* Especially when circuits are being designed for extreme constrains, e.g. very low-power (current), insufficient characterization of the devices usually results in inadequate predictions of circuit behavior by a simulator.

For a bipolar transistor three very illustrative plots can be generated:

- A plot of the current $I_c$ and the base current $I_b$, as a function of the base–emitter voltage, the "Gummel-plot".

- A plot of the current-gain factor $\beta_F$, as a function of the collector current $I_c$.

- A plot of the transit frequency $f_T$, as a function of the collector current $I_c$.

The results should look like the plots given in figure 1.9, figure 1.10 and figure 1.11, respectively. Figure 1.9 shows the Gummel-plot, in which clearly three area's can be distinguished. Area 2 is the "normal" region of operation for the bipolar transistor. Only the early effects at the base-collector and the base-emitter junctions affect the current-gain factor somewhat. This is modelled via parameters VAF and VAR, respectively. It is important to note that the parameter VAR, the "reverse" early voltage, is of influence in the forward mode just as well. In some circuits, like bandgap references, neglecting this parameter may even lead to considerable errors.
At low currents, in area 1, non-idealities of the base current become important.

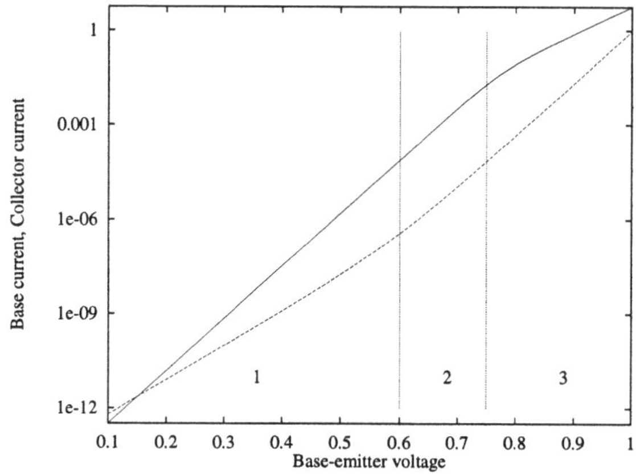


*Figure 1.9.* The Gummel-plot of a bipolar transistor.

In figure 1.9 it can be seen that this reduces the current-gain factor of the transistor at low currents. Parameters ISE and NE describe this behavior.
At high currents, in area 3, bulk resistors and high-injection effects reduce the current-gain factor. The important parameters are the bulk resistors RE, RC and RB and the "knee-current" at which high injection starts, IKF. Actually there are more than one parameters that describe the base resistance, because it is usually current dependent. A detailed description of this is, however, beyond the scope of this book. In the SPICE manual all parameters and equations are described in detail.
In figure 1.10 the current-gain factor as a function of the collector current is plotted. This can be derived straightforwardly from the Gummel-plot.

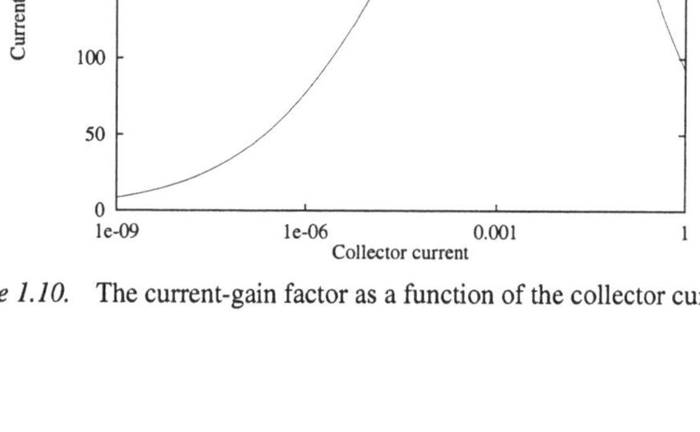


*Figure 1.10.* The current-gain factor as a function of the collector current $I_c$

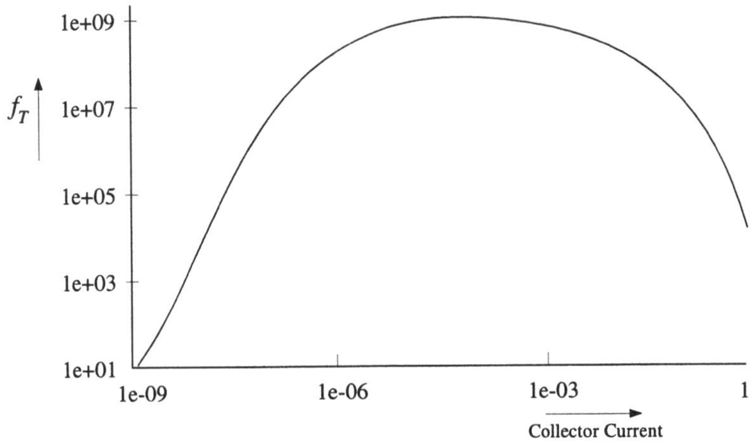


*Figure 1.11.* The transit frequency $f_T$ as a function of the collector current $I_c$

In figure 1.11 the $f_T$ of the transistor is shown as a function of the collector current. The transit frequency $f_T$ is the frequency at which the current-gain factor of the transistor becomes equal to unity. Several AC-analyses are necessary to obtain this plot. Figure 1.12 shows the $f_T$ measurement circuit. The collector-emitter voltage $V_{cc}$ must be chosen such that the transistor operates properly. For this analysis the transistor is biased at a constant collector voltage,

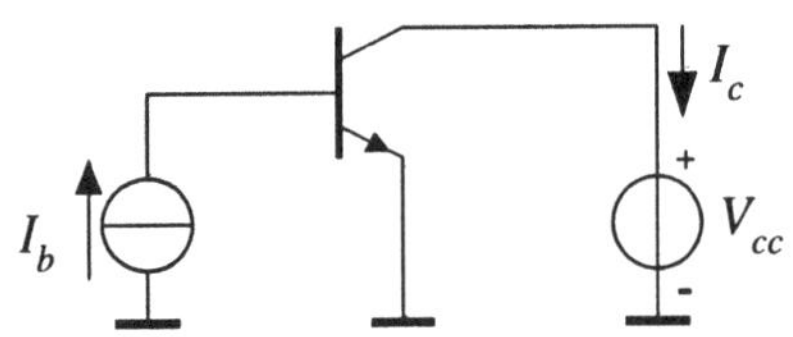


*Figure 1.12.* A circuit to find the current-gain factor and $f_T$.

because the output has to be short circuited for the signal for a correct measurement of the output *current*. The transistor is *current* driven. For a number of different values of the DC-bias current, the transit frequency is determined. The $f_T$ is mainly determined by parameters TF, CJE and to some extent CJC. At the flat part in figure 1.11, parameter TF is dominant in $f_T$ with:

$$f_T \approx \frac{1}{2\pi \mathrm{TF}} \tag{1.20}$$

Via parameter TF the "diffusion capacitor" of the forward biased emitter-base junction is modelled [21] . At high currents, parameter ITF is responsible for a reduction of $f_T$. At low currents, parameters CJE and CJC become important.

Parameters CJE and CJC model the "junction capacitors of the emitter-base and base-collector junction, respectively [21]. In this region the $f_T$ of the transistor becomes current dependent.

For FETs similar plots can be generated and similar areas can be distinguished. However, there are many different models for FETs. The selection of the appropriate model strongly depends on the application and also on the dimensions of the transistor itself. Therefore, it is not simple to present a short list of "essential parameters". The manual of the simulator should be consulted for this.

Figure 1.13 shows two simple circuits that can be used to find the small-signal parameters in a simulator like SPICE. For the bipolar transistor the base-

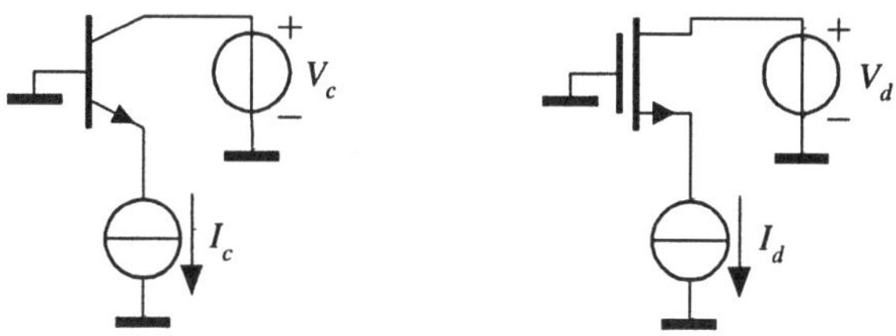


*Figure 1.13.* Example circuits that can be used to find the parameters in SPICE.

collector voltage is set by the voltage source $V_c$ and the collector current is set by the current source $I_c$. For the FET $V_d$ and $I_d$ are used for this. This is a biasing scheme that is also frequently used in measurement setups for parameter extraction[4]. When the desired biasing values are set, SPICE will produce a complete list of small-signal parameters for that operating point in its output.

## 1.6 The chain matrix

The synthesis of negative-feedback amplifiers presented in this book will be based on the interconnection of two-ports. There are many ways to describe a two-port, but in this book it has been chosen for the chain matrix. It is the most convenient way to describe a system with cascaded stages of which each stage is also described by a chain matrix. The chain matrix relates the input voltage and current of a two-port to the output voltage and current. The corresponding sign conventions are depicted in figure 1.14. It should be noted that the common sign convention for the output current is inward. The convention as depicted in figure 1.14 is more convenient when cascading two-ports as the outward directed output current directly matches the orientation of the inward directed

[4]When there is a separate substrate connection e.g. in the case of models for transistors on an integrated circuit, the biasing of this node should be taken care of too, of course.

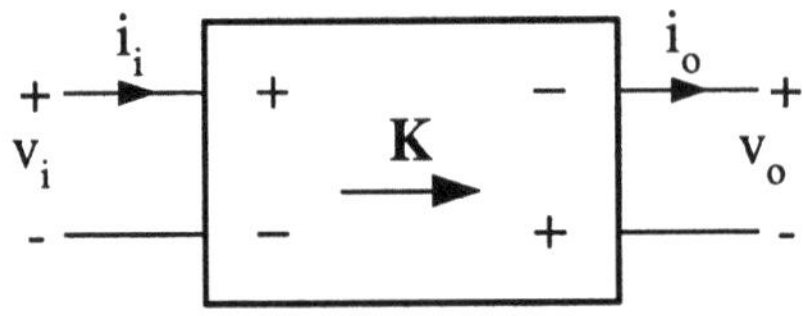


*Figure 1.14.* The sign conventions for a two-port with chain matrix **K**.

input current of the subsequent two-port. The chain matrix definition is given by:

$$\begin{pmatrix} v_i \\ i_i \end{pmatrix} = \begin{pmatrix} A & B \\ C & D \end{pmatrix} \begin{pmatrix} v_o \\ i_o \end{pmatrix} \tag{1.21}$$

with:

$$A = \frac{1}{\mu} = \left.\frac{v_i}{v_o}\right|_{i_o=0} \tag{1.22}$$

$$B = \frac{1}{\gamma} = \left.\frac{v_i}{i_o}\right|_{v_o=0} \tag{1.23}$$

$$C = \frac{1}{\zeta} = \left.\frac{i_i}{v_o}\right|_{i_o=0} \tag{1.24}$$

$$D = \frac{1}{\beta} = \left.\frac{i_i}{i_o}\right|_{v_o=0} \tag{1.25}$$

It is important to note the conditions $i_o = 0$ and $v_o = 0$ at the definitions. Frequently errors are made in this respect in practice.

### 1.6.1 Chain matrices of devices

The small-signal behavior of the available active and passive devices is described with chain matrices too. For the bipolar transistor used in common-emitter configuration (CE), it follows:

$$\mathbf{K_{bip}} = \begin{pmatrix} -\frac{1}{g_m r_o} & -\frac{1}{g_m} \\ -(\frac{1}{\beta_f} + \frac{j\omega}{\omega_T})\frac{1}{r_o} & -(\frac{1}{\beta_f} + \frac{j\omega}{\omega_T}) \end{pmatrix} \tag{1.26}$$

and for the FET used in common-source configuration (CS), it follows:

$$\mathbf{K_{FET}} = \begin{pmatrix} -\frac{1}{g_m r_d} & -\frac{1}{g_m} \\ -\frac{j\omega}{\omega_T}\frac{1}{r_d} & -\frac{j\omega}{\omega_T} \end{pmatrix} \tag{1.27}$$

The chain matrix of the gyrator with a transfer $Z$ is:

$$\mathbf{K_Z} = \begin{pmatrix} 0 & Z \\ \frac{1}{Z} & 0 \end{pmatrix} \tag{1.28}$$

For a transformer with a ratio $n$ the chain matrix is:

$$\mathbf{K_{tr}} = \begin{pmatrix} n & 0 \\ 0 & \frac{1}{n} \end{pmatrix} \tag{1.29}$$

Apart from chain matrices of single devices, it is very interesting to know the chain matrices of some frequently used device combinations, like the differential pair. When this chain matrix is expressed in terms of the matrix entries of a single CE-stage, it can be seen quickly what are the consequences of the replacement of a single CE-stage by a differential pair.
Starting from the chain matrix of a CE-stage given by equation (1.26), the chain matrix of a differential pair equals:

$$\mathbf{K_{CE,dif}} = \begin{pmatrix} A_{CE} & 2B_{CE} \\ \frac{C_{CE}}{2} & D_{CE} \end{pmatrix} \tag{1.30}$$

From this it can be seen that only $B$ and $C$ differ with just a factor of 2. This means, for example, that when in a later stage of the design it is decided that a CE-stage should be replaced by a differential pair, the calculations for the new circuit can be reduced to introducing a factor of 2 at some places.

## 1.7 Exercises

### Exercise 1.1

Fig.1.15 shows two two-ports. One with a parallel connected resistor, the other with a series connected resistor.

1. Calculate for each two-port the chain-matrix.

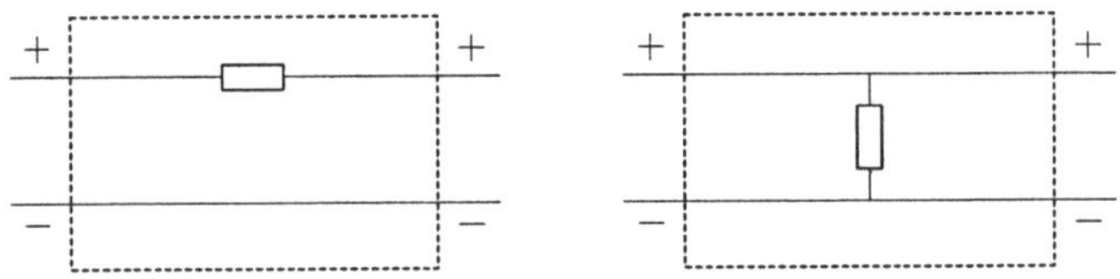

*Figure 1.15.* A parallel and a series resistor in a two-port

The chain-matrix of the amplifier depicted in fig.1.16 can be calculated in two ways. It can be done "directly" as it was done in the cases above, but it can also be done via a matrix multiplication of the chain-matrices of the two smaller two-ports that are indicated with dotted lines in the figure.

2. Calculate the chain-matrix for the amplifier shown in fig.1.16 via both methods.

3. Identify the matrix entry in this chain matrix that defines the desired transfer of the voltage amplifier.

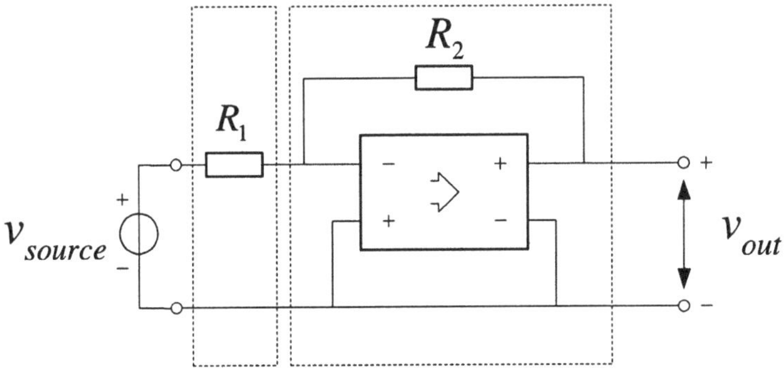


*Figure 1.16.* A voltage amplifier

4. Calculate the chain-matrix of the amplifier depicted in fig.1.17 on page 25 via a matrix multiplication of the chain-matrices of the two smaller two-ports that are indicated in the figure.

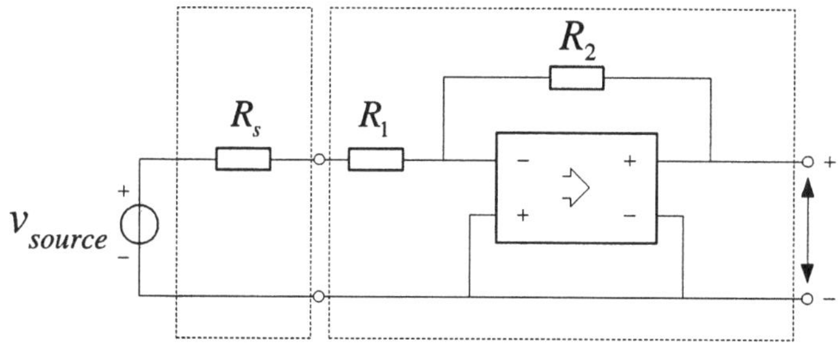


*Figure 1.17.* A voltage amplifier connected to a source

When the source impedance $R_s$ is inaccurate, the transfer of the amplifier becomes inaccurate.

5. Identify the entry in the chain matrix of the amplifier calculated in 2. that introduces the influence of the source impedance on the voltage-to-voltage transfer.

6. How can this entry be made zero?

## Exercise 1.2

Given the chain matrix of a CE-stage:

$$\begin{pmatrix} A_{CE} & B_{CE} \\ C_{CE} & D_{CE} \end{pmatrix} \tag{1.31}$$

1. Determine the elements of this matrix with the assumption that the frequency dependent elements can be ignored.

2. The stages depicted in figure 1.18 comprise one or more CE-stages. Determine for the depicted combinations the chain matrix as a function of $A_{CE}$,$B_{CE}$,$C_{CE}$ and $D_{CE}$. It can be assumed that all the CE-stages are identical.

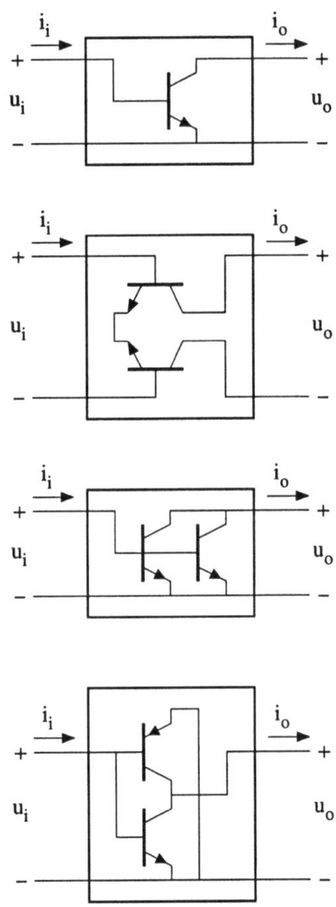


*Figure 1.18.* Various combinations of CE-stages

# 2

# SYNTHESIS OF ACCURATE AMPLIFIERS

## 2.1 Introduction

The design of high-performance amplifiers can be a very complicated task. There are many parameters that influence the performance. The complexity very often leads to a "chaos" in which the design procedure stalls. The main purpose of this book is to create order in this chaos. A design approach will be proposed in this chapter, strongly based on orthogonality, hierarchy and model simplification, as discussed in the previous chapter. The assumptions of orthogonality and hierarchy make it possible to quickly design an amplifier that is optimal within the constrains that are necessary to make the assumptions valid. Usually, the quality of this amplifier is so close the the absolute optimum, that further optimization is not worth the effort. But even when further optimization is still desired, the designer is usually well aware what to do, since the structure in the design approach has provided him with much insight in the behavior of that particular circuit. At the end it is "safe" to loosen the constraints, "Then the search space has become too small for chaos to fit into it..."

The amplifier-design methodology described in this book is based on the negative-feedback topology consisting of an active circuit (implementing a nullor) combined with a feedback network. This makes it possible to distinguish two large orthogonal design steps that can be assessed separately. These are:

- the design of the feedback network, while modelling the active circuit with a nullor (the "ideal active circuit");
- the design of an active circuit of which the properties approach that of the nullor as good as required for the application.

Within these two global steps, smaller orthogonal design steps can be distinguished. To complete the design, each of the smaller design steps can be assessed separately if the correct design sequence is used. It will be shown in this chapter that, to come as close as possible to true orthogonality, the design steps should be assessed in a specific order.

Six separate design steps can be distinguished within the two larger steps.

- Within the first global step:
  - Detailed source, load and transfer specification.
    This "trivial" step is a very important one. A wrong decision here will lead to an amplifier that is optimized for the wrong task. Performance will be poor despite all the design efforts.
  - Determination of the amplifier topology and dimensioning of the feedback network.
    Also this step is very critical. The amplifier performance will never be better than that of the feedback network. Optimizing a wrong amplifier topology often does lead to an acceptable amplifier, but never to the optimal one (see exercise 1).
- Within the second step, (the nullor design):
  - Noise
  - Distortion
  - Bandwidth
  - Biasing

Figure 2.1 shows a graphical representation of the design procedure. Each of these steps again consist of smaller orthogonal steps. The hierarchy that can be used in the design steps is a typical benefit of the orthogonality that makes the design clear and straightforward. Each step can be treated independently of what is to follow. Iterations will be scarcely necessary and if they are, the implications can been overseen easily. In each step, theory on a specific aspect of the performance is necessary. In this chapter it will only be discussed*what* is done and *why* it is done at that particular stage of the design. In the chapters to follow, it will be discussed *how* things are done. There the theory will be treated in full detail. Figure 2.1 shows a graphical representation of the design procedure.

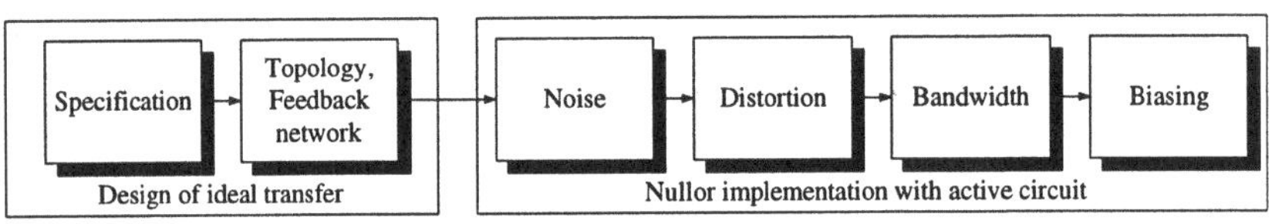


*Figure 2.1.* The design procedure.

## 2.2 The definition of an amplifier

An accurate definition of the function that has to be implemented should always be given first. This is to make sure that no design constraints "sneak into" the design procedure that could be dealt with more elegantly on a higher hierarchical (system) level.

An amplifier is a circuit that increases the power contents of an information carrying signal by multiplying it by an *accurate constant.*

This implies that the chain matrix of an amplifier has four entries that must be *constant* and *accurate.* The most simple solution to this design problem would be the use of a device with a chain matrix that meets with these demands. So, the first thing to do is to evaluate the chain matrices of various devices and rank them for these two properties. Two different classes of devices can be defined:

- devices with an *accurate* and *constant* transfer;
- the others.

Unfortunately, at present, the first class contains only passive devices that are unable to produce any power amplification. In the second class, there are many devices that can produce this amplification, but, unfortunately, non of them has sufficient accuracy. So a strategy has to be developed that is able to *combine* the best of both classes. Circuit topologies have to be found that make use of the accuracy of devices from the first class and of the power amplification of the second class. Of the many error correction techniques used in electronic design, negative feedback is the only one that is capable of doing this. In this chapter the design method to make optimal use of the feedback topology is introduced.

### 2.2.1 Configurations for high-performance amplifiers

The only transfer that can be made accurately with the devices from the first accurate class is a transfer that reduces the power content of the information it transfers. So, it is only possible to make very accurate attenuators. The only amplifier topology that makes use of an accurate attenuator to obtain accurate amplification is the feedback topology. This clearly distinguishes the negative-feedback topology from other error correcting topologies like error feed forward or compensation. In amplifiers using one of the latter two methods, the error at the output is not measured and corrected, but during design time the expected error of the active components is estimated and a correction circuit to remove this error is added. Since it is not possible to predict the errors perfectly, it is not possible with perfect passive devices from the first class to design a perfect amplifier. For the negative-feedback topology it is possible to design a perfect amplifier, so this is by far the preferred topology. The only reason for not using the feedback topology is an impossibility to generate adequate loop gain. For example, at very high frequencies the gain of the active devices may have reduced so much that it is difficult to generate sufficient loop gain. Of course, this frequency constraint is moving up swiftly with the ever growing speed of modern technologies[1]. Another reason might be the danger of instability of the feedback loop. The other two topologies mentioned above do not have an intentional loop and therefore are not likely to show this type of instability. However, a proper design of the frequency behavior of the feedback loop prevents this problem.

#### 2.2.1.1 The negative-feedback topology

Suppose there is a network of two resistors as shown in figure 2.2. This

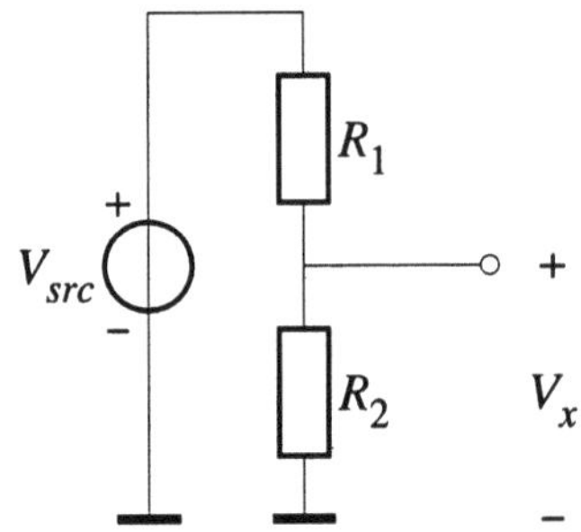


*Figure 2.2.* A voltage divider.

circuit is known as the voltage divider and its behavior is described with the following equation:

$$V_x = \frac{R_2}{R_1 + R_2} V_{src} \tag{2.1}$$

Since it merely consists of passive components, it can be used to generate a very accurately attenuated copy of the signal $V_{src}$ at its input. For a given value of the voltage source there is only one solution to this equation.

In a feedback amplifier, the output signal of the attenuator is actively made equal to an input signal, $V_{in}$, so:

$$V_x = V_{in} \tag{2.2}$$

In circuit theory, this equation is indicated with a *nullator*, as shown in figure 2.3. The nullator indicates that the voltage difference between the two nodes it connects is *defined* to equal zero. Also by definition, *no current* flows through the nullator. So, the nullator itself has no ability to force this equality. This is

[1]This is sometimes "missed" by circuit designers, of which some keep on using HF-techniques for circuit design even when sufficient loop gain could be obtained.

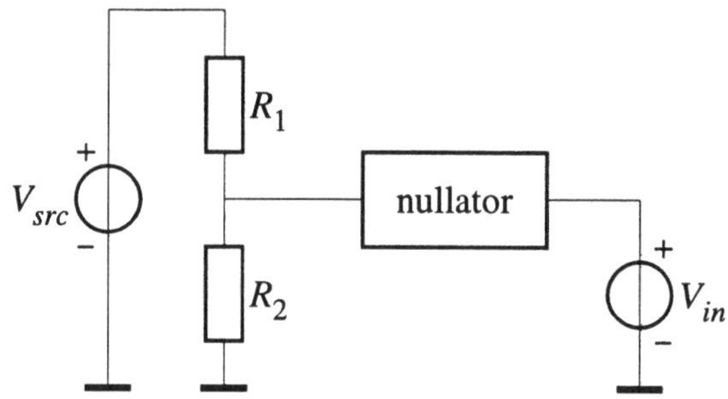


*Figure 2.3.* A voltage divider with a nullator.

done by a controlled source, $V_{src}$. The value of the source $V_{src}$ is controlled such that the conditions imposed on the circuit by the nullator are met. This variable source that adapts its output to fulfill the nullator conditions is called a *norator.*

The combination of a nullator and a norator makes it possible to have equation (2.2) met for every value of $V_{in}$.

*Though the output value of the norator is written as a voltage in this example, it is not defined whether the norator itself is a voltage source or a current source.* It is the nullator that gives the constraint and to fulfill this, the voltage across the norator and the current through the norator have to be adapted. The impedance at the terminals of the norator relates the norator current to the norator voltage, so when the norator produces the one, the other is automatically set via this impedance.

Later in the design, when the implementation of the norator is started, a choice has to be made whether the norator is implemented as a controlled voltage source or a controlled current source. A similar choice arises for the nullator which, when implemented, is replaced by either a current or a voltage sensor. This will be discussed later in this chapter.

With the voltage source $V_{src}$ acting as norator at the input of the voltage divider and the nullator at the output of the divider, as shown in figure 2.4, accurate gain is achieved, completely dominated by the value of the passive components. The voltage across the norator is given by:

$$V_{src} = \frac{R_1 + R_2}{R_2} V_{in} = \mu V_{in} \tag{2.3}$$

Note that the norator is directly at the output of the amplifier and is able to generate any voltage or current needed, so the amplifier is*completely independent of any load* that is connected to the amplifier. So, it can be concluded that it is possible to build amplifiers of which the transfer is completely dominated by accurate passive components using the negative-feedback topology. The only requirement for this is the availability of a nullator and a norator.

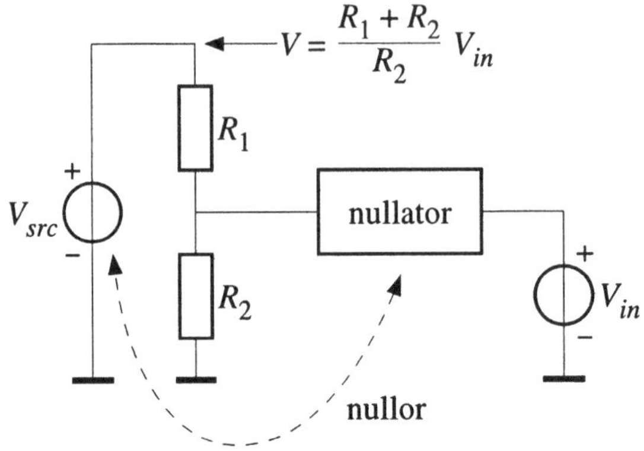


*Figure 2.4.* A voltage divider with a nullator and a norator ($V_{src}$).

### 2.2.2 The nullor

Since usually nullators and norators occur in pairs in circuits, it is convenient to use a network element that represents this combination. This element is called a *nullor*[2]. It is a two-port as depicted in figure 2.5. The relations between the

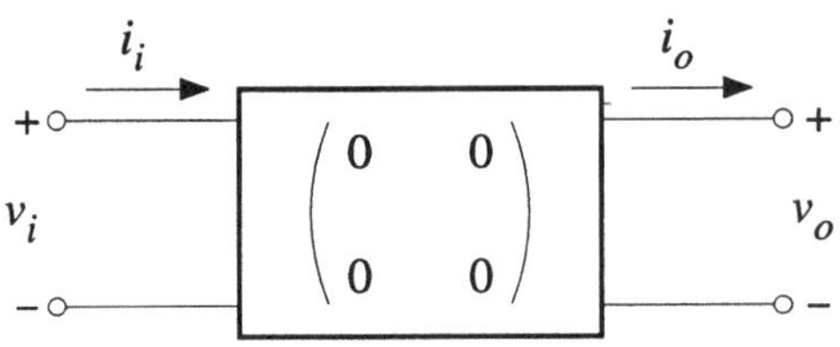


*Figure 2.5.* The nullor.

input and the output quantities are:

$$\begin{pmatrix} v_i \\ i_i \end{pmatrix} = \begin{pmatrix} 0 & 0 \\ 0 & 0 \end{pmatrix} \begin{pmatrix} v_o \\ i_o \end{pmatrix} \tag{2.4}$$

So, the chain matrix of a nullor contains only zero's. A nullor has an infinite voltage gain ($\mu$), current gain ($\alpha$), transconductance ($\gamma$) and trans-impedance

[2]There are a number of more complex amplifier circuits that can not be modelled efficiently with an equal number of nullators and norators. This may, for example, be the case when two controlled sources are assumed to produce correlated output signals. If this correlation is not explicitly introduced as a constraint (with a nullator) on the circuit, introducing nullors could introduce either too much equations or too much variables, resulting in circuits that do not work properly e.g. circuits that have multiple bias solutions (most of them being latch-up states). Working explicitly with nullators and norators prevents these problems in those cases. For this reason nullators and norators are mentioned in this chapter as the basic building block of the nullor. Problems in this respect are easily detected when at the start of the design it is made certain that *all* constrains imposed on a network are made explicit, especially required symmetry of signals e.g. in balanced differential amplifiers.

($\xi$). It adapts its output such that at its input terminals the voltage and current are equal to zero.
(From this it can already be seen intuitively that the design of amplifiers based on the nullor approach will make use of feedback configurations. For the nullor output to have influence on the nullor input, some of the nullor-output signal has to be fed back to the nullor input.)

Looking at the second class of components (the others), it can be seen that, though non of the devices have neither constant nor accurate entries in their chain matrices, most of them have *very small* entries. They have a large gain, which makes them to closely resemble a nullor. Cascading these devices increases the gain and therefore makes the resemblance to a nullor even better.

Evidently it is possible to implement nullors with the active components and, consequently, implement amplifiers that have the accuracy and constancy of the passive devices of the first class and use the gain of the devices in the second. The better the implementation of the nullor is, the better the behavior of the circuit is dominated by the passive components.

The negative-feedback topology makes it possible to split the design into two orthogonal parts:

1. The design of the feedback network, assuming a nullor is present.

2. The implementation of the nullor with suitable active devices.

From the first step, the fundamental limit to the performance can found. The nullor produces no noise, no distortion and does not limit the bandwidth of the circuit in any way: it is perfect. Practical implementations of the nullor never improve the performance found with the nullor. *If there is an improvement, the modelling of the circuit is incorrect or insufficient and should be corrected!*

After the first step is completed successfully and the performance is within specifications, the active circuit that fulfills the nullor function is implemented. When the resulting circuit is not performing according to the specifications anymore, it is just a matter of finding a better nullor implementation to bring the circuit within specifications. Usually it is very clear in what respect the active circuit is not performing sufficiently. Then it is fairly simple to determine what measures have to be taken to improve the circuit, or to decide that the circuit is not feasible.

A strategy will be described in this book which makes it simple to find directions for optimization. It consists of several clearly distinguishable design steps. It is of great importance, especially for automated design, that unfeasible circuits are recognized before excessive calculations have been performed on them. Therefore, for each step the models are kept as simple as possible. The predictions given by these models are "best case" predictions. When a circuit does not meet its specifications at a particular step it is certain the more detailed practical implementation will not meet its specs either. So, the circuit

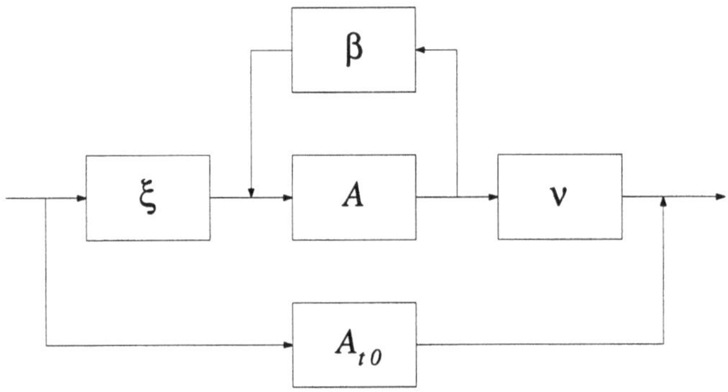


*Figure 2.6.* A negative feedback system.

can be rejected before extensive calculations have been performed. However, when it does meet its specifications at that step, there is no certainty that the specifications will be met in the end. The circuit may be rejected in a later stage. Therefore the steps have an increasing complexity. As a circuit passes for more tests, the risk of performing useless calculations on a non-solution reduces.

Also the steps are sorted for orthogonality. It is assumed that the three fundamental criteria noise, signal power and bandwidth are orthogonal, that a design can be optimized for one of them without affecting the performance for the others. In practice, it appears that the order in which the three criteria are dealt with affect the validity of the assumption of orthogonality very much. It will be shown later in this chapter that it is best to optimize for noise first, then for distortion and finally for bandwidth.

## 2.3 The asymptotic-gain model

The design of the amplifier consists of two major parts. The design of the feedback network and the design of the nullor. A model is needed that can be used to perform these two design steps independently. The asymptotic gain model is well suited for this purpose. In figure 2.6 a negative-feedback system has been depicted. $\beta$ is the transfer of the feedback network, $A$ is the transfer of the active circuit (nullor). Frequently a direct transfer exists between input and output, caused by various kinds of parasitic coupling. This is modelled via the direct transfer factor $A_{t0}$, i.e. the gain when the loop gain is zero. When the input of the active part is not connected directly to the source and the output of the feedback network, the extra transfer is modelled via $\xi$ and in a similar way $\nu$ models an extra transfer between between the norator and the amplifier output when they are not directly connected. Though $A_{t0}$ has been depicted as a separate signal path from the input to the output, generally it is a property of the active circuit itself. Also many feedback networks are not unilateral and therefore can produce a direct path from the input to the output. In a "proper design" $\xi$ and $\nu$ should both equal unity, though frequency compensation, components at the input or the output may even affect these parameters in this case.

The transfer of the system of figure 2.6 is:

$$A_t = \frac{A}{1 - A\beta} + A_{t0} \tag{2.5}$$

The product $A\beta$ is called the loop gain indicated with symbol $L$. When the amplifying part is a nullor, i.e. when $L = \infty$, the gain is:

$$\lim_{L \to \infty} A_t = -\frac{\xi\nu}{\beta} + A_{t0} = A_{t\infty} \tag{2.6}$$

The gain $A_t$ can be expressed in terms of $A_{t\infty}$ as:

$$A_t = A_{t\infty}\frac{-L}{1 - L} + \frac{A_{t0}}{1 - L} \tag{2.7}$$

The first factor of the first term ($A_{t\infty}$) is determined by the feedback network, its second factor is determined by the quality of the nullor implementation. It can be seen as the deviation from the ideal case (with the nullor). Ideally, it equals unity. In chapter 7, the details of this error term, like its frequency dependency will be discussed.

Generally the second term ($\frac{A_{t0}}{1-L}$) can be neglected for reasonable loop gains. However, this does not mean that $A_{t0}$ is of no importance! It is also a term in $A_{t\infty}$, so even when a nullor is present, a direct transfer caused by e.g. the feedback network is not removed. In some cases, $A_{t0}$ may even have a dominant influence on the transfer and large design errors are made when it is neglected. With the aid of this model it is possible to divide the synthesis of a negative-feedback amplifier into two orthogonal parts. The ideal transfer $A_{t\infty}$ is designed under the assumption that a nullor is present, it remains unchanged during the design of the nullor.

## 2.4 The synthesis of $A_{t\infty}$

The first design steps comprise the synthesis of the feedback network. The necessary gain is provided by a nullor.

### 2.4.1 Step 1: The determination of the input and output quantities

The amplifier has to match optimally to its environment, so apart from the amplification factor, also information about the source and the load is necessary. Therefore, to be able to start the synthesis of the feedback network, the following data have to be known:

- The source quantity and impedance
- The load quantity and impedance
- The bandwidth of the information
- The maximal signal level
- The dynamic range of the information
- The amplification factor

The source and load quantities are the quantities that are the*primary carriers of the information.* The source and load quantities can be:

- Voltage, ($V$)
- Current, ($I$)
- Power, ($P$) in which voltage and current are related via a characteristic impedance.

Usually, the source and load impedance are not known with sufficient accuracy, but when the topology is correctly chosen, these impedances have no influence on the quality of the transfer. This implies that the amplifier has to be sensitive to the primary information carrying quantity at its input and also has to deliver the primary carrier at its output.

Suppose, for example, a source of which the current is the primary information carrier, e.g. a piezo electric pressure sensor. It consists of a piezo electric crystal fitted between two metal plates serving as electrical connections. When pressure is put on the crystal its dipoles are deformed which produces charge built up over the crystal. The amount of charge is the primary measure for the pressure. When the two metal plates are short circuited and the current is measured, the amount of charge can be measured accurately.

$$i_{sensor} = \frac{dq_{plates}}{dt} \tag{2.8}$$

$q_{plates}$ is the charge that is generated by the crystal.

When the connections are left open, the charge creates a voltage difference across the plates of the sensor. The value of this voltage depends on the charge, but also on the capacitance that is between the plates.

$$v_{plates} = \frac{q_{plates}}{C} \tag{2.9}$$

The value of the capacitance between the plates "pollutes" the information with its inaccuracy. It may, for example, vary with the pressure, since the crystal is deformed under pressure, which causes non linearity. From this it can be seen that the choice of the appropriate input and output quantity is of primary importance.

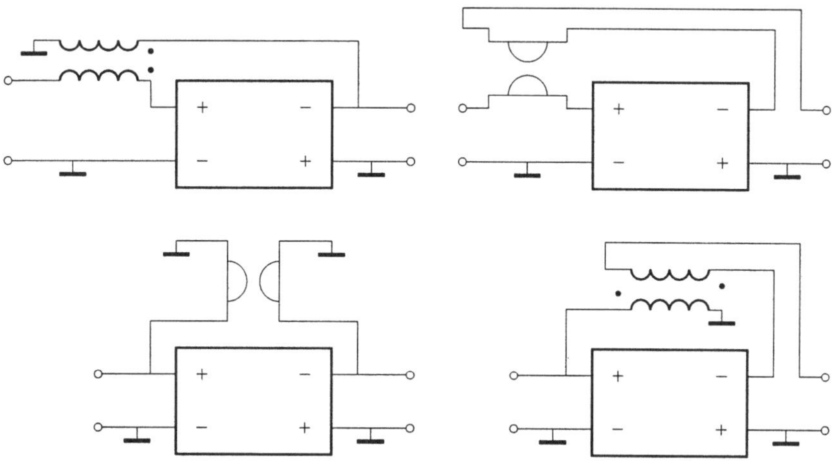

*Figure 2.7.* The four different negative-feedback amplifiers.

## 2.4.2 Step 2: The synthesis of the feedback network

The task of a feedback network is to take the output signal of the nullor, and reduce it into a signal that can be compared to the input signal by the nullator. Depending on the input and output quantities, four different feedback networks can be distinguished (see table 2.1)[3].

| | | | |
|---|---|---|---|
| 1 | $V_{fb,in} \rightarrow V_{fb,out}$ | transfer | $\frac{V_{fb,out}}{V_{fb,in}} = \frac{1}{A_{fb}} = \mu$ |
| 2 | $V_{fb,in} \rightarrow I_{fb,out}$ | transfer | $\frac{I_{fb,out}}{V_{fb,in}} = \frac{1}{B_{fb}} = \gamma$ |
| 3 | $I_{fb,in} \rightarrow V_{fb,out}$ | transfer | $\frac{V_{fb,out}}{I_{fb,in}} = \frac{1}{C_{fb}} = \zeta$ |
| 4 | $I_{fb,in} \rightarrow I_{fb,out}$ | transfer | $\frac{I_{fb,out}}{I_{fb,in}} = \frac{1}{D_{fb}} = \beta$ |

*Table 2.1.* Four different feedback-network transfers.

With these networks combined with a nullor, four different negative-feedback amplifiers can be made. In figure 2.7 these amplifiers are shown. Ideal non-energetic components (gyrator, transformer) are used as feedback network (The details can be found in chapter 4). The transfer of all amplifiers is dominated by the transfer of the network:

- When a voltage is fed back to the input of the nullor, the output of the feedback network is placed in series with the source.

[3]This book treats only amplifiers that have either current or voltage inputs and outputs. This implies they need only one feedback loop. For multi-loop amplifiers, having e.g. power inputs and outputs, the theory is similar, but calculations are a bit more elaborate. Details can be found in [1].

- For a current input, the output of the feedback network is placed in parallel to the source.

This is because in the first case the feedback network has to compensate the source *voltage* to obtain a zero input signal for the nullor, and in the latter case it has to compensate the source *current.*
The input of the *feedback network* is placed

- in parallel with the nullor output when the output voltage of the nullor is the primary information carrier
- in series with the nullor output when the current is the primary information carrier.

It can be easily seen that the amplifiers that have a current input, have a zero input impedance and that amplifiers that have a voltage input have an infinitely high input impedance. The amplifiers that have a current output, have a infinitely high output impedance and amplifiers that have a voltage output have a zero output impedance.

Ideally, the feedback network does not need to have any influence on the noise performance of the amplifier. This is the case when non-energetic components, like the transformer or the gyrator are used. In practice, these components are often not feasible. A gyrator can not be implemented without it producing at least the noise corresponding to its transfer and transformers are only of use in limited frequency ranges and certainly not suited for DC. Lacking gyrators and transformers of adequate performance, a network of impedances, like resistors, is used. In that case, the noise behavior is affected somewhat by the particular design of the network. In figure 2.8 amplifiers are shown that use impedances instead of non-energetic components as feedback network. In all cases, because of the presence of the nullor, the evaluation of the noise behavior of the amplifier is very simple. At this stage, noise contributions can be expected from the source and in some cases from the feedback network. This is the absolute best noise performance of the amplifier. After the nullor has been implemented, it will become worse. It can be seen that for the voltage and the current amplifier, the gain can be set by the *ratio* of two impedances in the feedback network. The noise contribution of the network is determined by the absolute value of the impedances (the *impedance level* of the network). Via the impedance level of the feedback network its noise contribution can be optimized without interfering with the gain. Unfortunately, the power consumption of the amplifier is also related to the impedance level of the network in an opposite way, so a compromise has to be found. Therefore, at this stage of the design it is known what the highest noise performance will be and also what the minimum output signal will be, that the nullor has to deliver. It has to supply the voltage and current for the load *and* for the feedback network. There are cases in which the feedback network loads the nullor output more than the actual load impedance!

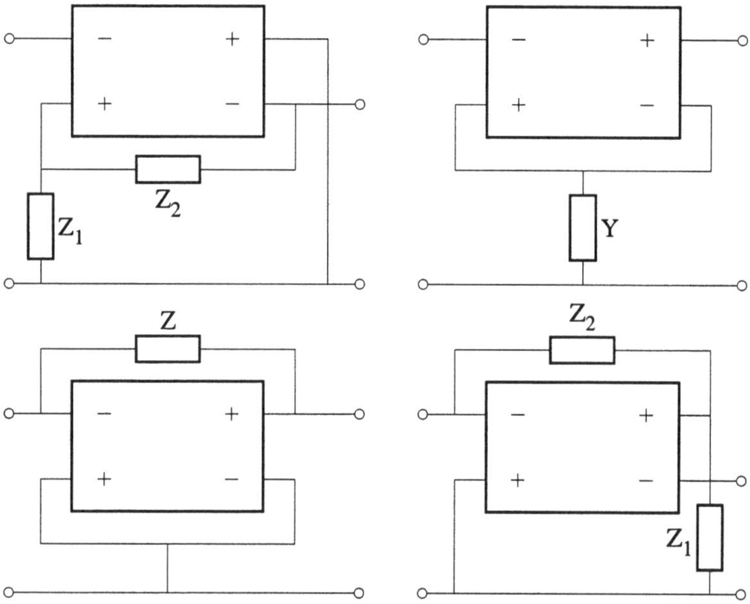


*Figure 2.8.* The four different negative-feedback amplifiers with impedances in the feedback network.

## 2.5 The next steps: Synthesis of the nullor

In design steps 1 and 2 the feedback network ($A_{t\infty}$) has been designed. At this point in the design, the nullor will be implemented in several steps with practical components, thereby taking care that the degradation of the performance that can be expected is identified and kept within tolerable limits.

In a properly designed amplifier, in which every stage offers maximum gain, only the noise behavior of the first stage of the active circuit is of importance.

*Property noise is localized at the input.*

Clipping, the most aggressive way of distortion is likely to happen at the output, since in a properly designed amplifier, the signals are the largest in the output stage.

*Property distortion is localized at the output.*

The fact that these two properties can be assigned to different parts of the amplifier makes the assumption of orthogonality for these two valid. The third property, bandwidth cannot be assigned to any particular part of the amplifier. Both the input stage and the output stage contribute to the bandwidth performance of the complete amplifier too. Therefore noise and clipping-distortion optimization always interfere with bandwidth optimization. For this reason noise and clipping-distortion optimization should be performed *before* bandwidth optimization. During these optimizations bandwidth is not taken into account. It is assumed that the bandwidth can be repaired later outside the first and last stage. The contributions of the first and last stage to the bandwidth are taken for granted during bandwidth optimization. They are taken into account, but in principle not changed.

Bandwidth calculations can be very tedious. Therefore, before the actual bandwidth of the amplifier is determined, first a prediction of the bandwidth is made by way of a fairly simple calculation, using simplified models. When the circuit passes this test, the actual bandwidth calculations are performed using fully detailed models. So, especially bandwidth optimization consists of a number of (orthogonal) stages:

- bandwidth estimation, based on simplified models
- bandwidth optimization or *frequency compensation*, still based on simplified models
- model refinement combined with circuit measures to maintain the performance found with the simplified models

The reason for using very simple transistor models at first, and refining the model later after each successful calculation, is that this enables to detect unfeasible solutions before extensive calculations are performed. Also this gradual refinement of models provides a lot of insight for the designer in the exact cause of appearing problems in the amplifier.

After the bandwidth optimization is completed successfully, the biasing circuits are introduced. First as ideal voltage and current sources. At this stage a first verification with a circuit simulator like SPICE can be performed. After this, gradually the ideal biasing sources are replaced by practical implementations. In this way, the influence of each source on the performance of the amplifier can be evaluated and if necessary corrected.

## 2.6 Topology synthesis of the nullor

### 2.6.1 The cascade topology

Before starting the synthesis of the various nullor stages and optimizing the performance of each stage, it has to be decided first what kind of stages will be used. Since the nullor is the part of the negative-feedback amplifier that produces the gain, it should only consist of devices that can deliver this gain. *So, only active devices are used to implement the nullor.* Passive devices can only reduce a signal (especially resistors that dissipate signal), so their presence in the nullor circuit would only degrade its performance. Only during the bandwidth optimization step, sometimes passive compensation components may be added to the nullor circuit to influence its frequency behavior. So, after the frequency compensation passive components may be found in the nullor circuit that are "legal", of course only if the better places, like the feedback network do not offer an opportunity to do the frequency compensation there.

The active devices are seen as two-ports with a chain matrix with small (but unfortunately not zero) entries. The configuration consisting of more than one device that produces the most gain is the cascade of devices. It is not the only "correct" configuration, but it is the correct "first guess", and usually it is the best choice. A typical nullor circuit, consisting of three stages, is shown in figure 2.9. To each stage one of the properties of the negative-feedback

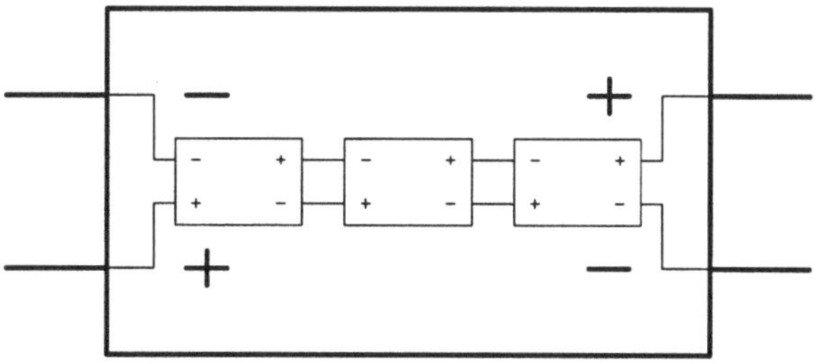

*Figure 2.9.* The nullor as a cascade of three two-ports

amplifier can be assigned as the property to be optimized. Noise to the first stage, clipping to the last stage and frequency behavior to the middle stage(s). Each two-port contains active devices only.

## 2.6.2 Two-ports with active devices

The active devices that can, for example, be used in the nullor are the bipolar transistor and the MOS-transistor. For simplicity, in the figures only bipolar transistors will be shown, but at all times MOS-transistor or even vacuum tubes could replace them. Figure 2.10 shows the 6 ways a transistor can be used to implement a two-port. Note that the three single-transistor two-ports have one terminal of the input port directly connected to one terminal of the output port because actually the transistor is a tree-terminal device. This limits the number of configurations in which they can be used as two-port. It is important to understand that although transistors have been drawn inside the two-ports in figure 2.10, actually small-signal models are used during this design step as shown in figure 2.11. Initially, even more simplified models are used as shown in figure 2.18 on page 52 and explained in section 2.9.3.

*Unfortunately, no special symbol has been defined that represents a small-signal model, so the standard symbol for a transistor is used both in small-signal diagrams and circuit diagrams representing the actual fully biased circuit.* This often leads to confusion in interpreting a circuit diagram.

The connection between one input and one output terminal in the case of the single-transistor cases has been drawn explicitly outside the two-port in order

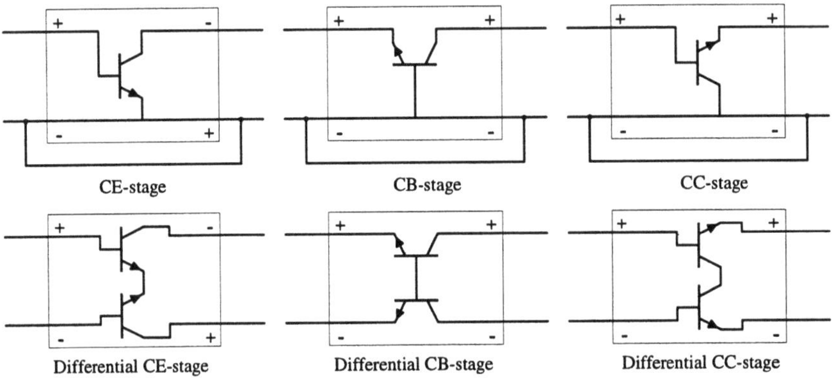


*Figure 2.10.* Six ways to use transistors in a two-port.

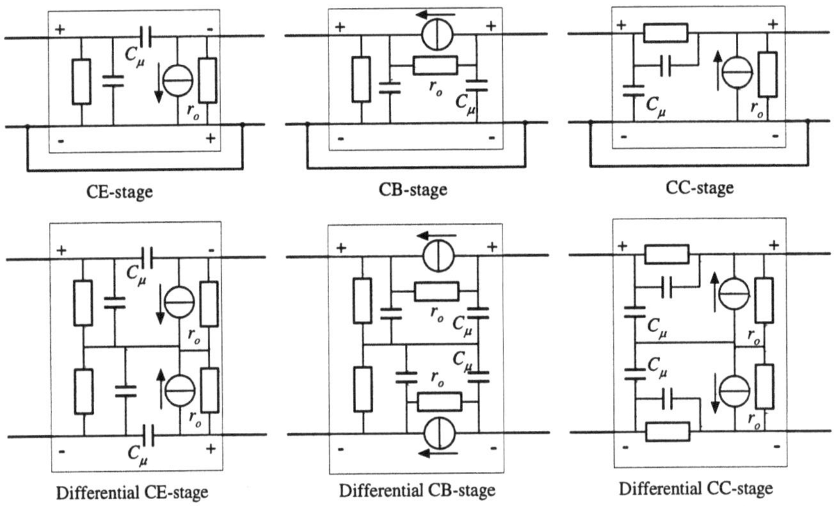


*Figure 2.11.* Six ways to use transistors in a two-port, using the small-signal models.

to make it visible at this synthesis level. If the connection can be tolerated, a single-transistor stage can be used. If not, a stage containing a differential stage must be used. Using the two-ports shown in figure 2.11, it can be easily found what type of two-ports is necessary to implement negative feed-back amplifiers as shown in figure 2.8. In figure 2.12, the current amplifier implemented with two CE-stages is shown. Only one CE-stage can be used because of the direct connection between one of the input terminal and one of the output terminals. With two CE-stages, it is not possible to have the correct sign of the loop gain. One of the stages has to be a differential stage, but there is freedom to choose where to put it. Of course, it is always possible to use two differential stages.

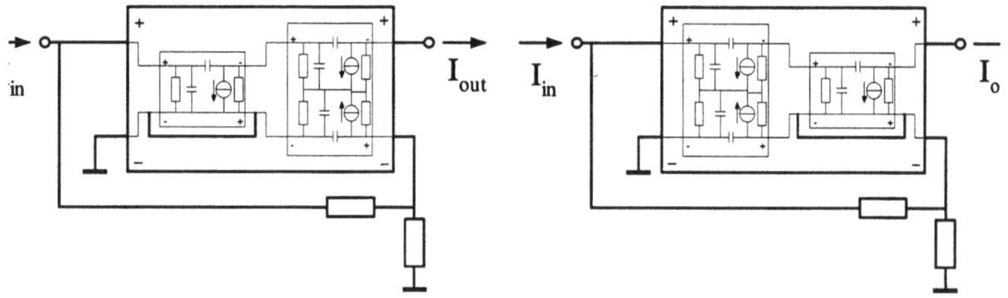


*Figure 2.12.* Two two-stage current amplifiers.

### 2.6.3 Replacing the nullator and the norator

The nullor has a nullator at its input and a norator at its output. The nullator represents an element that is described as just one point in the origin of the impedance $V$–$I$ plane, i.e. no current and no voltage. *There is no practical element with this property.* But it is possible to create a set of nodes in a circuit that meet the specs, i.e. no voltage between the nodes and no current from one node to the other. Negative feedback is needed to do this. To be able to realize this with negative feedback, a measurement has to be done at the nodes involved. There are two measurements that can be done:

1. A measurement of the current from one node to another, where the negative-feedback loop nullifies this current and via the impedances across the nodes, consequently, also nullifies the voltage.
2. A measurement of the voltage difference between the two nodes, where the negative-feedback loop nullifies this voltage difference and via the impedances across the nodes, consequently, also nullifies the current from the one node to the other.

*This implies that the nullator is replaced by either a devices that measures current or a device that measures voltage.* This is a choice that has to be made explicitly, because it may have a considerable impact on the performance of the amplifier.

A similar thing holds for the norator. The norator is an element that relates the voltage across it to the current through it in such a way that the set of equations that describe the circuit has a solution. So, the output signal of the norator will be such that for the nullator the zero current and voltage conditions are met. The norator is a source, in this respect that it can supply power to the circuit, but it is not explicitly a voltage source or a current source. So, for the practical situation in which only voltage and current sources are available, a choice has to be made: *the norator has to be replaced by either a voltage source or a current source.* Also this choice has to be made explicitly because of the impact it has on the performance of the amplifier.

### 2.6.4 The ideal substitution and the practical situation

Looking at the four configurations shown in figure 2.7, it can be seen that for each one a different choice should be made for the substitution of the nullator and the norator. This will be explained below.

#### 2.6.4.1 The voltage amplifier

In the voltage amplifier, the feedback network generates a voltage at its output. This voltage is made equal to the input signal. An error would generate a voltage difference, so the primary quantity that should be observed by the nullator-substitute is *voltage*.

At the output of the amplifier, the feedback network senses the voltage. This implies that the norator should be replaced by a voltage source.

When the nullator-substitute is chosen to be an element that observes current, it is of course a zero(low)-impedance element. Then it can easily be seen that a current flows through it that depends on the difference between the input voltage and the voltage returned by the feedback network, converted into a current via the source impedance. So, then the loop gain also depends on the source impedance which makes it less accurate. In a similar way, the load impedance starts having influence when at the output the current is sensed instead of the voltage.

This shows that only when the proper substitutions have been done, the source and load impedances have absolutely no influence on the transfer, even if the loop gain is not infinite any longer. To have the best independence of source and load impedance, in order to have the most accurate voltage amplifier, the nullator should be substituted by a voltage sensor and the norator by a *voltage* source in this case.

#### 2.6.4.2 The voltage-to-current amplifier.

In the voltage-to-current amplifier, at the input the same situation exists as for the voltage amplifier. So, also in this case the primary quantity that should be observed by the nullator-substitute is *voltage*.

At the output of the amplifier, the feedback network senses the current. This implies that the norator should be replaced by a *current* source.

Also in this case it can be easily seen that source and load impedances have absolutely no influence on the transfer, even if the loop gain is not infinite anymore. So, in order to have the most accurate voltage-to-current amplifier, the nullator should be substituted by a voltage sensor and the norator by a current source.

#### 2.6.4.3 The current-to-voltage amplifier

In the current-to-voltage amplifier, the feedback network generates a current at its output. This current is made equal to the input signal. An error would generate a current difference. So, the primary quantity that should be observed by the nullator-substitute is *current.*

At the output of the amplifier, the feedback network senses the voltage. This implies that the norator should be replaced by a *voltage* source.

So, in order to have the most accurate current-to-voltage amplifier, the nullator should be substituted by a current sensor and the norator by a voltage source.

#### 2.6.4.4 The current amplifier

In a similar way it can be found that in order to have the most accurate current amplifier, the nullator should be substituted by a current sensor and the norator by a *current* source.

#### 2.6.4.5 The practical situation

Figure 2.13 shows the amplifiers of figure 2.7 with their ideal nullator and norator substitutes. To have perfect independence of source and load, *even when*

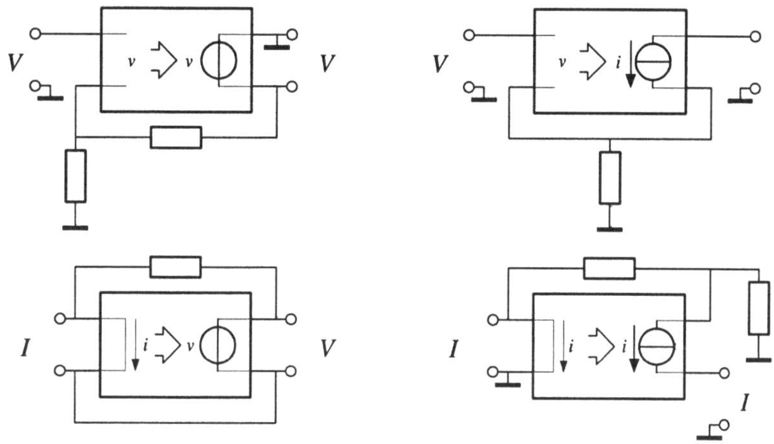


*Figure 2.13.* The four different single-loop negative-feedback amplifiers with the ideal substitute for the nullator and the norator.

*the gain of the nullor circuit is not infinite anymore*, the nullator and norator should be replaced by the elements shown in this figure. Looking at figure 2.11, it can be seen that the current technology (with transistors) does not provide these elements in a perfect way. Perhaps with the CE-stage the technology matches the best to the requirements for the voltage-to-current amplifier, but the same stage matches very poorly to the requirements for the voltage-to-voltage amplifier. So, for many configurations the nullator and norator are replaced by elements of the "incorrect" type.

When the nullator and the norator are not replaced by elements of the correct type, the source and the load impedances influence both the transfer of the amplifier and the loop gain. This influence, basically, can be separated into two effects:

1. A reduction of the loop-gain, making the factor $\frac{-L}{1-L}$ in equation 2.7 deviate more from unity. This makes the influence of the feedback network on the transfer less dominant. The result is less accuracy.

2. An increased unpredictability of the loop gain, since both the source and load impedance might be not very accurately known.

Problems caused by the first effect can be simply reduced by increasing the loop gain, for instance, by adding an extra stage.
The latter effect can cause the most serious problems. When the source and load impedance are very unpredictable—this is for example the case for operational amplifier for which the source and load impedance are not known during design time—the value of the loop gain also is very unpredictable and it may be impossible to do a reliable frequency compensation.
*In this case,* **and only in this case,** *it is necessary to give the most priority to the substitution of the nullator or norator by the best matching transistor stage concerning its input or output impedance. In all other cases priority should be given to the maximum contribution of the substituting stage to the loop gain.*
If a problem can be solved by increasing the loop gain, this always is the best method. Only if this is no option, other methods may be considered.

Selecting the proper stage, two criteria are used. In order of priority, the available stages can be sorted as:

1. stages that offer the largest gain, in other words, the stages that have the smallest entries at every location of the chain matrix are preferred;

2. stages that match best to the source and load impedance of the amplifier as described above.

Both the CE-stage and the differential CE-stage always match best to the first criterion, but not always to the second one. From table 2.2 it can be seen that for the CB-stage chain-matrix parameter $D$ equals unity and for the CC-stage this holds for parameter $A$. So, the contribution of these stages to the loop gain is lower than that of a CE-stage. Also the noise and the distortion behavior is usually negatively affected when these stages are used at the input or the output of the nullor circuit. However, the CB-stage has an input impedance that comes closest to the "current-sensor substitution" for the nullator. Therefore if the

| | A | B | C | D | sense type | source type |
|---|---|---|---|---|---|---|
| CE-stage | small | small | small | small | I/V | I |
| CB-stage | small | small | small | **1** | I | I |
| CC-stage | **1** | small | small | small | V | V |

*Table 2.2.* Typical entries of the chain matrix of the three transistor stages and their typical input and output behavior.

current-source impedance is very unpredictable, at the expense of a decreased noise performance and reduced loop gain, the influence of the source can be reduced, making a predictable design possible.
In a similar way, the CC-stage has an output impedance that comes closest to the "voltage source substitute" for the norator. Therefore if the load impedance is very unpredictable, at the expense of the distortion behavior and the loop gain, the influence of the load impedance can be reduced, making a predictable voltage-output design possible.[4]

From all this it can be concluded that, except for some very rare cases, the CE-stage and the differential CE-stage should be the building blocks to implement the nullor circuit. Later it will be shown that sometimes CB-stages could occur as a cascode stage to improve the behavior of a CE-stage.

### 2.6.5 The current mirror as a stage in the nullor circuit

The current mirror is a "strange" translinear stage that does not fit very well in the classification of stages that has been made so far. Still they are often used, so there must be a reason to use stages like this, even when their use is not indicated by the theory described above (yet?).

In figure 2.14 a circuit diagram is drawn that shows the behavior of a current mirror when a signal current $i$ is supplied at its input. It just multiplies the current at its input by a factor 2. With one of its output terminals connected to the supply or the ground it always injects signal into the supply environment. Since only one terminal of the output port is freely available, the current mirror cannot be used to process differential signals. Sometimes it is used to increase the gain of a nullor circuit by a factor 2 as can be seen in figure 2.15. Without the current mirror, this nullor circuit would have had half the current gain it has now. The price to pay for this is the connection of one of the terminals

[4]This is generally the reason to design operational amplifiers (OPAMPS) with an output stage like this. The load impedance is not known. Especially a capacitive load could cause stability problems when this capacitance can generate an unpredictable dominant pole. Usually the CC-stage at the output shifts this pole to the non-dominant group. For "OPAMPS" that are used internally in integrated circuits, where the source and load impedance are known, there is no reason to use the CC-stage. Giving in on the reflex of using a CC-stage in this case means giving in on the over-all performance. But designers do think....

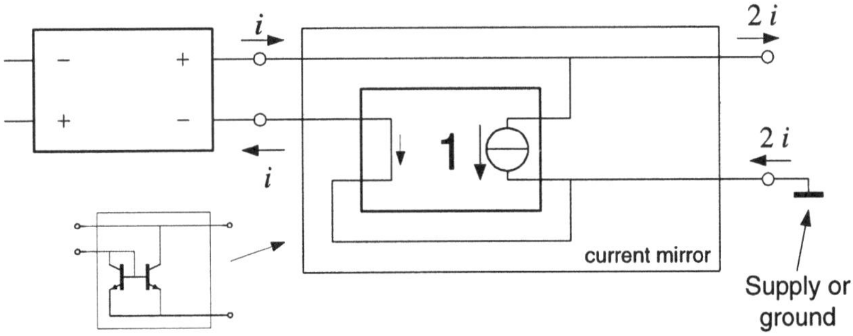


*Figure 2.14.* The signal behavior of a current mirror on the "system level".

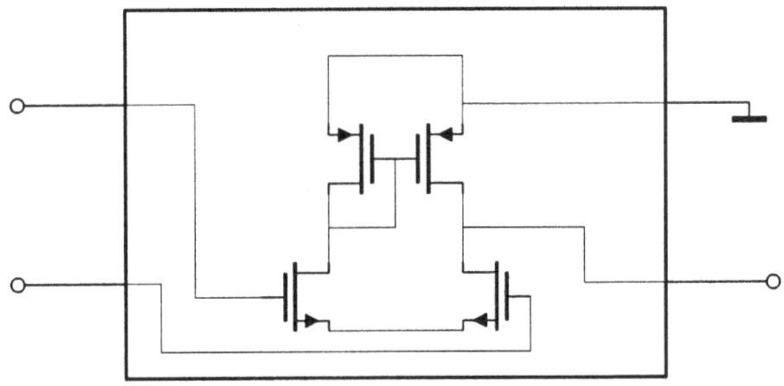

*Figure 2.15.* A current mirror used to increase the gain of a nullor circuit by a factor 2.

to the signal ground (which can also be the supply voltage). Even when the current mirror is used in an intermediate stage the ground connection occurs, thus creating a leakage path for the signal to ground. Used as a differential pair or as a cascoded CE-stage instead of as a current mirror, the two transistors could have contributed much more to the gain of the nullor circuit and no ground connection would occur.

The input impedance of the current mirror is low. In this respect it is very similar to the CB-stage. So for instance it could be used as a cascode stage, to improve the properties of a CE-stage. However the CB-stage performs better in this case.

Figure 2.16 shows the current mirror in a circuit with a differential pair in which the bias currents for the differential pair are shown too. It can be seen that in this case the current mirror can be used effectively to separate bias current from signal current. At the free output terminal, the signal current is available without bias currents and the differential pair is nicely symmetrically biased following the tail current that sets the value of the bias current. In section 8.12.3 this will be discussed. From all this it could[5] be concluded that it only makes

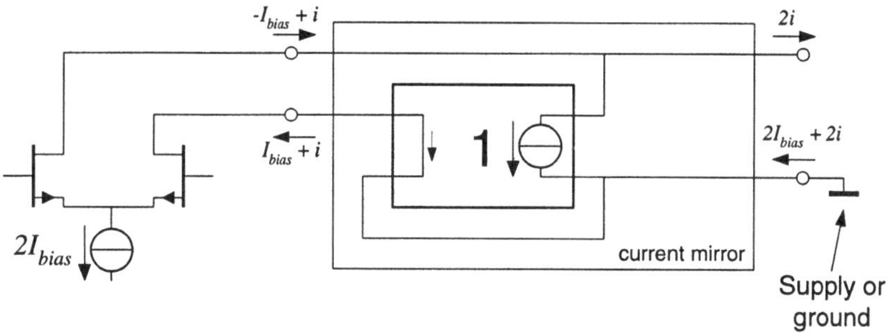


*Figure 2.16.* The current mirror as a bias device.

sense to introduce a current mirror in a circuit for biasing reasons....

## 2.7 Step 3: Design of the first nullor stage: noise

The first property that is optimized in the nullor design, is the noise performance. In figure 2.17 the nullor configuration is shown that is used for this. The first stage is put in front of the nullor. When the gain of the first stage

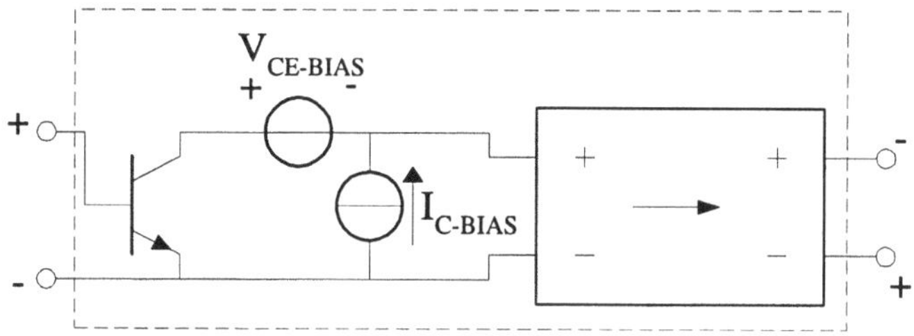


*Figure 2.17.* The model for the active circuit used for noise optimization.

is sufficient, the noise contribution of the rest of the circuit can be neglected. Therefore, as a first stage a CE or a CS stage should be used, because they offer the most gain. Basically, three different ways of noise optimization can used at the input:

- noise matching to the source via a transformer;
- optimization of the bias current of the input stage;
- connecting several input transistors in series/parallel.

All of them can essentially be viewed as variants of the first type. The difference is the number of noise sources which are taken into account in the optimization. Details will be described in chapter 4.

[5]It will take some more research before the word "could" can be replaced by the word "can".

The advantage of starting with the noise optimization is the fact that it is possible to model the circuitry following the first stage by a nullor. In this way, the other two criteria, bandwidth and distortion, remain ideal. The nullor will supply infinite power if necessary, its infinite gain results in infinite bandwidth and zero distortion. The designer only has to be concerned with noise in the input stage. As long as the implementation of the nullor is good enough, the performance of the first stage with respect to bandwidth and distortion can be of no importance. An orthogonal design of the noise performance is therefore possible.

## 2.8 Step 4: Design of the last nullor stage: distortion

The clipping distortion, though also localized at a specific place in the amplifier, is better not taken as the first aspect to be optimized. This is because it is not possible, like it was for noise, to take the stage concerned and cascade it with a nullor to make the optimization independent of the noise and bandwidth aspects. The nullor would be placed in front of the stage concerned and with its infinite gain, it would reduce the distortion caused by this last stage to zero, irrespective of the implementation of that stage. So, to keep calculations simple as long as possible, by keeping a nullor in the design as long as possible, the design should start with noise optimization.

During most of the design small-signal models are used. As already mentioned in section 1.5.2 and section 1.5.3, these models are only valid for small signals, for which the higher-order terms in equation 1.17 are negligible. This is the criterion that sets the validity range of the model. Outside this validity range, the behavior of the circuit differs from the model to a certain extent. The distortion that is caused then can be divided into two classes:

- weak distortion;
- clipping distortion.

When the signal becomes so large that the higher-order terms are not negligible anymore, but are not causing a complete change in the signal behavior of the amplifier this is called "weak distortion". Usually an increase of the loop gain can reduce it.

When the signal becomes so large that an additional increase of the input signal does not lead to an increase in the output signal, "clipping" occurs. A direct consequence of the clipping of an amplifier is that the feedback loop is broken. This tends to result in a violent and unpredictable behavior of the amplifier and should be prevented at all times. When an amplifier is close to clipping, the small-signal parameters become strongly signal dependent yielding an considerable increase of the nonlinear distortion. Clipping occurs, for example, in the circuit with transfer $\tilde{g}_(x)$ shown in figure 1.7 on page 16, when a signal with magnitude $X_Q$ is applied to the input. Then the small-signal transfer is zero; the loop is broken.

Clipping occurs when the signal has a large amplitude. When every stage in the nullor circuit is designed for maximum gain as is stated in section 2.6, the largest signals can be expected in the output stage. So, as the first stage is optimized for noise since a that place the signal is the smallest, the last stage is designed for preventing clipping. Bandwidth can be manipulated anywhere in the amplifier, so any restriction on bandwidth caused by the last stage, can be corrected elsewhere in the amplifier.

For the best orthogonality, the gain of the last stage should be as large as possible. The magnitude of the signals in the preceding stage is reduced by the gain of the last stage. Therefore it is unlikely that this preceding stage will cause clipping distortion problems if the last one does not. So, also at the output a CE or a CS stage is preferred.

The dominant contribution to the weak distortion, caused by the non-linearity of the devices, may not come from the last stage. In a lot of cases it appears to be the first stage that causes this type of distortion. However, this type of distortion can be reduced by increasing the loop gain, a measure that can be taken anywhere in the circuit and that is not in conflict with bandwidth or noise optimization. Details will be described in chapter 5.

## 2.9 Step 5: Bandwidth optimization

Bandwidth optimization comprises three steps. Firstly, an estimation of the bandwidth capabilities of the amplifier obtained so far. When this estimation is large enough, the second step can be done, the actual frequency compensation. These first two steps use simple models yielding simple (and fast) calculations. The third step in the bandwidth optimization is to check whether these simple models are valid or not. If not, countermeasures should be taken to make these simple models valid again.

### 2.9.1 Bandwidth estimation

The exact calculation of the bandwidth of an amplifier is very complicated and not necessary to assess the feasibility. Therefore, first a prediction is made on the achievable bandwidth of a solution that is not based on complicated calculations. By model simplification, calculations are reduced even further.

In figure 2.18 the models that will be used for the CE and the CS stage are depicted. The output impedance is made infinite. The capacitances$C_\mu$ and $C_{gd}$ from base to collector and from gate to drain, respectively, are not included in the model.

The two stages that have resulted from the noise and the distortion optimization are now cascaded to form the first guess for the complete active circuit. The

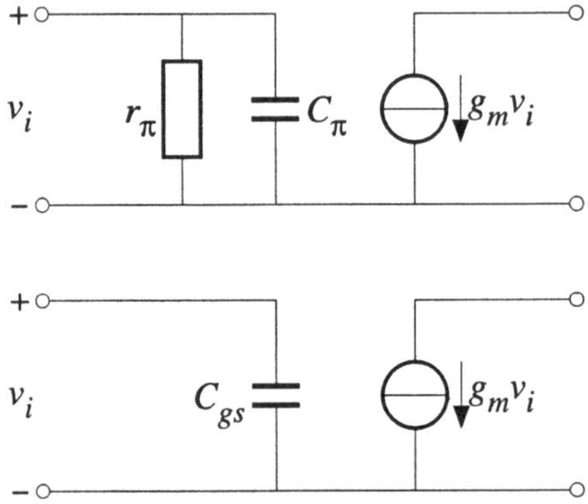


*Figure 2.18.* The simplified model for the CE and the CS stage for used for bandwidth estimation.

simple models are used. In some cases differential pairs may be used instead of single transistors. Details about this are found in the appropriate chapters. A negative-feedback amplifier results that at least meets the noise and the clipping specs. Now the product of the poles and the DC-loop gain is calculated. Because of the simple models this is a very simple job. The result is called the *loop–gain–poles product* (LP-product). The following prediction about the bandwidth of the amplifier can be made with the use of this LP-product:

*When n poles of the amplifier can be forced into Butterworth position during compensation, the bandwidth of the amplifier will be:*

$$B_w = \sqrt[n]{LP_n} \tag{2.10}$$

It is a necessary, but not sufficient requirement on the amplifier. When the LP-product is not sufficient, the amplifier will never meet its bandwidth specs. When it is sufficient, there is a good chance that an amplifier results after frequency compensation that meets the bandwidth specs.

When the LP-product is not sufficient an extra stage is added to increase it. Every time a CE or a CS stage is added, a factor $f_T$ is contributed to the LP-product. Other than these stages contribute less than $f_T$, so they are not favored. At this design stage, only extra intermediate CE or CS stages are added until the LP-product predicts sufficient bandwidth.

Not all poles can be forced into Butterworth position. Only so-called "dominant poles" can. A method to distinguish dominant poles from non-dominant poles will be described in chapter 6. There also the biasing requirements for the intermediate stages and slewing effects will be discussed.

### 2.9.2 The actual frequency compensation

When the LP-product is sufficient, the poles of the amplifier have to be forced to the required position. Several methods exist that can be used. They are phantom-zero compensation, pole-splitting, pole-zero cancellation and resistive broad banding. These are treated in detail in chapter 7. It is important to note that this frequency compensation is performed using the simplified models!

### 2.9.3 Model refinement

model, refinement Frequency compensation is at first performed with the simple models. When a full compensation is established, the models are refined step by step. Each time an output impedance or a Miller capacitance is added and the pole-zero pattern is evaluated. In the cases where the pole-zero pattern is seriously affected, an ideal current follower is cascaded to the stage that causes the problems. In figure 2.19b this is shown for a stage in which a Miller capacitance is added. The ideal follower short circuits the output impedance, so

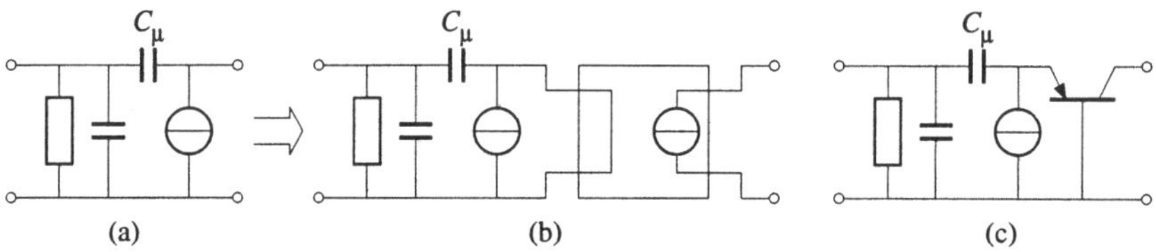


*Figure 2.19.* Gradual correction of errors that occur during model refinement.

any problems that occur due to this should disappear. The collector–base ($C_\mu$) capacitance is shorted to ground also and thus the pole of the stage is back at its original place again. Only the *right-half-plane zero* that is introduced by $C_\mu$ is not removed. When the pole-zero pattern is not restored by the introduction of the follower, it is certain the problems are caused by this zero. Then for a bipolar transistor it can sometimes be forced to a higher frequency by proper voltage biasing of a collector. Also a device with a smaller Miller capacitance could be selected. In all other cases a more complicated compensation, taking the zero into account, has to be performed.

In case the ideal follower restores the original pole-zero pattern, it is replaced by a practical implementation, a CB or CG stage. In figure 2.19c this is shown. In case the practical follower does not restore the pattern while the ideal follower does, it is certain the right-half-plane zero does not cause the problems—for this reason the step with the ideal follower was introduced—but the insufficient performance of the practical follower. Appropriate measures can then be taken.

## 2.10 Step 6: Biasing

The biasing step starts when the small-signal design is completed. This means that until now, the small-signal models have been used for the devices and the deviations from this model caused by the non-linearity of active devices was dealt with as distortion or clipping. Biasing values like a collector current have been used already in the small-circuit design, but not as actual bias currents and voltages actually occurring in the circuit. They appeared as parameters that set the value of the small-signal parameters. For example, the small-signal transconductance $g_m$ of a bipolar transistor is related to a "parameter" $I_c$ with the name collector current. In the small-signal model no collector current$I_c$ is found. During the small-signal design the only thing that matters is to know that when the bias parameters have the correct value, the small-signal model is valid, in principle only in the bias-point itself, but allowing for acceptable errors (distortion) the model is valid in a certain range around this bias point. During the biasing step, the actual transistor, being a non-linear device, is combined with bias sources in such a way that in a specified signal range it behaves sufficiently similar to the small-signal model used in the previous design steps.

In figure 2.20 the small-signal model of a MOSFET has been depicted. Apart

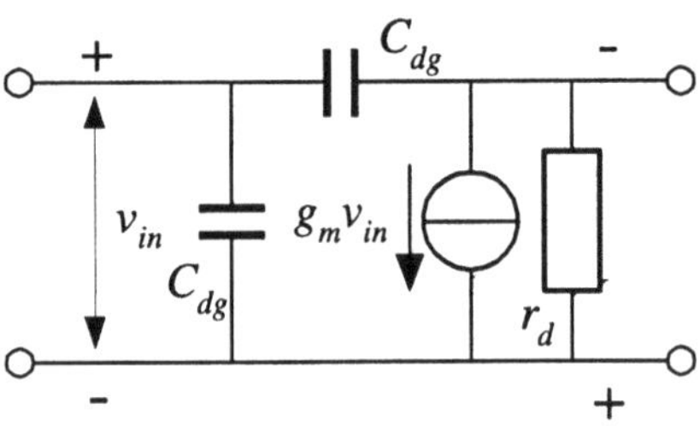


*Figure 2.20.* The small-signal model for a MOSFET .

from some passive components, there also is a voltage-controlled current source. This source can generate power at any frequency completely under control of the signal at the controlling port. This can be a completely linear circuit.

When precisely compared to the behavior of a practical device it becomes clear that this small-signal model cannot be the model for just the MOSFET. A practical MOSFET contains no sources. It cannot generate power at any frequency under control of the gate voltage. Also, the most dominant brand of sources available in practice are uncontrollable DC sources. So, a direct practical implementation of a controlled source is not feasible. To generate power at any desired frequency with only DC sources available requires a non-linear device, because only with a non-linear system energy can be exchanged between different frequencies. A practical circuit that can approximate the behavior of the small-signal model must comprise at least one non-linear device and one DC-source. As mentioned in section 1.5.2, also an offset makes a system non-linear, so to make the desired practical circuit meet the small-signal requirement, also off-set sources are needed. Figure 2.21 shows the complete practical circuit that has a behavior close enough to the behavior of the small-signal model shown in figure 2.20 in a limited signal range. The components are:

1. the non-linear device: the MOSFET;
2. the DC-source that supplies the power;
3. the DC-sources that remove the offset and define and maintain the operating point.

Two of the DC sources are controlled. Note that even a MOSFET has a controlled DC current source at its input. Even when a nominal value of zero is expected, this does not mean that the control can be deleted. The details on this topic are found in chapter 8.

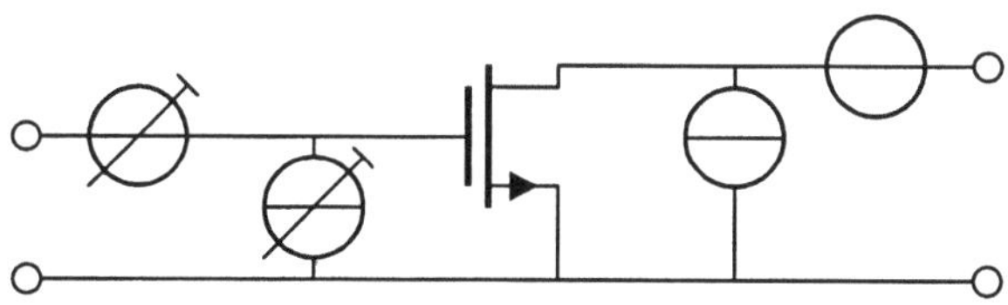

*Figure 2.21.* A biased MOSFET .

When all small-signal models have been replaced by transistor-source combinations, the number of DC-sources can be reduced by shifting them through the network and merging them. Voltage sources that can be shifted into grounded branches form the supply voltage. *So the supply voltage is not given as a specification, but found from the required performance.* When only a lower a supply voltage is allowed, it is easy to understand that then the required performance can not be obtained. Still, since it is known where the voltage sources originated, it is possible to select which performance aspect will be reduced to be able to meet with the supply voltage constraints. Sometimes it is also possible to reduces the necessary supply voltage by changing the type of some of the devices, e.g. by replacing some N-MOST transistors for P-MOST transistors.

Of course, first ideal voltage and current sources are used for biasing. During this design stage, the functioning of bias loops can be evaluated and frequency compensation of each bias loop can be performed. After this, one by one the ideal sources are replaced by practical implementations, to see what influence each source has on the performance.

Chapter 8 will deal with the complete biasing theory.

## 2.11 A note on current conveyors

In this book, apart from this section, no current conveyors will be found. Current conveyors can be seen as basic building blocks (atoms) that are used to construct circuits. The same holds for nullors. In this chapter, nullors have been introduced as the power and gain providing atoms for the synthesis of amplifiers. Both the nullor and the current conveyor are sufficient by themselves to cover the complete synthesis space of electronic circuits. It is possible to build a synthesis method based on either one of them. When one of them is chosen as the atom, the other can be constructed with it. In figure 2.22a, it can be seen how a current conveyor is constructed from a nullor by applying the appropriate feedback. In figure 2.22b, it is shown how a nullor is constructed from two current conveyors. It is obvious that it is not necessary to introduce both of them as an atom in a design theory.

It depends on the demands of a designer which element is chosen as an atom. Surely both elements have their advantages and their disadvantages, and it is not easy to rank them. A big advantage of a current conveyor is that the most common active devices presently available in technology all resemble current conveyors. Both bipolar transistors and MOSFETs easily match to the current conveyor. When a circuit has been synthesized with current conveyors, it therefore seems to be not too difficult to translate this circuit into hardware. When a circuit has been designed with nullors as atoms, often constructions like that given in figure 2.22b are used before the circuit can be translated into hardware. This implies that the center node, which connects the "X" terminals of the current conveyors in figure 2.22b, remains unnoticed during the first stages of the design. It is hidden within the nullor. Still, this node can be "an input for trouble" in the practical circuit. This node is usually involved in the biasing of a circuit, making it a signal node too, especially when there is a special biasing loop. Choosing the nullor as the atom inevitably introduces this problem, and one might think it is best to forget about nullors and to start designing with current conveyors as atoms. Also the fact that the present devices behave like current conveyors might suggest into this direction. However, appearances are deceptive. The devices are poor performing current conveyors and for high performance applications they will not be able to do the job on their own. It is not wise to mix different hierarchical levels, the system level and the device level. To find new circuits, new applications and to drive technology, the choice for an atom on the system level should never be dictated by technology. The most convenient atom for the system level design should be chosen and this is beyond any doubt the nullor.
Therefore, in this book the nullor has been chosen as the atom. It has been chosen because of its largest advantage over the current conveyor: it is more general and therefore matches better to the creativity of the designer.

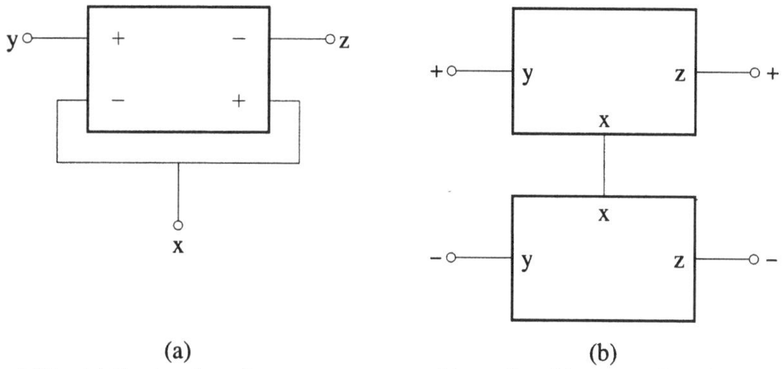


*Figure 2.22.* (a) Construction of a current conveyor with a nullor, (b) construction of a nullor with two current conveyors.

Looking at figure 2.22a, it can be seen that a current conveyor can be constructed from a nullor, by applying feedback to it. And this is an important difference between the two elements. The current conveyor can be seen as an element with an internal feedback loop and consequently an internal loop gain. And this is where the problem pops up in practice. As long as the loop gain in a device that is used to implement a current conveyors is sufficient, there is no problem in translating designs into hardware. The problem occurs when the loop gain is not sufficient, which tends to be in high-performance designs. It is not a big problem to increase the loop gain in the case of figure 2.22a, since only the gain of the nullor implementation has to be increased. As always, when an implementation of a nullor fails to do the job, only its gain needs to be adapted and the problem is solved. However, when a device does not operate as a current conveyor within specifications, there is no simple way to increase its internal loop gain. It is given by technology. Then more devices are needed to do the job. When this happens, the distinction between "true" current conveyors and constructions shown in figure 2.22a, becomes very vague. Then the nullor approach is "loosely" followed to construct better current conveyors. Then these conveyors are used to constructs circuits. There is a great risk this results in sub optimal circuit! In this situation, it is of course better to directly construct the circuit with the nullor approach. Also, when the current conveyors are implemented with more than one device, internal nodes are created, similar to the internal X-node of figure 2.22b.

Considering these practical conditions, it is obvious that the nullor is the best choice as the atom for a clear, insight-providing design strategy. It will be shown later that the awareness of a potential "X-node problem", makes it, if it does occur, easier detectable and solvable.

## 2.12 Exercises

### Exercise 2.1

In the figure given below, fig.2.23, the gain A of an opamp (without feedback) is depicted. The gain equals 140 dB ($A_0$) for the relatively low frequencies,

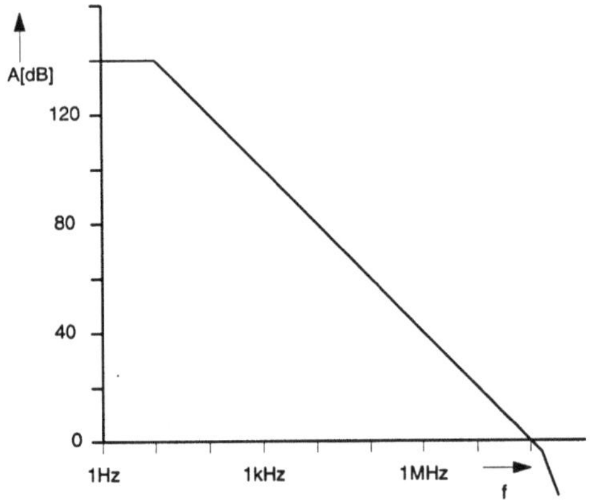


*Figure 2.23.* The gain of an operational amplifier as a function of frequency

whereas the gain drops with 20 dB/dec beyond 10 Hz. The gain is reduced to 1 at 100 MHz ($f_T$, the transit frequency). When this operational amplifier is used in a negative-feedback structure, the figure can be used to determine the bandwidth of the closed-loop amplifier.

With this operational amplifier a negative-feedback amplifier is made with a gain of 10.

1. Determine with the help of this plot the bandwidth of the negative-feedback amplifier, using the operational amplifier as the active part, for a 90% accurate transfer.

2. Repeat this, but now for a 99% accurate transfer.

3. When you, as a designer, have the ability to increase one of the two variables, $f_T$ or $A_0$, in order to increase the accuracy of the feedback amplifier, which one should you choose and why?

## Exercise 2.2

A transducer supplies a signal current $i_s$ equal to:

$$i_s = X \cdot 10\,\text{pA}, \tag{2.11}$$

in which $X$ is the information carrying quantity. The output impedance of the transducer is approximately 100 k$\Omega$.
This signal has to be amplified to a signal, $v_s$:

$$v_s = X \cdot 1\,\mu\text{V} \tag{2.12}$$

The load of the amplifier is approximately 1 k$\Omega$.
Below, 6 basic configurations are depicted which are proposed for the required amplifier. For 3 configurations the Norton-Thevenin transformation is applied to the source.
Motivate for each of these 6 configurations whether it is a good choice for the required signal processing function, or not.

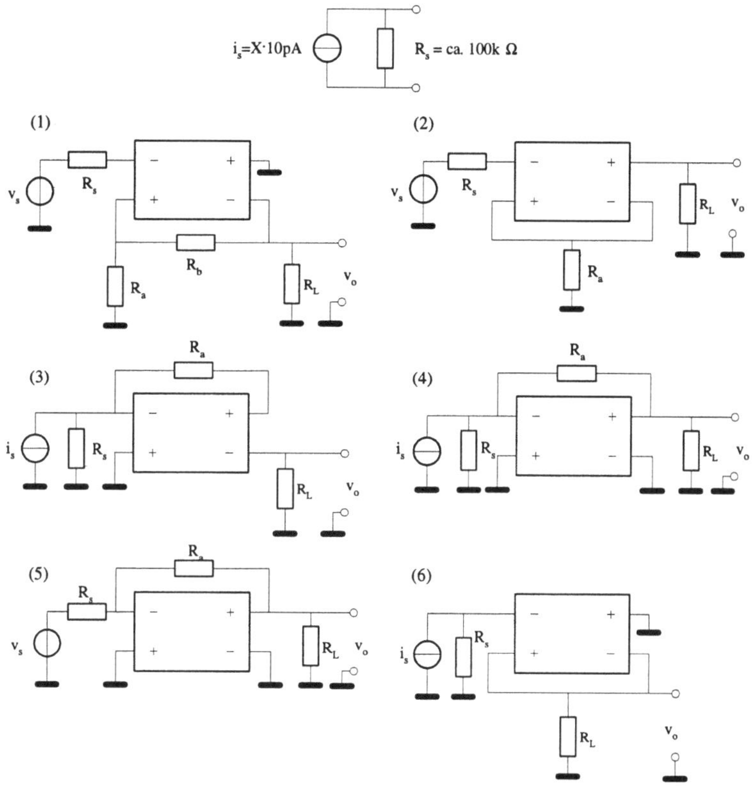

## Exercise 2.3

1. Using a nullor, design a current follower ($D$=1) and a voltage follower ($A = 1$) by applying unity feedback to this nullor.

2. Use one CE-stage, as shown in fig.2.24, as a nullor implementation. Compare the resulting circuits to the CB and the CC stage. What conclusions can be drawn from this?

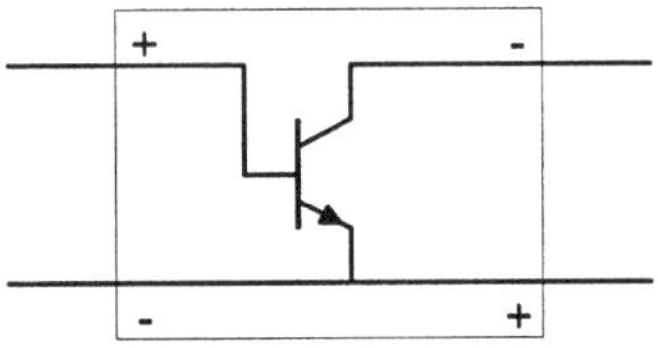


*Figure 2.24.* A single transistor nullor.

## Exercise 2.4

Small-signal diagrams are used to analyze the behavior of, for instance, nonlinear transistors when excited by a relatively small signal.

1. Derive for relatively low frequencies (DC) the small-signal diagram of a PNP bipolar transistor.
2. Compare it to the low-frequency (DC) small-signal diagram of a NPN bipolar transistor.
3. What is your conclusion? Explain your findings.

## Exercise 2.5

Figure 2.25 depicts three nullor implementations.

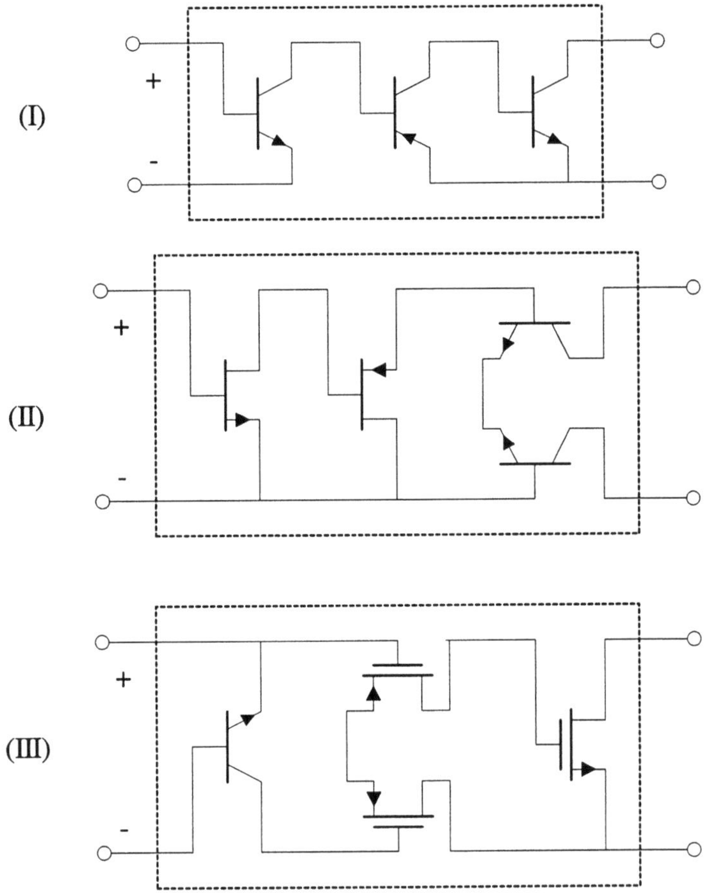


*Figure 2.25.* Three arbitrary nullor implementations.

1. Which of the implementations of figure 2.25 are correct and which incorrect? Motivate your choices.
2. Indicate in the figures the intended output polarities. A reference polarity for the input is given in the figure.

## Exercise 2.6

Figure 2.26 depicts two, two-stage nullor implementations realizing an inversion between input and output, as indicated.

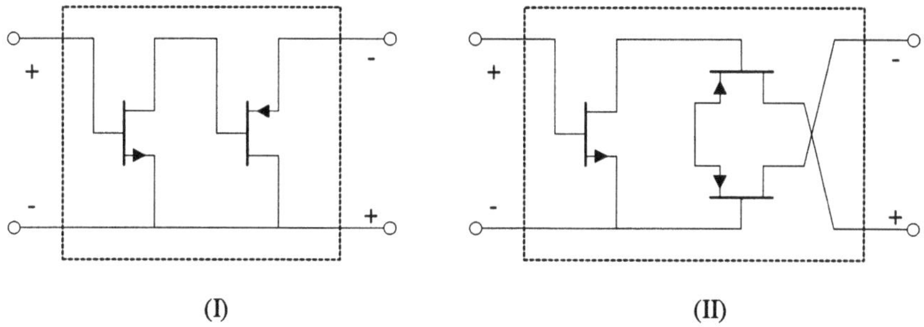


*Figure 2.26.* Two two-stage nullor implementations realizing inversion between input and output.

1. Are both implementations correct implementations?
2. No: describe what is wrong and indicate what the consequence is on the performance of the nullor implementation.
3. Yes: which one do you prefer? Motivate!

## Exercise 2.7

Given the four amplifiers in figure 2.27.

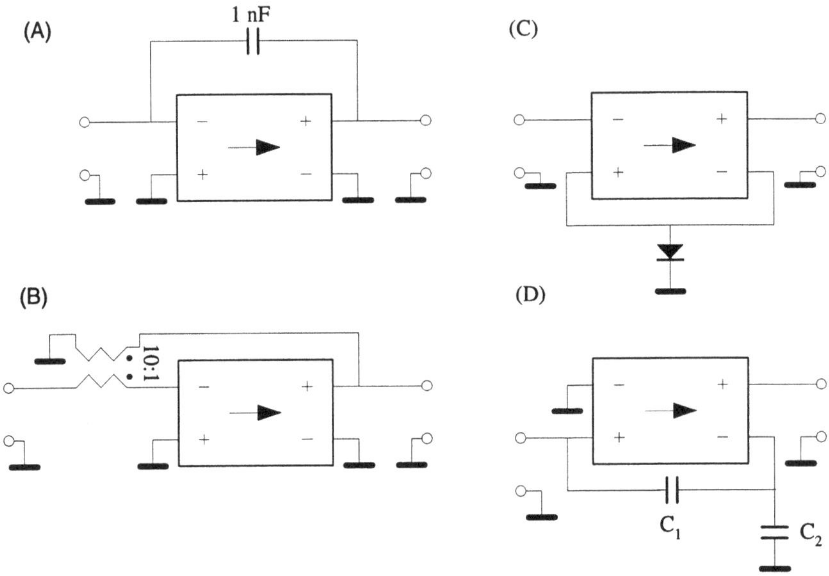


*Figure 2.27.* Four feedback amplifiers.

1. Determine the ideal transfer, $A_{t\infty}$, of each of these amplifiers by using the nullor constraints.
2. Give for each of the nullors a two-stage implementation using MOSFETs
3. Derive for each of the implementations the loop gain, $L(s)$, and the direct transfer, $A_{t0}$.

## Exercise 2.8

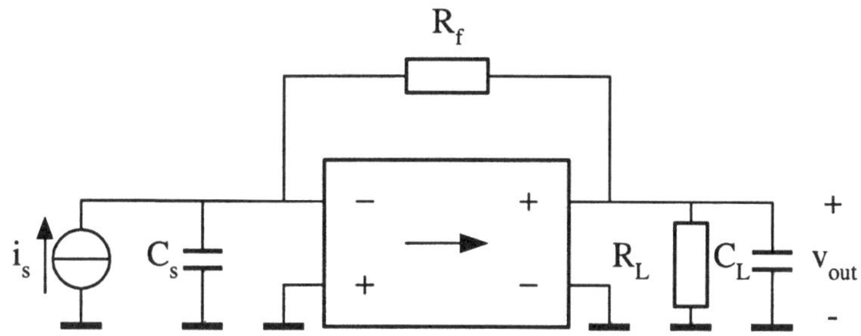


*Figure 2.28.* A transimpedance amplifier.

The above depicted transimpedance amplifier needs to be connected to a current signal source. The output impedance of the signal source is capacitive and not accurately known. This is because the amplifier is intended to be connected via a long shielded wire to a sensor. The exact length of the wire is determined only at the moment of the application of the amplifier. Thus the value of $C_s$ is not known.

1. What is the effect of the uncertainty in $C_s$ on the transfer of the amplifier when the active part is a nullor?
2. What are the consequences of the uncertainty in $C_s$ when the nullor is implemented by means of transistors?
3. What type of transistor stage (CE, CB or CC) is the most appropriate one, in order to minimize the effect of the uncertainty in $C_s$ the most?
4. What is the effect on the three quality aspects: noise; distortion and bandwidth for the chosen stage?

# 3

# NEGATIVE FEEDBACK

## 3.1 Introduction

The goal in electronic circuit design is to realize a specified transfer function. Three fundamental signal-processing quality aspects are of importance:

- noise
- bandwidth
- distortion

In order to attain an implementation that is optimal to all three aspects quickly, a systematic design method has to be followed. According to chapters 1 and 2, this method consists of a top-down method, in which the three design aspects are orthogonalized. In this chapter negative feedback is dealt with as a method for realizing the accurate transfer. The asymptotic-gain model is discussed, which models the amplifier as a combination of a nullor, or a circuit implementation of it, and an accurate passive feedback network. This division orthogonalizes the design of the active circuit and the design of the feedback network, that realizes the accurate transfer. This makes the asymptotic-gain model well suited for synthesis.

Ideally, the gain part of a negative-feedback amplifier is a nullor, so it has infinite amplification. In practice, when the nullor is implemented with transistors, an error is made. Then the loop gain is finite and is likely to show some non-linearity. However, when the loop gain is still high enough, the dominance of the feedback network in determining the transfer may still be sufficient.

Ideally, the transfer of the overall amplifier is fully determined by the feedback network. The feedback network may consist of ideal elements: gyrators and transformers. They consume no power and produce no noise. With combination of these elements all possible single-loop negative-feedback amplifier

configurations can be realized as can be seen in figure 2.7. The transfer can be both inverting and non-inverting, depending on the connection of the feedback network. So, totally eight configurations exist.
By using the more practical passive components, not all amplifier combinations are possible with only one nullor. As can be seen in figure 2.8, the voltage amplifier and the current amplifier have to be non-inverting and the other two must be inverting. The use of "indirect feedback" and "active feedback" may yield feasible results but this is beyond the scope of this book. If only passive feedback networks and one nullor are to be used, only four amplifiers out of eight types can be realized. And, of course, the impedances in the network load the nullor circuit and may introduce noise or increase the noise contribution of existing sources.

## 3.2 Accurate transfer

In chapter 2 it has already been shown that negative feedback is the only strategy that has the potential of synthesizing amplifiers of which the accuracy is completely dominated by passive components. This chapter will describe in more detail the procedure that makes use of the asymptotic-gain model to do this.

### 3.2.1 Negative feedback

Negative feedback has best accuracy, because an accurately reduced copy of the output signal is compared with the original input signal. Any deviation from the intended output signal is found at the input of the nullor circuit, which adapts its output signal to the required value. The feedback network determines the transfer function completely when the nullor circuit has a perfect behavior. It only delivers the required signal power. Therefore, the nullor circuit needs not to be accurate, but only needs to have large gain, ideally an infinite gain. The feedback network consists of passive elements only, which must have high accuracy.

### 3.2.2 Adaptation to source and load

In electronic systems, the information can be coded in three domains:

- current
- voltage
- power

Both the source and load have the signal coded in one of these domains, resulting in nine different types of amplifiers (18 when the inverting and non-inverting possibilities are taken into account). All these types can be realized by negative-feedback amplifiers. The feedback network of the amplifier must be chosen such that the amplifier has an ideal match with respect to the source and load conditions, as explained in the following.

#### 3.2.2.1 Source condition

It is important to prevent the source impedance from having influence on the transfer. So, a voltage source requires an amplifier with an infinite input impedance, while a current source requires an amplifier with a zero input impedance. Via these ideal terminations of the sources, the influence of the source impedance is prevented. When the source is not terminated correctly, the source impedance is somehow found in the transfer of the amplifier. Both impedances can be inaccurate or even nonlinear, thus resulting in performance degradation, *even when a nullor is used as active element*. This phenomenon can be used to do a quick and simple evaluation to see if the correct amplifier configuration has been chosen. A nullor should make a circuit perfect. If in a circuit the active part intended for amplification is replaced by a nullor and the circuit does not become perfect, the circuit is not correct for the application[1].

Suppose the signal of a voltage source with a source impedance of 1 kΩ is amplified by an ideal current amplifier (with nullor). The input impedance of the current amplifier behaves like a short circuit, so a signal current flows with a magnitude that is set by the source impedance. When this source impedance is inaccurate or non-linear, the resulting current is a much worse representation of the information that was originally contained in the signal voltage of the source. The lost information cannot be recovered, not even by a perfect current amplifier.

#### 3.2.2.2 Load condition

The load condition for amplifiers is similar to the source condition. The signal to be delivered must be coded in the required domain, otherwise the non-ideal load impedance determines the signal transfer too, even when the amplifier itself is a perfect one.

## 3.3 The Asymptotic-Gain model

Several models exist for describing feedback systems, of which Blacks model is the best known. However, Blacks model assumes ideal unilateral transfers and no loading effects at the input and the output, which is far from practice in electronics. A better model for describing electronic feedback systems is the asymptotic-gain model. The starting point to come to the asymptotic-gain model is the superposition model, shown in figure 3.1. The signals in this model

[1]This is a general feature of the introduction of a nullor and not limited to just amplifiers.

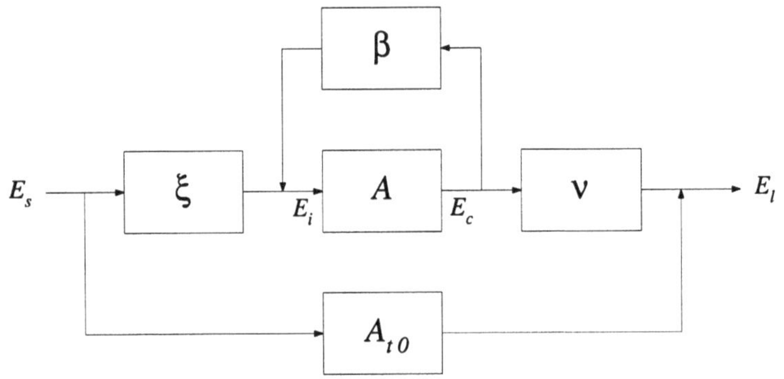


*Figure 3.1.* The superposition model.

are linear combinations of the signal from the source $E_s$ and an arbitrarily chosen controlled source $E_c$ [2]. As it is not important whether the signals are voltages or currents, they are denoted by $E$. For the model the following holds:

$$E_l = A_{t0}E_s + \nu E_c, \quad (3.1)$$

$$E_i = \xi E_s + \beta E_c, \quad (3.2)$$

in which:
- $E_l$ is the signal at the amplifier output
- $E_i$ is the control input of the controlled source
- $E_s$ is the signal at the amplifier input
- $E_c$ is the signal from the controlled source

The output of the controlled source depends on its control signal according to:

$$E_c = AE_i \quad (3.3)$$

$A$ is the gain of the controlled source, $\beta$ represents the feedback.

The transfer $A_t$ of the amplifier is defined as the ratio between the load and source signal:

$$A_t = \frac{E_l}{E_s} = A_{t0} + \nu\xi\frac{A}{1 - A\beta} \quad (3.4)$$

The product

$$L = A\beta \quad (3.5)$$

is called the *loop gain.* The higher the loop gain, the more dominant the feedback $\beta$ becomes on the overall transfer. For the loop gain approaching infinity the asymptotic gain, $A_{t\infty}$, is found as:

$$\lim_{L\to\infty} A_t = A_{t\infty} = A_{t0} - \frac{\nu\xi}{\beta}. \quad (3.6)$$

[2]Of course, we are thinking of a nullor here.

The factors $\nu$ and $\xi$ represent a non-ideal coupling between the input source and the control input of the controlled source and a non-ideal coupling between the controlled source and the output, respectively. Normally they equal unity. When an incorrect amplifier type is chosen as discussed before, source or load impedances may cause them to have a different value. Usually that results in inaccuracy[3]. Therefore, to keep the equations simple, the substitutions $\nu = 1$ and $\xi = 1$ are made. Then the expression for the transfer $A_t$ yields:

$$A_t = \frac{E_l}{E_s} = A_{t\infty}\frac{-L}{1-L} + \frac{A_{t0}}{1-L}. \tag{3.7}$$

For an infinite loop gain, the second term disappears and even in practical amplifiers the first term tends to dominate the second one. Note that $A_{t0}$ is also in the expression for $A_{t\infty}$, so loop gain does not remove the influence of a direct transfer! Still, for high loop gains the transfer can be approximated by:

$$A_t = A_{t\infty}\frac{-L}{1-L}. \tag{3.8}$$

So, when there is infinite loop gain—which would be there if the controlled source was a nullor—the transfer of the amplifier is:

$$A_{t\infty} = -\frac{1}{\beta}, \tag{3.9}$$

an expression that can realized with just passive components, since $\beta$ is smaller than one for an amplifier with a gain larger than one. It can be seen that the design of amplifiers can be separated into two orthogonal design steps:

- design of the asymptotic gain $A_{t\infty}$;
- realization of a large (infinite) loop gain $L$,

or in circuit terms:

- design of the feedback network, assuming the presence of a nullor;
- design of a circuit with a behavior that approximates the nullor sufficiently.

The model that makes use of the asymptotic gain, $A_{t\infty}$, is called the asymptotic-gain model. The model takes implicitly the source and load impedances into account. Their influence on $A_{t\infty}$ is modelled via $\xi$ and $\nu$, respectively. By choosing the appropriate feedback network they can be made one and then the source and load impedance do not have an effect on the ideal transfer. This

[3]There are also negative-feedback amplifiers with indirect feedback, that may have a $\nu$ or $\xi$ not equal to unity by nature. So not in all cases inequality to unity indicates an incorrect choice of the amplifier type.

means that the amplifier implicitly has the correct input and output impedances. The influence of the source and load impedance on the accuracy is taken into account via the loop gain. When the loop gain is sufficiently large (infinite in the ideal case) the transfer function is equal to $A_{t\infty}$: the signal is transferred according to specifications.

## 3.4 Transfer of an amplifier

The transfer of an amplifier is determined by the feedback network. The feedback network and the nullor implement a strict relation between the currents and voltages at the input and output ports of the amplifier. The behavior of the amplifier is fully determined by four (anti causal) transfers, given by the chain matrix:

$$\begin{pmatrix} v_i \\ i_i \end{pmatrix} = \begin{pmatrix} A & B \\ C & D \end{pmatrix} \begin{pmatrix} v_o \\ i_o \end{pmatrix} \tag{3.10}$$

in which the chain parameters are given by:

$$A = \left.\frac{v_i}{v_o}\right|_{i_o=0}, B = \left.\frac{v_i}{i_o}\right|_{v_o=0}, C = \left.\frac{i_i}{v_o}\right|_{i_o=0}, D = \left.\frac{i_i}{i_o}\right|_{v_o=0}. \tag{3.11}$$

The reciprocal values of these parameters are the transfer parameters, and give the transfer from the input to the output of the two port.

$$voltage - gain\ factor \quad = \quad \mu = \frac{1}{A} = \left.\frac{v_o}{v_i}\right|_{i_o=0} \tag{3.12}$$

$$transadmittance\ factor \quad = \quad \gamma = \frac{1}{B} = \left.\frac{i_o}{v_i}\right|_{v_o=0} \tag{3.13}$$

$$transimpedance\ factor \quad = \quad \zeta = \frac{1}{C} = \left.\frac{v_o}{i_i}\right|_{i_o=0} \tag{3.14}$$

$$current - gain\ factor \quad = \quad \alpha = \frac{1}{D} = \left.\frac{i_o}{i_i}\right|_{v_o=0} \tag{3.15}$$

By applying negative feedback, it is possible to accurately determine one of these transfer parameters.

## 3.5 Nullor feedback networks

The asymptotic-gain model shows the possibility for accurate amplification, if a high-gain stage and an accurate passive feedback network are available. In that case the transfer function approaches $A_{t\infty}$. The ideal circuit-theoretical element that fulfills the high-gain requirement is the nullor as discussed in section 2.2.2. The nullor is an ideal element with infinite gain, infinite bandwidth and infinite output capability, without addition of any noise. When a design, in which a nullor is incorporated, is not functioning according to specifications already, it is sure that it will never meet the specifications irrespective of the successive design steps. It is sure that the behavior of any practical circuit that is designed to substitute the nullor is worse than the ideal nullor behavior. If it happens that a practical circuit yields better results than a nullor would, probably a serious error has been made in the choice of the circuit topology. The correct choice would give a better ideal circuit with the nullor and also a better "practical" circuit.

## 3.6 Feedback networks

The source and load representation of the information can be voltage, current or power. In case the information is coded in power, the voltage and the current have a strict relation to each other. To obtain an accurate relation between the source signal and the load signal, an accurately reduced copy of the output signal must be compared to the input signal. Output series feedback is used when the output of the amplifier is a current. Output shunt feedback is used when the output of the amplifier is a voltage. To allow the nullator to make a comparison between the output signal and the input signal, the input shunt feedback is used for current comparison and the input series feedback is used for voltage comparison.

### 3.6.1 Transformer and gyrator networks

The use of gyrators and transformers constitutes a fully ideal feedback network. No noise is introduced, no bandwidth limitations imposed and no distortion is present. They seem to be the ideal candidates for implementing the feedback network. And indeed, transformers nowadays can be attractive feedback components, though integrated transformers only become feasible above a few hundreds of MHz. At low frequencies they are frequently considered to be too bulky. Unfortunately gyrators are just ideal circuit-theoretical elements that can not exist in practice. The gyrator function can only be simulated by electronic components, thereby fundamentally introducing noise and a limited dynamic range.

Still, transformers and gyrators as ideal circuit-theoretical non-energetic two-ports can be used to design a desired transfer function. In figure 3.2, two gyrators and two transformers are used to realize all possible transfer functions, yielding a relation between current, voltage or power at the input and output of the amplifier. The designer can choose one, two, three or four feedback loops to realize the required signal transfer accurately, according to the table 3.2. Single loop amplifiers –the main topic of this book– only allow for current and voltage signals; no power transfer is possible, because for this at least two loops are necessary.

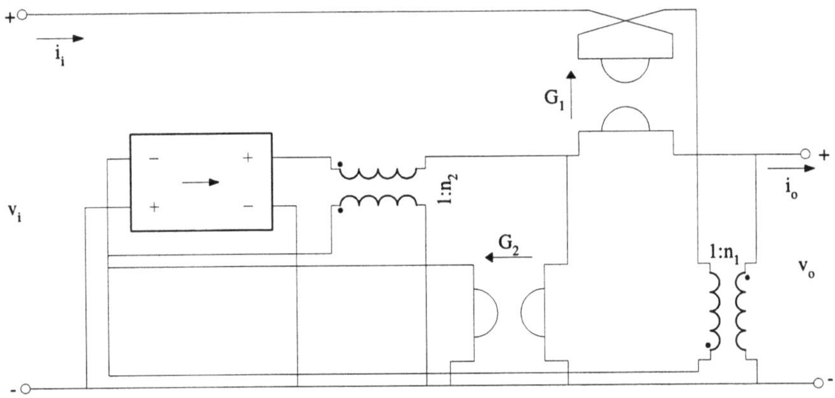


*Figure 3.2.* All possible ideal feedback networks.

| | output voltage | output current | output power |
|---|---|---|---|
| input voltage | $n_1$ | $G_1$ | $n_1$ and $G_1$ |
| input current | $G_2$ | $n_2$ | $n_2$ and $G_2$ |
| input power | $n_1$ and $G_2$ | $G_1$ and $n_2$ | $n_1, n_2, G_1$ and $G_2$ |

*Table 3.1.* Required feedback loops as a function of input and output quantities.

### 3.6.2 Passive single-loop feedback by one port elements

When transformers and gyrators are not available, the feedback network can employ resistors, inductors and capacitors to determine the overall transfer function. The configurations that are possible are given in figure 3.3. According to its definition (section 1.4.2) the transfer function of an amplifier is frequency independent, so the transfer of the feedback network must also be frequency independent. When the transfer depends on the ratio of two impedances, the two impedances should have an equal frequency behavior. When resistors are used, they introduce noise, which have to be taken into account when optimizing the noise performance of the amplifier.

Of course, when using only resistors, capacitors and inductors it is not possible to realize each type of transfer both inverting and non-inverting, but this hardly causes a real problem. Still, when it does, using indirect feedback or using active feedback it is possible to realize both inverting and non-inverting transfer functions for every amplifier type. Both methods will only be treated briefly in the following.

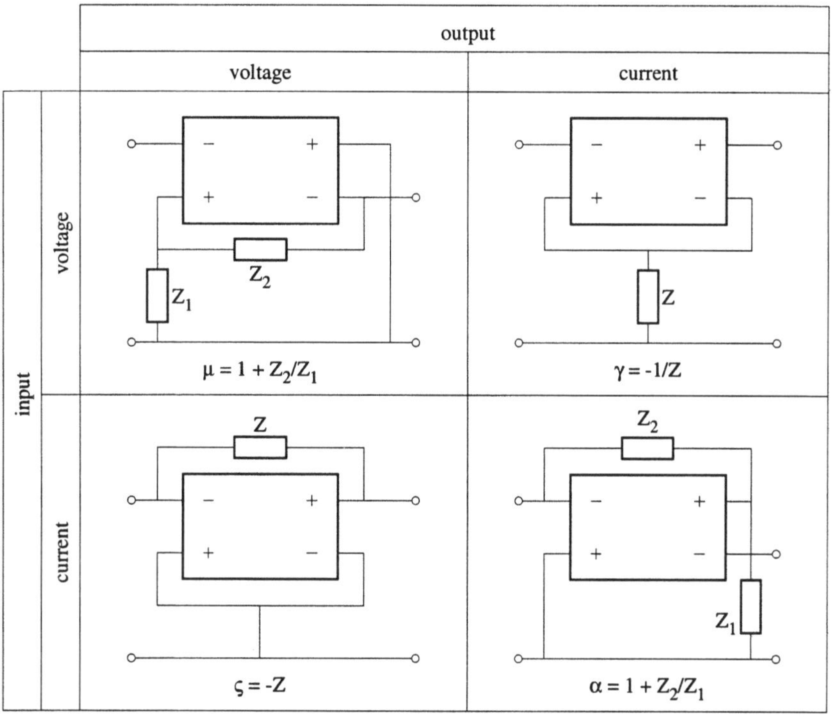


*Figure 3.3.* The four possible feedback configurations that can be realized with a single feedback network using impedances

### 3.6.3 Indirect feedback

For indirect feedback the load signal is not directly sensed, or the input signal is not directly compared to the signal coming from the feedback network. For indirect feedback at the output *a copy* of the load signal is used as input for the feedback network. Especially in low voltage applications (and) when the output signal is a current this method could be applied. Because the feedback signal is only a copy of the load signal, the load-signal accuracy is limited by the accuracy of the copying. The copy can be an inverted version of the original signal, e.g. via the use of a current mirror, thus making it possible to design both inverting and non-inverting amplifiers of every type. Similarly, indirect feedback at the input uses a copy of the input signal to compare with the signal which comes from the feedback network.

### 3.6.4 Active feedback

Active feedback amplifiers make use of active components in the feedback network to realize a specified transfer function. The active element can be inverting. This again makes it possible to design both inverting and non-inverting amplifiers of every type.

## 3.7 Example: asymptotic-gain model

In this example it is shown how the asymptotic gain and the loop gain can be calculated using the asymptotic-gain model. The amplifier depicted in figure 3.4 is studied. The gain of the amplifier is set by resistor $R_{feedback}$. The

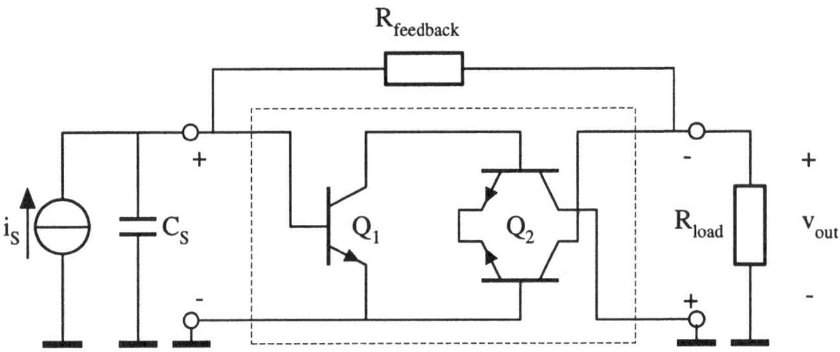


*Figure 3.4.* The signal diagram of a transimpedance amplifier with an two-stage nullor implementation.

input signal is a current $i_s$ and the output signal is a voltage $v_{out}$. The nullor is implemented with two stages. The first stage is a CE-stage and the second stage is a differential pair in order to guarantee a negative loop gain. The detailed small-signal circuit of this amplifier is given in figure 3.5. For the differential

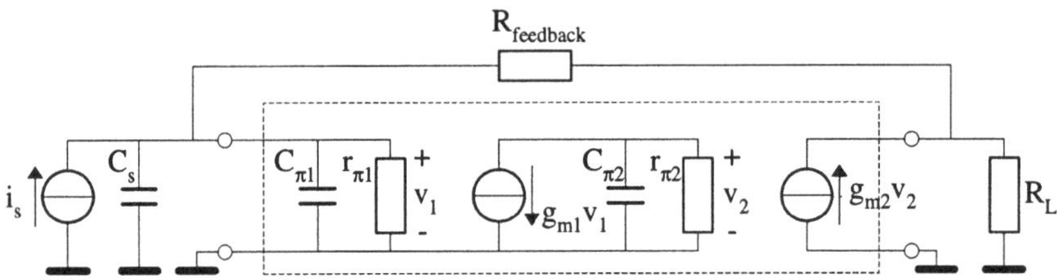


*Figure 3.5.* The small-signal diagram of the amplifier of figure 3.4.

pair, a simplified small-signal model is used. In figure 3.6 the relation between this simplified model and the two-transistor model is shown. For the case that both base currents are equal in the differential pair, no current flows through the branch indicated with X. So, it can be removed from the diagram. The circuit that remains is easily reduced to the diagram at the right side of figure 3.6. For the resulting diagram, the input and output port do not have a terminal in common and thus an inverting as well as a non-inverting transfer can be realized

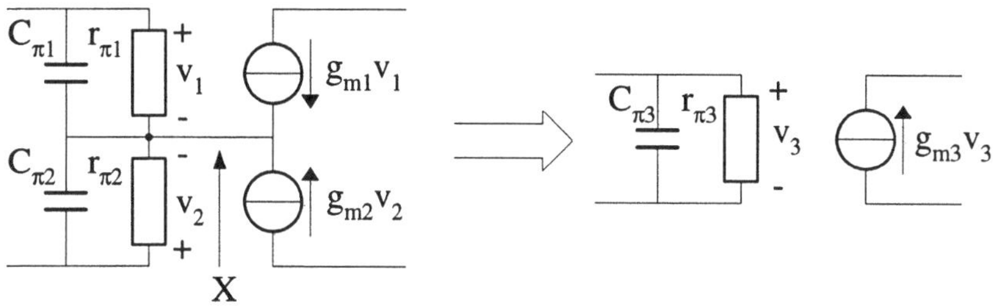


*Figure 3.6.* The small-signal diagram for a differential pair.

with this stage. The relations between the elements are:

$$r_{\pi 3} = r_{\pi 1} + r_{\pi 2} = 2r_{\pi} \tag{3.16}$$

$$C_{\pi 3} = \frac{C_{\pi 1} C_{\pi 2}}{C_{\pi 1} + C_{\pi 2}} = \frac{C_{\pi}}{2} \tag{3.17}$$

$$g_{m3} = \frac{g_{m1} g_{m2}}{g_{m1} + g_{m2}} = \frac{g_m}{2} \tag{3.18}$$

in which the last equalities hold for the case of two identical transistors biased at the same current.

The model shown in figure 3.5 is used to calculate the asymptotic gain and the loop gain.

The asymptotic gain is found for infinite loop gain, i.e. the active part has nullor properties. The nullator constraints are imposed on the input port of the active circuit. This implies that the input current of the active part is zero and thus all the signal current flows through the feedback resistor. As also the input voltage of the active part is zero, the output voltage is found to be $-R_{feedback} i_s$. Note that the only correct location in the network to impose the nullator conditions on is the input of the active circuit. Other nodes inside the active circuit do not exist on a higher hierarchical level, since the nullor at that level only has an input and an output port defined. Any loop gain calculated via another set of nodes in the active circuit would not be defined at the higher hierarchical levels. Obviously the loop gain should be defined in the same way at all hierarchical levels. *So there is only one loop gain defined for every feedback circuit.*[4] Imposing the nullator constraints on the input port of the active circuit, the asymptotic gain is easily found to be:

$$A_{t\infty} = -R_{feedback} \tag{3.19}$$

[4]Sometimes, textbooks state that depending on the choice of ports, the expression for the loop gain could differ, so different loop gains exist for the same circuit. This is an incorrect statement based on an improper definition of the loop gain.

For calculating the loop gain, $L$, the loop has to be "broken" somewhere, and the gain should be calculated between the two open ends, assuming that the input signal ($i_s$) is zero. Of course, care should be taken that by breaking the loop the impedance at non of the nodes should change. For instance, when the loop would be broken by cutting $R_{feedback}$ from the input, the impedance seen at $R_L$ is changed as $R_{feedback}$ is floating now. In contrast, when a controlled source is assumed to be uncontrolled, the loop is also broken. In this case the topology is not changed and thus the loop gain is calculated exactly. As example, when the controlled current source with current $g_{m1}v_1$ is assumed to be uncontrolled with output current $i_x$ (see figure 3.7), the loop is broken. The loop gain is found easily now. First the transfer from $i_x$ to $v_1$ needs to be calculated. As in the real amplifier the relation between "$i_x$" and $v_1$ is $g_{m1}$, multiplying by $g_{m1}$ yields the loop gain:

$$L = \frac{v_1}{i_x} \cdot g_{m1} \tag{3.20}$$

It should be noted that the controlled source which is assumed to be uncontrolled for calculating the loop gain, cannot be chosen arbitrarily. The criterion is that by making the controlled source uncontrolled, the overall loop is really broken. This can easily be checked by making the corresponding control variable infinite. In that case the loop gain should become infinite. It may be safest to select the controlled source that is at the nullator position to be independent. This is not as strict a rule as the rule to select the location to impose the nullator constraints on, because the loop gain is a loop property that is not concentrated in one stage or location, as the nullator property is. In a cascade topology this would make any controlled source useable and the one that yields the most simple calculations can be chosen.

An incorrect choice easily made, for instance, can occur dealing with a differential pair. This is when the small-signal diagram on the left side of figure 3.6 is used and one of the two controlled sources is assumed to be uncontrolled. The two sources are correlated and this correlation must be maintained. The correct choice is using $g_{m3}$ of the diagram at the right side of figure 3.6 in which the correlation is taken into account. Errors like this are easily prevented when the circuit is inspected on a higher hierarchical level, that shows the two-ports, that are used to implement the nullor (see for example figure 2.12). All controlled sources within such a two-port are correlated.

In the example, the voltage-controlled current source of the input transistor is made uncontrolled. Figure 3.7 shows the new small-signal diagram. The first step is to calculate the transfer from $i_x$ to $v_1$. This can be done by the MNA method but also by inspection: use current division at the several nodes and the

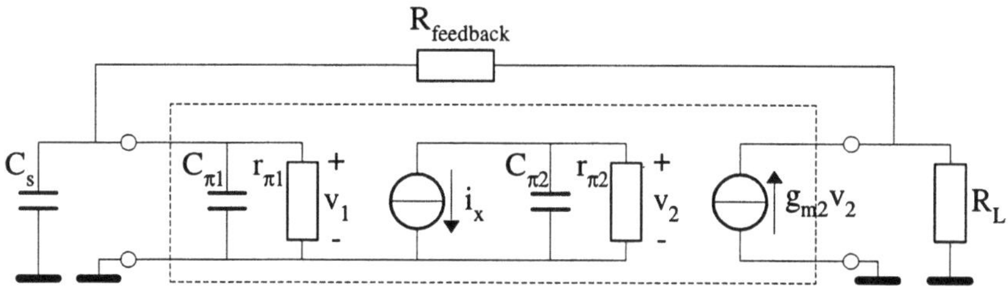


*Figure 3.7.* Small-signal diagram for calculating the loop gain; the voltage-controlled current source of the input transistor is assumed to be uncontrolled and the input signal is made zero.

gain of the transconductances:

$$V_1 = -I_x \cdot \frac{r_{\pi 2}}{1 + s r_{\pi 2} C_{\pi 2}} \cdot g_{m2} \times \frac{R_L}{R_L + R_{feedback} + \frac{r_{\pi 1}}{1 + s r_{\pi 1}(C_s + C_{\pi 1})}} \cdot \frac{r_{\pi 1}}{1 + s r_{\pi 1}(C_s + C_{\pi 1})} \tag{3.21}$$

in which $s$ is the Laplace variable. The loop gain is found by multiplying this expression by $g_{m1}$. Simplifying the resulting expression yields:

$$L(s) = \frac{-\beta_1 \beta_2 R_L}{r_{\pi 1} + R_L + R_{feedback}} \times \frac{1}{1 + s r_{\pi 2} C_{\pi 2}} \cdot \frac{1}{1 + s \frac{r_{\pi 1}(R_L + R_{feedback})}{r_{\pi 1} + R_L + R_{feedback}} \cdot (C_s + C_{\pi 1})} \tag{3.22}$$

in which $\beta = g_m r_\pi$. From this expression the DC loop gain [$L(0)$] and the poles ($p_1, p_2$) are readily found to be:

$$L(0) = \frac{-\beta_1 \beta_2 R_L}{r_{\pi 1} + R_L + R_{feedback}} \tag{3.23}$$

$$p_1 = \frac{-1}{2\pi r_{\pi 2} C_{\pi 2}} \tag{3.24}$$

$$p_2 = \frac{-(r_{\pi 1} + R_L + R_{feedback})}{2\pi r_{\pi 1}(R_L + R_{feedback})(C_s + C_{\pi 1})} \tag{3.25}$$

A detailed procedure to calculate the poles and the DC loop gain and to use them to predict and later to optimize the frequency behavior of the amplifier will be discussed in chapter 6 (the predictions) and chapter 7 (the optimization).

## 3.8 Exercises

### Exercise 3.1

With negative feedback, the quality of passive elements (accuracy) and the quality of active elements (gain) can be combined.

1. Discuss the problems of error feed forward when trying to design accurate amplification with passive and active elements.
2. Discuss the problems that arise when error-compensation techniques are used for realizing active amplification.
3. What is the essential difference between negative feedback and the two previous discussed methods?

## Exercise 3.2

Assume that the inaccuracy of the passive elements that can be used in the feedback network is 1%. The active part is implemented such that it approximates the nullor *well enough.*

1. What would you consider in this case a reasonable minimum loop gain? Motivate your selection.

2. What is the required bandwidth for the amplifier when the inaccuracy should be below 1% up to 1 MHz? Assume a second-order behavior for the amplifier.

## Exercise 3.3

Given the three amplifiers in figures 3.8 and 3.9. For each of the amplifiers,

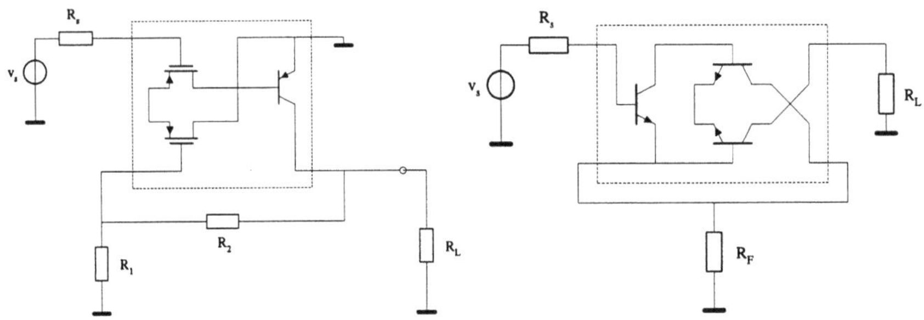


*Figure 3.8.* Amplifier 1 and 2.

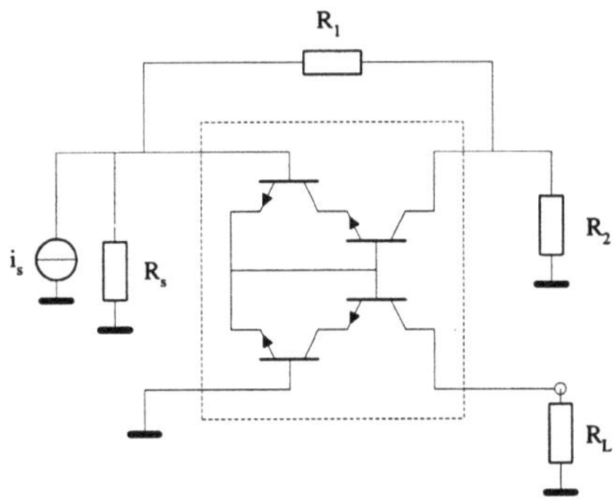


*Figure 3.9.* Amplifier 3.

calculate (using simple small-signal models, i.e. $g_m$, $r_\pi$, $c_\pi$, $c_{gs}$):

1. the Asymptotic-gain for the transfer from source to the load quantity;
2. the loop gain in terms of the small-signal parameters, source, load and feedback impedances.

## Exercise 3.4

Consider the transconductance amplifier with a capacitive load, as depicted in figure 3.10.

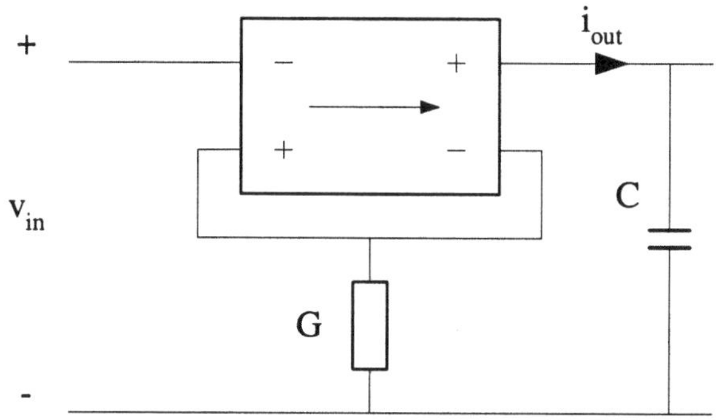


*Figure 3.10.* A transconductance amplifier.

1. Determine $A_{t\infty}$ for the amplifier.

Subsequently, assume that the nullor is approximated by a three stage implementation. The corresponding small-signal diagram is depicted in figure 3.11. For this situation the loop gain is studied.

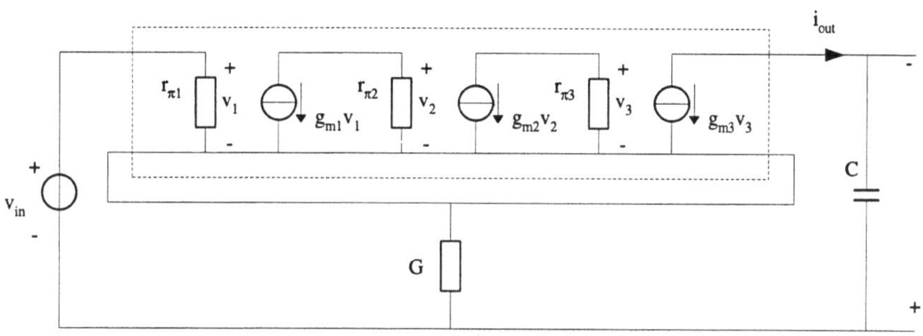


*Figure 3.11.* The small-signal diagram for the amplifier when the nullor is implemented by three stages.

2. For calculating the loop gain, the loop needs to be broken somewhere. How and where can you break the loop for the amplifier of figure 3.11?
3. What should be the polarity of a correct loop gain? Motivate!
4. What is the dimension of a correct loop gain? Motivate!
5. What is the effect of the integrator capacitor, C, on the loop gain?
6. Determine the expression for the loop gain.

## Exercise 3.5

Given a transimpedance amplifier as depicted in figure 3.12.

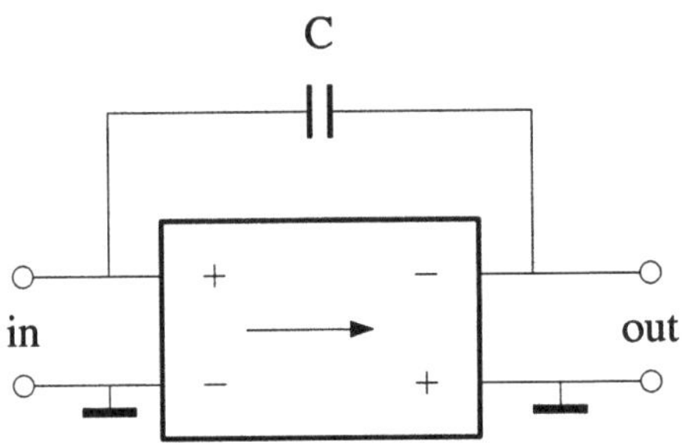


*Figure 3.12.* A transimpedance amplifier.

1. For what source and load type is this amplifier optimal? Motivate your choices.

2. Determine $A_{t\infty}$ for this amplifier.

The nullor is implemented by means of two amplifying stages. The corresponding small-signal diagram is depicted in figure 3.13. In the figure a load resistor, *R*, is added.

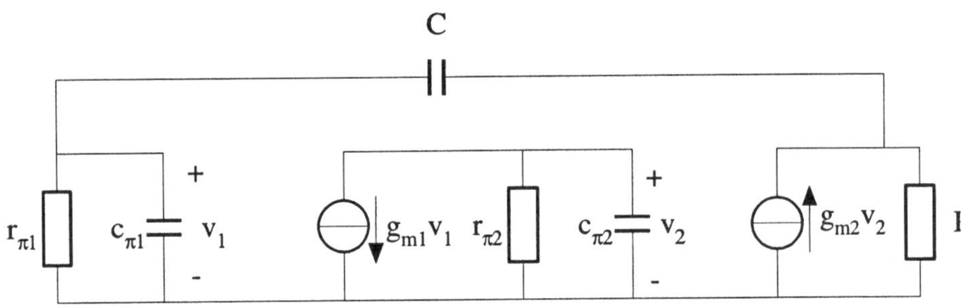


*Figure 3.13.* The small-signal diagram of the transimpedance amplifier when the nullor is approximated by two amplifying stages.

2. What is the DC loop gain of the integrator ?

3. Calculate the loop gain as a function of the frequency.

# 4

# NOISE

Noise is one of the phenomena that is ubiquitously present throughout physical mechanisms and physical systems. Such systems obey the well-predictable, deterministic rules only to a certain extend; the noise introduces slight non-predictable perturbations from it. The hissing sounds produced e.g. by radios, televisions, and telephones between messages is perhaps the most widely known example of noise. In engineering, however, the term noise is used to refer to a much wider class of very diverse random phenomena. As you may have expected already, electronic components are subjected to noise as well; their behavior is also impaired by slight non-predictable perturbations.

Returning to our amplifier design procedure, what is the effect of electronic circuit noise on the operation of electronic amplifiers? The most general answer, given in chapter 2, is that it limits the amplifier information handling capacity. More specifically, this means that it defines a lower bound on the range of signal magnitudes that can be reliably processed; only signals larger than the noise can be reliably distinguished.

What measures can we take during amplifier design to minimize the amplifier noise, such that very weak signals can processed reliably? The answer on that is the subject of this chapter. You will learn the techniques to design amplifiers with an optimal noise performance (= minimal noise), and be able to predict the minimum achievable noise level in advance. Altogether, this constitutes the first step to the implementation of the nullor by transistors.

The first thing to asses is the way noise manifests itself in the amplifier. There are many noise sources and it would be very inconvenient to deal with every source separately. Fortunately, the effect of all noise sources can be modelled in one "equivalent noise source". In section 4.2 this will be discussed, together with the various noise transformations that can be used to find the equivalent noise source. When the contributions of various noise sources are added, it is very important to take into account whether noise sources are correlated or uncorrelated. This difference leads to different ways in which the sources are added. In section 4.3 this will be discussed. The noise sources related to the components used in the amplifier will be discussed in section 4.4. Subsequently, section 4.5 provides you some short-cut tools to analyze the noise behavior of a given circuit. Section 4.6 derives criteria for the input stage of the nullor implementation for optimum noise performance. Finally, section 4.7 provides the techniques to design amplifiers with an optimal noise performance.

## 4.1 Measure for the Amplifier Noise Performance

The first thing we have to do to realize an amplifier with an optimal noise performance, is specifying in detail what we actually mean by a "good" and a "bad" noise performance. That is, we need a criterion in order to optimize the noise performance, i.e. find the best possible match to it.

This section will select a suitable criterion for the amplifier noise performance, that will be used throughout this chapter. As shown below, this criterion is the result of considerations from information and circuit theory.

### 4.1.1 Signal-to-Noise Ratio

To obtain a suitable measure for the amplifier noise performance, we return to the principle concept of the design procedure; the idea that an electronic amplifier can be considered as a realization of (a part of) an information channel. The speed at which the information transfer through the channel takes place cannot exceed the *capacity* of the information channel. Any information supplied to the channel in excess of its capacity will be lost during transmission.

Noise is one of the factors that limits the information handling capacity of the channel; it reduces the accuracy by which one signal value can be distinguished from another (resolution). Shannon's information theory, which in its simplest form is described by equation (1.1), states that the channel capacity increases logarithmically with the Signal-to-Noise Ratio (SNR) in the channel. Amplifier noise reduces the (maximum possible) SNR in the channel, and thereby also decreases the channel capacity. As a result, the information rate at the destination decreases; the noise destroys part of the transmitted information.

The amplifier design strategy has to make sure that an as small as possible amount of information is destroyed by circuit noise. In other words, it should maximize the amplifier channel capacity by maximizing the (maximum possible) signal to noise ratio in the amplifier. Consequently, the maximum possible SNR in the amplifier is a suitable measure for its noise performance.

The (maximum possible) signal-to-noise ratio in the amplifier can be maximized through separate optimization of the maximum tolerable signal power, and the generated noise power. This is due to the fact that both characteristics can be made orthogonal, by concentrating them in different parts of the amplifier design. The maximum tolerable signal power is determined by the distortion, which is mainly generated in the output part of the amplifier. The weakest signals in the amplifier, encountered mainly in the input part, are most vulnerable to noise. Consequently, the design strategy has to minimize the amplifier noise production through proper design of the amplifier input part.

### 4.1.2 Equivalent Input and Output Signal-to-Noise Ratio

Although we know right now that the maximum possible signal-to-noise ratio is a suitable measure for the amplifier noise performance, we are not ready yet to apply it in our design strategy. We haven't determined at which position in the amplifier circuit this SNR, or the signal power and noise power, should be observed.

According to the channel model from information theory we have been using so far, the most suitable position would be the amplifier output. It is the output of the information channel, and the information contained in the signal at this position is used as input for further information processing.

From the viewpoint of the amplifier design strategy, however, it would be more appropriate to associate the maximum possible SNR to the input of the amplifier; this is the part to be optimized with respect to noise. Furthermore, as shown in the remainder of this chapter, an input referred SNR is easier to calculate, and is better suited to explain the various noise optimization techniques.

In section 4.3 it will be shown that an input referred SNR, the so called equivalent input SNR, can, and preferably should be used in design and analysis of the noise performance if (and only if) the amplifier transfer is frequency independent. This statement can be made plausible at this point already using figure 4.1. When the output signal is a voltage, the amplifier output as observed

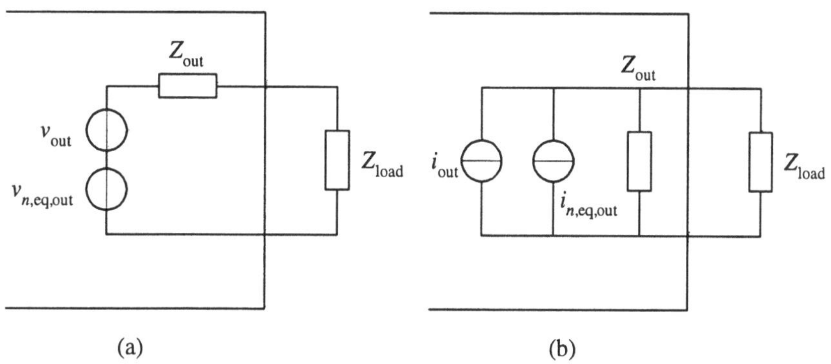


*Figure 4.1.* Thevenin equivalent (a) and Norton equivalent (b) of the amplifier output in the presence of noise.

from the load can be represented by the Thevenin equivalent of figure 4.1a. Likewise, if it is a current, it can be represented by the Norton equivalent of figure 4.1b. Both equivalents consist of the amplifier output impedance $Z_{out}$, and two independent sources. One of these, of course, represents the intended amplifier output information signal. The other one, the so called *equivalent output noise source* represents the combined contribution of all noise processes in the amplifier to the output signal. It is an ordinary independent[1] source, of which the value is a stochastic signal. In fact, the equivalent noise source replaces all other noise in the amplifier; after insertion of $v_{n,eq,out}$ or $i_{n,eq,out}$, the amplifier itself can be considered 'noise free'. The amplifier output SNR equals the ratio of the power content of $v_{out}$ and $v_{n,eq,out}$, or $i_{out}$ and $i_{n,eq,out}$.

The input referred SNR, or equivalent input SNR, is easily related to the output signal and noise. Assume that the amplifier gain $G$ is frequency independent. Then, the amplifier noise performance can also be modelled by the combination of the input information signal and equivalent input noise. Both quantities, which are also represented by independent current or voltages sources, are related to the output signal and equivalent output noise through:

$$e_{out} = Ge_{in}, \tag{4.1}$$

$$e_{n,out,eq} = Ge_{n,in,eq}, \tag{4.2}$$

where $e$ denotes either current or voltage. Let $P_{s,out}$ and $P_{n,eq,out}$ represent the output signal and equivalent output noise power, respectively, and $P_{s,in}$ and $P_{n,eq,in}$ the corresponding input quantities. Then, it follows that the equivalent input- and output SNR are related through:

$$\mathrm{SNR}_{eq,out} = \frac{P_{s,out}}{P_{n,eq,out}} = \frac{G^2 P_{s,in}}{G^2 P_{n,eq,in}} = \mathrm{SNR}_{eq,in}. \tag{4.3}$$

Consequently, both SNRs are equivalent and the input SNR can be used in the design process instead of the output SNR, as stated before.

Thus, in order to assess the amplifier noise performance, the equivalent input noise power and input signal power have to be determined. Techniques to determine the equivalent input noise source and its associated power are considered in detail in section 4.2 and section 4.3.

### 4.1.3 Available Input Signal-to-Noise Ratio

Although we have determined by now that the equivalent input SNR is a suitable measure for the amplifier noise performance, there is one slight problem

[1]The word 'independent' should be interpreted in the circuit theoretical sense here; the current/voltage impressed upon the circuit by an independent source is not affected by any other current or voltage in the network. Independence in the stochastic sense has a different interpretation in circuits, as discussed later on.

looming: both the signal power and (equivalent) noise power delivered to input of a single-loop amplifier equal zero! Therefore, it seems that formally, the equivalent amplifier input SNR is not well defined. An elegant solution to this problem is provided by the the concept of an 'available Signal-to-Noise Ratio', as shown below.

Figure 4.2a and figure 4.2b represent the input part of an amplifier with a voltage input and a current input, respectively. In order to ensure accurate

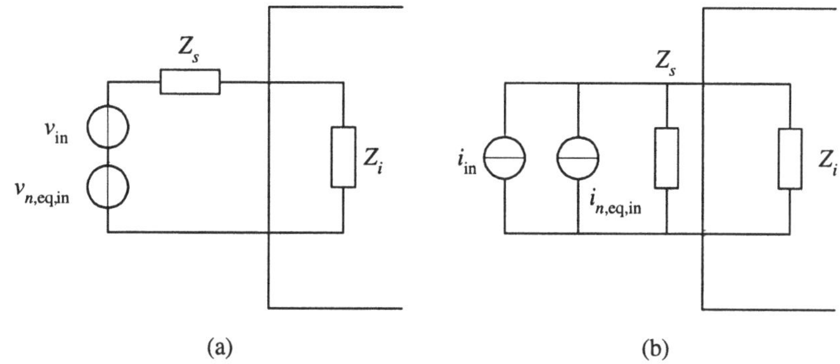


*Figure 4.2.* Amplifier with a voltage input (a) and a current input (b).

amplification of the information signal, we concluded in chapter 2 that the usually inaccurate source impedance should not appear in the amplifier transfer, which means that its power dissipation should equal zero. In figure 4.2a, this is achieved when the amplifier input impedance $Z_i$ is infinite, and in figure 4.2b when $Z_i$ equals zero. The consequence of this is that also the power delivered to $Z_i$ becomes zero.

The power delivered to the amplifier is some fraction of the maximum power that the source can deliver, the so called *available power.* As is well-known from circuit theory, the complete available power is delivered to $Z_i$ when it equals the complex conjugate of the source impedance $Z_s$:

$$Z_i = Z_s^* = R_s - jX_s. \tag{4.4}$$

This yields an available input signal power $P_{s,a,in}$ and available equivalent input noise power $P_{n,a,in}$ for figure 4.2a equal to:

$$P_{s,a,\text{in}} = \int \frac{S_{v_s}(f)}{4\Re\{Z_s(f)\}} \text{d}f, \tag{4.5}$$

$$P_{n,a,\text{in}} = \int \frac{S_{v_{n,\text{eq,in}}}(f)}{4\Re\{Z_s(f)\}} \text{d}f, \tag{4.6}$$

Where $S_{v_s}(f)$ and $S_{v_{n,\text{eq,in}}}(f)$ represent the power spectral densities (see section 4.3) associated to $v_s$ and $v_{n,\text{eq,in}}$, respectively. As can be observed from the equations above, a major advantage of $P_{s,a,\mathrm{in}}$ and $P_{n,a,\mathrm{in}}$ is that they are independent of the amplifier input impedance $Z_i$; it is an 'intrinsic' property of the source.

In circuit theory, it is usually assumed that the source still provides the maximum available power when $Z_i$ is not equal to $Z_s^*$, i.e. when there is no power match. In that case, $Z_i$ absorbs only part of this, and reflects the remainder back into the source. The power absorbed by $Z_i$ can, after transmission-line theory, be viewed as the 'transmitted power', and the part that is not absorbed as the 'reflected power'. The associated transmission coefficient $T(f)$ and reflection coefficient $R(f)$ are customarily defined per unit of frequency:

$$T(f) \triangleq \frac{\text{Spectral density of transmitted power}}{\text{Spectral density of available power}} \tag{4.7}$$

$$R(f) \triangleq \frac{\text{Spectral density of reflected power}}{\text{Spectral density of available power}} \tag{4.8}$$

Further, based on the conservation of energy, we know that:

$$T(f) + R(f) = 1, \tag{4.9}$$

$$0 \leq T(f), R(f) \leq 1. \tag{4.10}$$

A graphical representation of $T(f)$ and $R(f)$ is depicted in figure 4.3. For

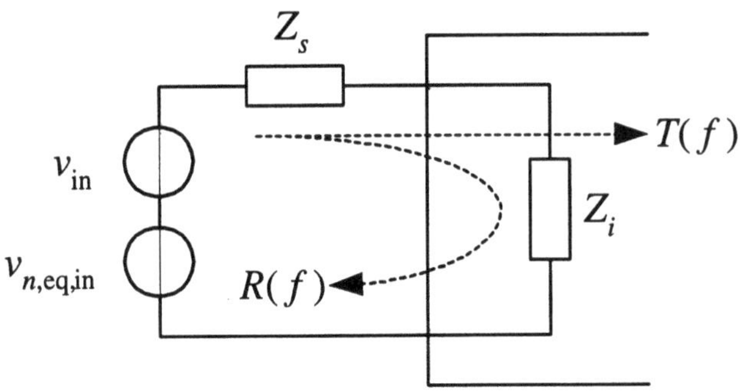


*Figure 4.3.* Power transmission and reflection at the amplifier input.

figure 4.2a, the equivalent input SNR can be expressed as:

$$\mathrm{SNR}_{\mathrm{eq,in}} = \frac{\int \frac{S_{v_s}(f)}{4\Re\{Z_s(f)\}} T(f) \mathrm{d}f}{\int \frac{S_{v_{n,\mathrm{eq,in}}}(f)}{4\Re\{Z_s(f)\}} T(f) \mathrm{d}f}. \tag{4.11}$$

This expression looks rather complicated. Fortunately, however, we know that a well-designed single-loop amplifier doesn't load the signal source; all delivered signal power is reflected, such that $T(f) = 0$ and $R(f) = 1$. In other words: for a well-designed amplifier, both $T(f)$ and $R(f)$ are frequency independent. In that case, the equivalent input SNR becomes equal to the available, equivalent input SNR, $\text{SNR}_{a,\text{in}}$:

$$\text{SNR}_{\text{eq,in}} = \frac{TP_{s,a,\text{in}}}{TP_{n,a,\text{in}}} = \frac{P_{s,a,\text{in}}}{P_{n,a,\text{in}}} = \text{SNR}_{a,\text{in}}. \tag{4.12}$$

The available input SNR is well-defined, and therefore very suited for noise optimization purposes, to be discussed later on. When necessary, all noise calculations will therefore be referred to this SNR.

### 4.1.4 Summary

According to information theory, which is the underlying principle for the entire design strategy, the Signal-to-Noise Ratio (SNR) is the best suited measure for the amplifier noise performance. The signal power and noise power can be separately optimized, since they are determined by different parts of the amplifier; the noise by the input part, and the maximum allowable signal power by the output part. Optimization of the noise performance is therefore equivalent to minimization of the generated circuit noise power.

Although formally the amplifier output SNR is of most importance (the output signal is the wanted signal), an input referred SNR is better suited for design purposes, since the amplifier noise behavior is determined by the input circuitry. It appears possible to use such an SNR for almost all types of amplifiers. The amplifier circuit noise can be represented by a single, independent voltage or current source (depending on the nature of the input information signal) that adds to the input signal source. The input referred SNR, the so called equivalent input SNR, equals the ratio of the power contained in the signal source and the equivalent input noise source.

The equivalent input SNR is not directly suited for noise optimization purposes, since both the signal power and equivalent noise power absorbed by single-loop amplifiers equals zero; their power gain is infinite. As shown, the available equivalent input SNR is better suited for this purpose; it is equal to the equivalent input SNR for well-designed single-loop amplifiers, but is associated to the ratio of the nonzero available signal power and the nonzero available equivalent noise power, instead of the power absorbed by the amplifier.

## 4.2 Equivalent Input Noise Source

As one of the first steps in a noise analysis, we have to determine the so called equivalent (input) noise source. This source models the noise experienced at the amplifier output, due to internal circuit noise production. It concentrates the entire circuit noise production into one circuit branch, and thereby provides straight-forward calculation of the experienced noise power.

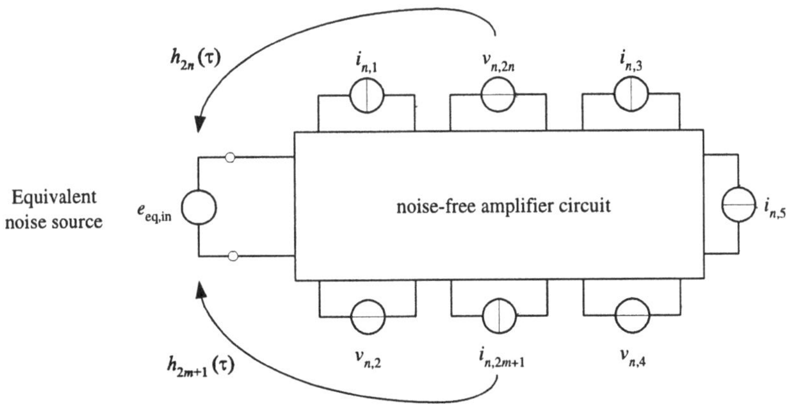


*Figure 4.4.* Representation of the amplifier circuit as a noise-free multi-port network, connected to multiple noise sources.

To determine the contributions of all internal noise sources to the equivalent noise sources, a combination of various transformations, originating from circuit theory, is needed. In contrast, no use of any statistical property of the noise sources is required to obtain the equivalent noise source; these are required only to determine the equivalent noise power, as explained in section 4.3.

In this section, you will learn how to determine the equivalent input noise source of an amplifier efficiently, using the appropriate circuit transforms. We start with a rather general description of the relation between the equivalent source and the various (original) noise sources in the circuit in section 4.2.1. Subsequently, section 4.2.2 through section 4.2.5 focus on four transforms that enable you to perform all necessary noise source manipulations. Section 4.2.6 gives an example calculation.

### 4.2.1 The Equivalent Input Noise Source

The equivalent input noise source replaces all other noise sources in the amplifier circuit. Determination of this source may therefore be viewed as 'wiping' all circuit noise to the amplifier input, resulting in one noise source, and a noise-free amplifier.

This 'wiping' or transformation of circuit noise to the input is schematically represented by figure 4.4. The circuit noise processes are distributed all over the amplifier circuit, and can each be represented by an independent current or voltage source. These sources can be considered as external inputs to the otherwise noise-free amplifier, which is represented as a multi-port network.

As far as the noise is concerned, the amplifier network behaves linear, such that we can describe the contribution of each noise source to the equivalent input noise source as a (time-domain) convolution with an impulse response$h_k(\tau)$, as depicted for $v_{n,2n}$ and $i_{2m+1}$ in figure 4.4. If we reserve the odd indices for noise current sources and the even ones for noise voltage sources, we can formally express the equivalent input noise source (either current or voltage) as:

$$e_{n,\mathrm{eq,in}}(t) = \sum_{i=1}^{n} h_{2i} * v_{n,2i}(t) + \sum_{k=0}^{m} h_{2k+1} * i_{n,2k+1}(t), \tag{4.13}$$

where '$*$' denotes convolution. When $e_{n,\mathrm{eq,in}}(t)$ is a voltage the impulse responses $h_{2i}(\tau)$ are dimensionless (voltage-to-voltage), while $h_{2i+1}(\tau)$ are (trans-)impedances. Likewise, when $e_{n,\mathrm{eq,in}}(t)$ is a current, $h_{2i}(\tau)$ are (trans-)conductances, while $h_{2i+1}(\tau)$ are dimensionless (current-to-current).

The above shows that in order to obtain the equivalent noise source, we basically have to determine the impulse responses $h_{2i}(\tau)$, or the associated transfer functions $H_{2i}(j2\pi f)$. We could do this in a straight-forward way, by evaluating the circuit equations of the amplifier. Such a procedure, however, is rather involved, and not very suited for design purposes. It doesn't yield much insight into the origin of dominant noise contributions, and hides the key design parameters that are essential to optimization of the noise behavior.

Instead of such a 'global' one-step transform, we will apply another approach in this book, one that breaks the transformation of a noise source to the equivalent input noise into several consecutive steps. In this way, the approach exploits knowledge about the topology of the network, which is very similar for the various amplifier types, and directly shows the dominant factors and key parameters determining the amplifier noise behavior. The approach uses combinations of four different types of transformations, each of which is separately discussed below.

### 4.2.2 Transform-I: Voltage Source Shift

The voltage source shift (V-shift) is a transform that enables to move (noise) voltage sources through the amplifier network. Generally, it is used to shift these sources to the amplifier input, the output, or a branch where it can be subjected to another type of transform.

The major constraint to be posed on any source transform is that the transform itself does not change the noise current/voltage experienced at the circuit input (or output); it should not be noticeable to an observer measuring the noise voltage/current at these points whether or not noise sources have been transformed/shifted internally in the amplifier network. For the V-shift, this means that it must not change the Kirchhoff Voltage Law (KVL) of any mesh in the circuit.

The V-shift obeying this constraint is visualized in figure 4.5. The original (noise) source $v_n$ is shifted out of the branch between the nodes 1,4 into the two other branches connected to node 4; the ones between 2,4 and 3,4. In order

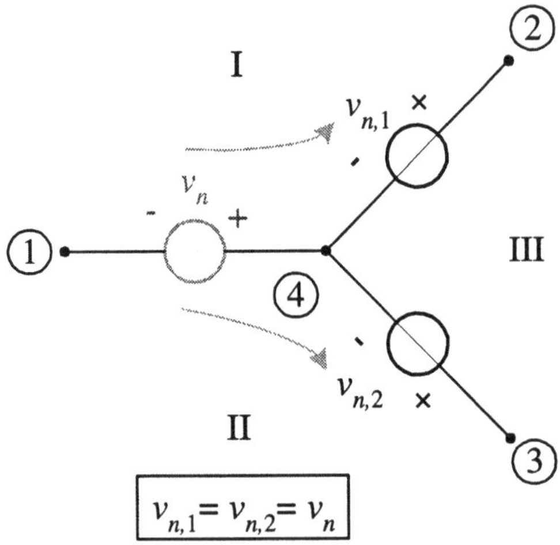


*Figure 4.5.* V-shift transform.

to guarantee that the KVLs of the meshes I,II,III, associated to node 4 remain unchanged, the new sources $v_{n1}$ and $v_{n2}$ have to be exactly equal to each other and to the original source $v_n$.

As we will see later on, this means that the stochastic processes $v_n$, $v_{n1}$ and $v_{n2}$ are fully correlated, and possess identical stochastic properties. Observe what happens with the V-shift in the general case, when node 4 is connected to $n$ branches ($n > 2$). Then, shifting $v_n$ through node 4 from its original branch to the $n-1$ other branches yields $n-1$ identical sources, instead of just 2. Further, note that the V-shift allows to move $v_n$ around mesh I and II, but not out of it; this would change the KVL.

### 4.2.3 Transform-II: Current Source Shift

The dual transform of the V-shift is the I-shift, which allows to move current (noise) sources through the amplifier network. It is generally used to shift these sources to the amplifier input port, output port, or intermediate nodes that allow another type of transform.

Whereas the V-shift is not allowed to affect the KVL of any circuit mesh, the I-shift is not allowed to change the Kirchhoff Current Law (KCL) of any circuit node, for the same reason. The transform obeying this constraint is depicted in figure 4.6. The original (noise) current source $i_n$ is redirected from the branch between nodes 1,2 through the sources $i_{n,1}$ and $i_{n,2}$ between nodes 1,3 and 2,3. In order to keep the KCLs of the nodes 1,2, and 3 unchanged, $i_{n,1}$ and $i_{n,2}$ have to be exactly equal to each other and to $i_n$, and also have to be directed as illustrated.

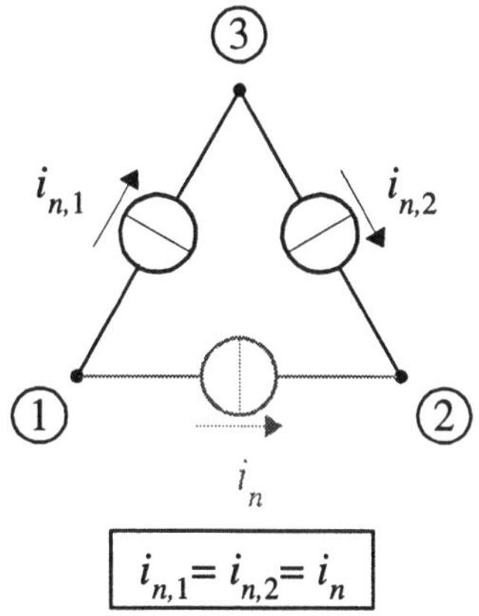


*Figure 4.6.* The I-shift transform.

Similar to the V-shift, this means that the original source $i_n$ and the transformed sources $i_{n,1}$ and $i_{n,2}$ have to be fully correlated and possess identical statistical properties. We further notice that, similar to the V-shift, the I-shift allows to move a current source around the network, but cannot be used to disconnect it from the original nodes; this would change the KCL.

### 4.2.4 Transform-III: Norton-Thevenin Transform

The equivalence of the well-known theorems of Norton and Thevenin can be used to transform a (noise) current source into a (noise) voltage source and vice versa. This type of transform does essentially not move sources through the amplifier network, but is used to switch between the V-shift and I-shift.

The Norton-Thevenin transform, which automatically obeys the constraint that it does not affect the observed output noise (why ?), is depicted in figure 4.7.

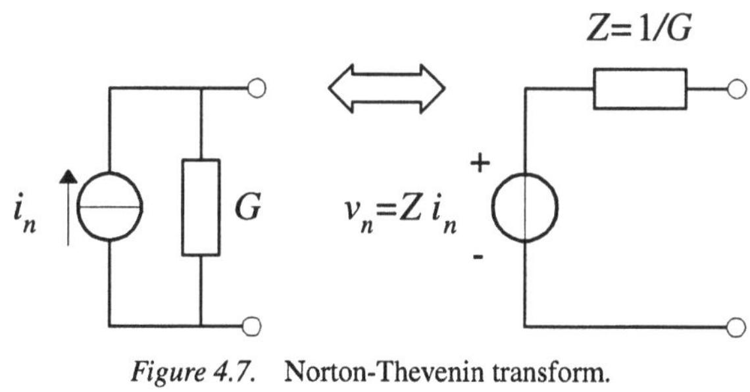


*Figure 4.7.* Norton-Thevenin transform.

The current source $i_n$ and the voltage source $v_n$ have a one to one correspondence through the impedance Z. The stochastic processes $i_n$ and $v_n$ are therefore fully correlated, and possess similar, though not identical stochastic properties. Note that this transformation does change the KVLs and KCLs of the circuit; it exchanges a circuit branch for a circuit node, and vice versa. For this reason, the Norton-Thevenin transform, in combination with the V-shift and I-shift can be used to eliminate a voltage source from a mesh, or a current source from a node.

### 4.2.5 Transform-IV: Shift through Two-ports

The three transformations we have considered so far are all concerned with two-terminal elements (one-ports) only. If a network consists entirely of such elements these three transforms are all we need to determine the equivalent input noise.

An amplifier network, however, will always contain elementary two-ports, controlled sources (due to transistors) or a nullor, that cannot be replaced by any combination of one-ports. For such networks, we need an additional transform: the two-port shift. This transformation is illustrated by figure 4.8. The output

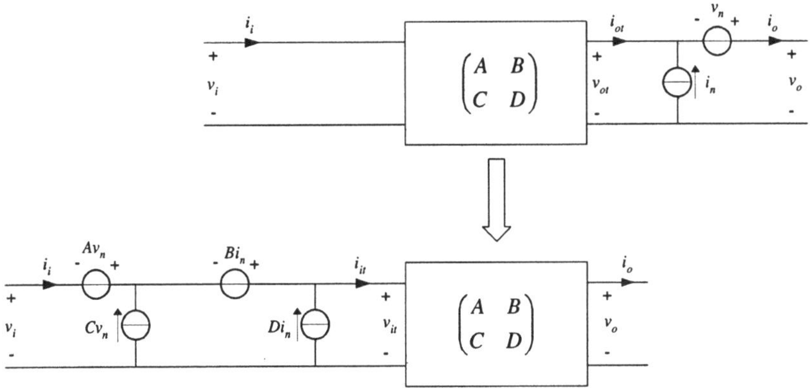


*Figure 4.8.* Shifting a noise current and voltage source through a two-port.

voltage $v_o$ and the output current $i_o$ of the two-port are mutilated by a noise voltage $v_n$ and noise current $i_n$, respectively, as depicted in the upper part of the figure. The purpose of the two-port shift transform is to obtain the equivalent input noise sources that yield this output noise current and voltage, as depicted in the lower part of figure 4.8.

The transformation is established using the chain matrix of the two-port, that relates the two-port input voltage $v_{it}$ and input current $i_{it}$, which in the upper part equal $v_i$ and $i_i$, respectively, to the two-port output voltage $v_{ot}$ and output current $i_{ot}$. In the upper part of the figure, the output voltage $v_o$ and output current $i_o$ are related to the two-port output voltage and current as:

$$v_o = v_{ot} + v_n, \tag{4.14}$$

$$i_o = i_{ot} + i_n. \tag{4.15}$$

In the lower part of the figure, $v_n$ and $i_n$ have been transformed to the input, such that $v_o$ and $i_o$ have become the output current and voltage of the two-port. Substitution for $v_{ot}$ and $i_{ot}$ into the chain matrix equation then yields:

$$\underbrace{\begin{pmatrix} v_i \\ i_i \end{pmatrix} + \begin{pmatrix} A & B \\ C & D \end{pmatrix} \begin{pmatrix} v_n \\ i_n \end{pmatrix}}_{\begin{pmatrix} v_{it} \\ i_{it} \end{pmatrix}} = \begin{pmatrix} A & B \\ C & D \end{pmatrix} \begin{pmatrix} v_o \\ i_o \end{pmatrix}. \tag{4.16}$$

As indicated, the left hand side of this equation denotes the two-port input voltage $v_{it}$ and input current $i_{it}$, shown in the lower part of figure 4.8.

We observe that the output noise voltage source $v_n$ is transformed into an input noise voltage source $Av_n$ and a noise current source $Cv_n$, which, as we will see, are fully correlated (originate from the same source), and possess similar statistical properties. Likewise, the output noise current source $i_n$ transforms into an input noise voltage source $Bi_n$ and an input noise current source $Di_n$.

### 4.2.6 Example 1

In order to illustrate the use of the four previously discussed transforms, we will now determine the equivalent input noise voltage source of the voltage amplifier depicted in figure 4.9. The positions of the noise sources match the practical situation, as will be explained in section 4.4. For the time being, however this is not important, nor is it to know the origin of the sources. In accordance with figure 4.4 and equation (4.13), all noise voltage sources have been labelled with even indices, while the current source has an odd index. Further, to emphasize its presence, the ground terminal has been explicitly plotted as a node.

Unfortunately, no explicit rules are available (yet) that show us which sequence of transforms most efficiently leads us to the equivalent input noise source. In many cases, several different sequences lead to the same result with a comparable efficiency. However, it is generally observed that the most efficient transform sequences always shift all noise sources out of the amplifier core, i.e. towards the source, the load (amplifier output) or the nullor output. The reason for this is that circuit nodes and meshes in the amplifier core are not connected to the amplifier input or output signal, and therefore eventually have to become 'noise-free'. In the final step, the sources shifted towards the output are transformed to the input by means of the two-port shift.

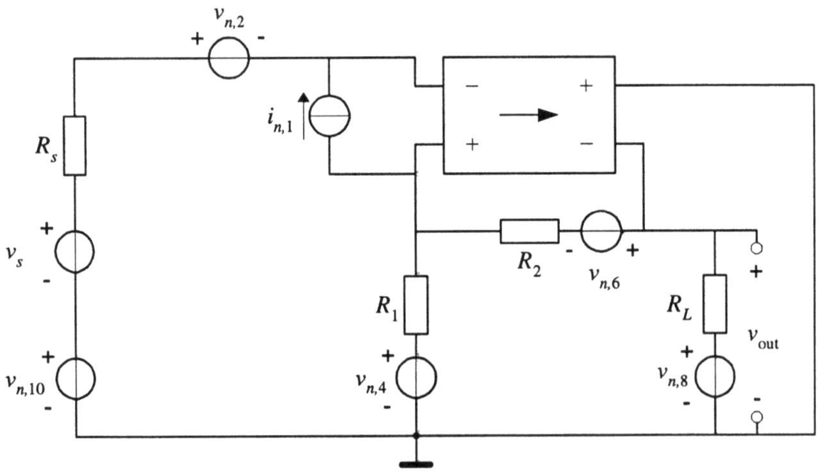


*Figure 4.9.* Voltage amplifier with noise.

Using this information, we proceed with the determination of the equivalent noise source of the voltage amplifier in figure 4.9 from the edged towards its core, starting with the sources located nearest to the signal source and the load..

**Step 1: $v_{n,2}$ and $v_{n,10}$**

The voltage sources, $v_{n,2}$ and $v_{n,10}$ are located already in series with the signal source $v_s$, and therefore can be included directly into the equivalent noise voltage source. No further transforms on these sources are required.

**Step 2: $v_{n,8}$**

The source $v_{n,8}$ in series with the load impedance is a tricky one, that will easily lead to mistakes. We will show that, although a different conclusion might be drawn at first sight, this source does *not* contribute to the equivalent input noise. The reason for this is as follows.

In the first place, notice that the combination of $v_{n,8}$ and $R_L$ represents the Thevenin equivalent of the noisy load. This means that it describes the behavior of the load correctly only between its terminals, i.e. the minus output terminal of the nullor, and the ground node. The node between $R_L$ and $v_{n,8}$ is an internal node of the Thevenin equivalent, that is not physically present in the circuit. The output voltage $v_{out}$ can therefore not be measured across the load $R_L$ alone, but instead only over the combination of $R_L$ and $v_{n,8}$, as shown in figure 4.9.

Secondly, notice that the voltage amplifier impresses its output voltage directly across the Thevenin equivalent of the load. Since the output impedance of this (ideal) amplifier equals zero, the Thevenin equivalent is short-circuited such that $v_{n,8}$ does not contribute to $v_{out}$. Figure 4.10 illustrates the situation in two alternative ways, where the amplifier output port is modelled by its Thevenin equivalent. In figure 4.10(a), the V-shift transform has been applied

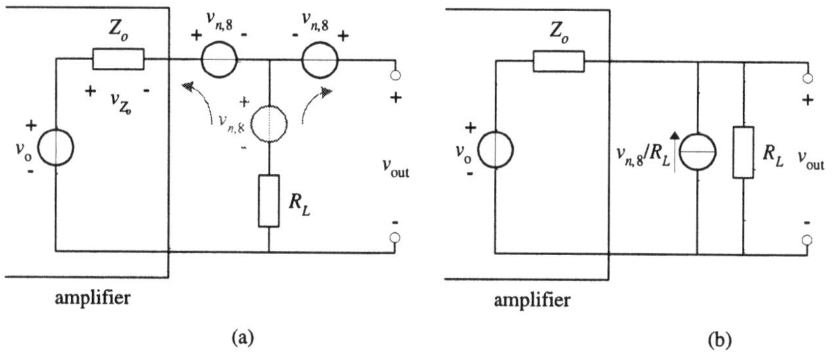


*Figure 4.10.* Contribution of $v_{n,8}$ to the output voltage $v_{out}$.

to $v_{n,8}$, in order to shift it in series with the output voltage $v_{out}$. From the KVL of the outer mesh of this circuit, we obtain:

$$v_{\text{out}} = v_o - v_{Z_o}. \tag{4.17}$$

Consequently, $v_{n,8}$ only contributes to the noise in $v_{out}$ when it induces a voltage across the amplifier output impedance $Z_o$. However, since $Z_o$ of the ideal voltage amplifier equals zero, the voltage across it equals zero, and $v_{n,8}$ has no effect on $v_{out}$. For arbitrary values of $Z_o$, one obtains:

$$v_{out} = \frac{R_L}{R_L + Z_o} v_o + \frac{Z_o}{R_L + Z_o} v_{n,8}. \tag{4.18}$$

This confirms our statement: $v_{n,8}$ vanishes if (and only if) $Z_o = 0$. In practice, however, $Z_o$ will be much smaller than $R_L$, such that the contribution of $v_{n,8}$ is negligible anyway.

Figure 4.10(b) depicts the Norton equivalent of the load. From this figure it is easily seen that $v_{n,8}$ transforms into a voltage source in series with $v_o$ only when $Z_o \neq 0$, which is consistent with our previous observation. Alternatively, application of the two-port transform to the noise current source, shows that the corresponding equivalent input noise source equals zero, since the chain parameters $B$ and $D$ of the amplifier are zero (see figure 4.8).

### Step 3: $i_{n,1}$

The source $i_{n,1}$ is also located in the input circuitry and, according to the previously mentioned guidelines, should be shifted away from the input of the nullor towards the source. It is clear that only the I-shift can be used to perform the first step in the transform sequence; to shift $i_{n,1}$ towards the signal source, it should become connected between the amplifier input terminals, i.e. the minus-terminal of the nullor and the ground node. The required I-shift is depicted in figure 4.11. The leftmost of the resulting current sources is easily transformed

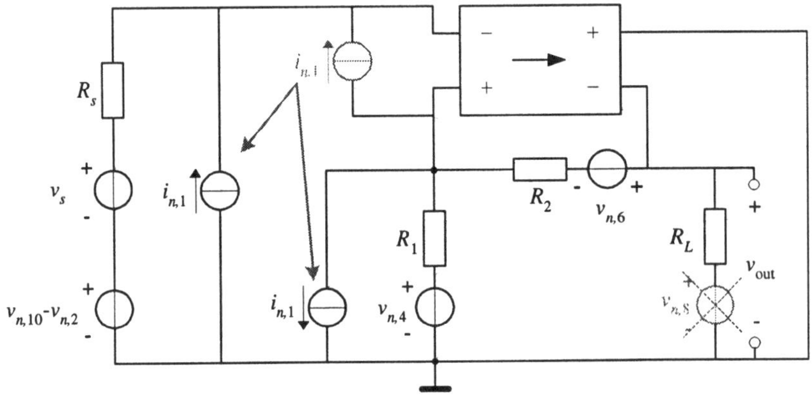


*Figure 4.11.* Application of the I-shift transform to $i_{n,1}$.

into a voltage source in series with $v_s$ using the Norton-Thevenin transform across the source impedance $R_s$. The other one still has to be removed from the internal amplifier node, connected between ground and the plus-terminal of the nullor. Since we have used the I-shift already, the only suitable transform for this purpose is a Norton-Thevenin transform across $R_1$. The resulting voltage source is connected in series with $v_{n,4}$. Since the same transforms will apply to both sources from this point on, they will be combined into a single source $v'_{n,4}$, as depicted in figure 4.12.

### Step 4: $v'_{n,4}$ and $v_{n,6}$

The sources $v'_{n,4}$ and $v_{n,6}$ should be shifted away from the amplifier core, according to our guidelines.

The source $v'_{n,4}$ should thus be moved downward through the ground node by use of the V-shift. Since six[2] branches are connected to this node, the transformation yields five copies of $v'_{n,4}$, as indicated in figure 4.13. The source $v_{n,6}$ should be moved towards the load $R_L$, through the node that connects the feedback network to the load and the nullor. Since this node has four branches,

[2]The ground symbol doesn't count as a branch; it is connected only to one node in the circuit, to mark it as the datum. A voltage source in series with it doesn't appear in any KVL of the circuit, since it is not included in any mesh.

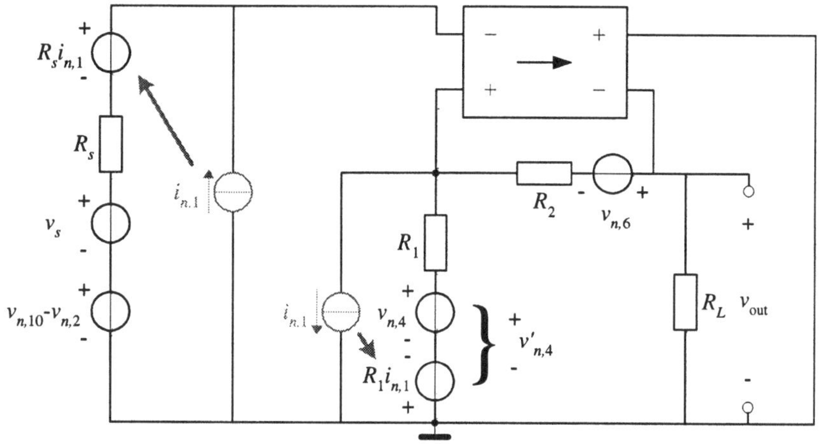


*Figure 4.12.* Application of the Norton-Thevenin transform to both current-sources $i_{n,1}$.

the applied V-shift will result in three copies of $v_{n,6}$ (see figure 4.13). The

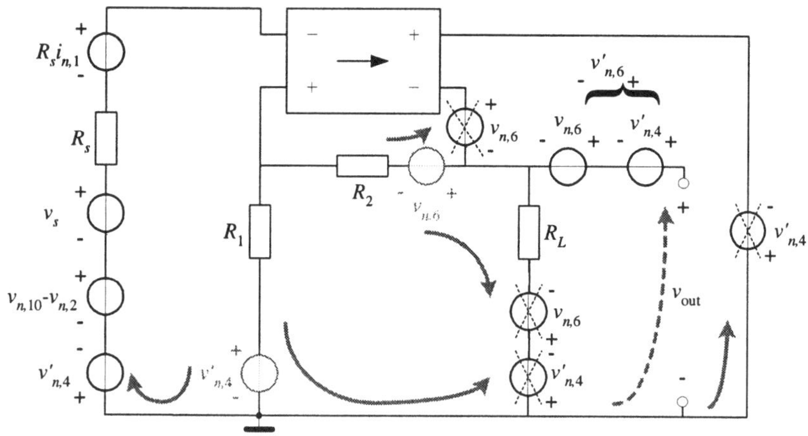


*Figure 4.13.* Application of the V-shift transform to $v'_{n,4}$ and $v_{n,6}$.

sources in series with the load $R_L$ vanish, for the same reason that $v_{n,8}$ vanishes. The sources in series with the output are combined into one voltage source $v'_{n,6}$.

One copy of $v'_{n,4}$ and $v_{n,6}$ is located directly in series with the output port of the nullor. The two-port transform, applied to the nullor shows us directly that these sources will not yield any equivalent input noise source, since all chain parameters of the nullor equal zero. Consequently, these sources vanish too.

**Step 5: $v'_{n,6}$**

The final step consists of the transformation of $v'_{n,6}$ to the amplifier input. Since $v'_{n,6}$ is connected in series with the output port of the complete voltage amplifier, this transformation can be established by application of the two-port shift to the complete voltage amplifier. As is known from chapter 3, only the chain parameter $A$ of this configuration is nonzero, and equals the reciproke of the voltage gain. $v'_{n,6}$ therefore results only in one equivalent input voltage source, and no current source. The result of this transformation is depicted in figure 4.14. From this figure, we observe that the equivalent input voltage noise

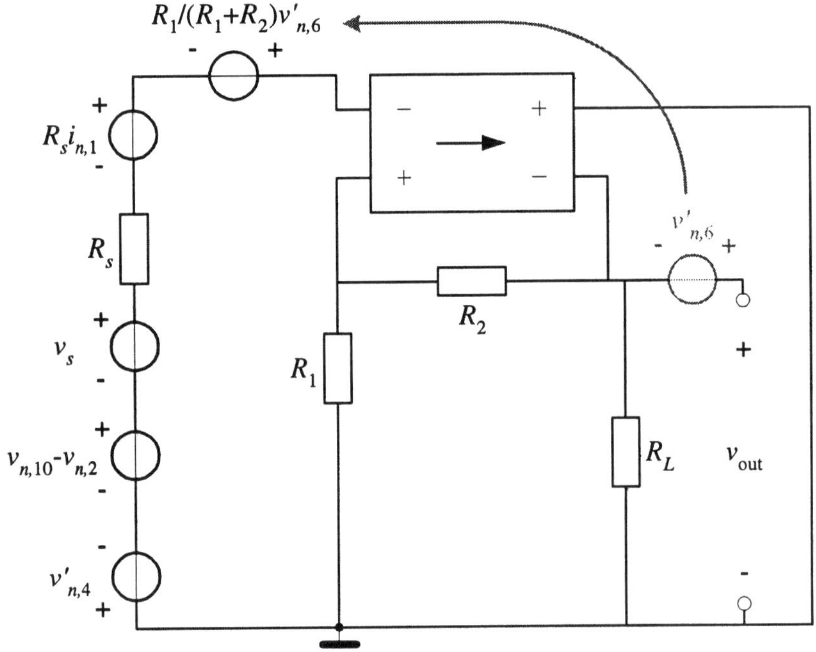


*Figure 4.14.* Application of the two-port transform to $v'_{n,6}$.

source of the amplifier, $v_{n,\text{eq,in}}$ equals:

$$v_{n,\text{eq,in}} = \left(R_s + \frac{R_1 R_2}{R_1 + R_2}\right) i_{n,1} - v_{n,2} - \left(\frac{R_2}{R_1 + R_2}\right) v_{n,4} + \left(\frac{R_1}{R_1 + R_2}\right) v_{n,6} + v_{n,10}, \quad (4.19)$$

where the polarity of $v_{n,\text{eq,in}}$ has been chosen the same as the polarity of $v_s$.

### 4.2.7 Noise transformations at the output port

It is easy to make an error while transforming noise sources via the output of an amplifier. For example in figure 4.14, the source $v'_{n,6}$ at the output is easily forgotten when the output terminal is not explicitly drawn. There would be no branch for the source to exist in. This situation is shown in figure 4.15. The left hand side shows the output port of an amplifier with a voltage output and the right hand side shows the output of an amplifier with a current output. In this situation also errors may be made with the noise source of the load resistor itself. This is shown in figure 4.16. When the noise sources are put into the

*Figure 4.15.* Common ways to indicate the output quantity.

circuit schematic, an output indicator easily ends up at the wrong position. In both cases the indicated signal contains a noise contribution of the noise source, but the *true* output signal of the amplifiers does *not*.

*Figure 4.16.* Output indicators at the wrong position in the schematic.

A good method to make sure that the output indicators are at the right position and no branches are "forgotten", is to insert a voltage or a current meter in the schematic that measures the amplifier output signal, like a for example a multi meter would be used in a practical situation. Only signals that show up in the read-out of the meter are of interest. This is shown in figure 4.17. From this figure it can be easily seen that in both cases the noise source of the load resistor does not produce a signal in the meter, so there is no contribution of the noise of the load resistor in the amplifier output signal.

In the left hand situation in figure 4.17 the noise of the load resistor only manifests itself as a noise current through the resistor. The voltage across the resistor, which is the designed output quantity of the amplifier is completely determined by the amplifier output voltage and does not contain noise caused by the load resistor itself.

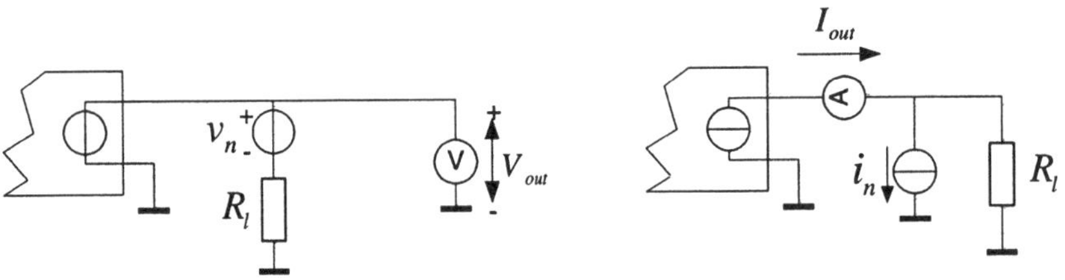


*Figure 4.17.* Explicit measurement of the output quantity with a volt or a current meter.

In the right hand situation in figure 4.17, the noise of the load resistor only manifests itself as a noise voltage across the resistor. The current through the resistor which is the designed output quantity of the amplifier is completely determined by the amplifier output current and does not contain noise caused by the load resistor itself.

## 4.3 Equivalent Input Noise Power

Once we have determined the equivalent input noise source, using circuit transforms, the equivalent input noise power and with that the signal-to-noise ratio can be determined. For this purpose, we need to use some of the statistical characteristics of the noise.

This section will acquaint you with the techniques required to determine the equivalent input noise power. In order to demonstrate the significance of the various quantities and measures involved, we traverse the procedure in reverse order. First, section 4.3.1 considers the relation between the equivalent input noise power and its distribution over frequency, leading to the so called power spectral density. Section 4.3.2 and section 4.3.3 subsequently relate this spectral density to the equivalent noise source. As an example, section 4.3.4 applies the procedure to the equivalent input noise of the amplifier considered in section 4.2.6. Section 4.3.5 summarizes the results.

### 4.3.1 Definition of the Equivalent Input Noise Power

It will be clear from our previous discussions that the equivalent input noise power, which will be referred to as $P_{n,\text{eq,in}}$, is tightly related to the power produced by the equivalent input noise source. What has been less clear so far, is that the equivalent input noise power is *not* exactly equal to the power produced by that source; some of the power produced by the source is not included into $P_{n,\text{eq,in}}$. The purpose of this section is to explain the difference and relation between both power measures.

The difference between $P_{n,\text{eq,in}}$ and the power produced by the equivalent input noise source is related to the distribution of these powers over frequency.

As we know, the distribution of the energy of a (deterministic) signal over frequency can be made visible through application of the Fourier transform. The spectrum resulting from this transform is a complex function describing the magnitude and phase of the contributing frequencies. In a similar way, we can visualize the distribution of signal power over frequency by means of a so called *power spectral density.* This function, referred to as $S(f)$, describes the power contained in the signal per unit of frequency; the power contained in a very small (infinitesimal) frequency band between $f$ and $f+df$ equals $S(f)df$. For the moment, this definition of $S(f)$ is sufficient; a more accurate definition, and methods to calculate such a density are given later on.

We can assign a power spectral density to both the amplifier input signal source, denoted by $S_{s,\text{in}}(f)$ and the equivalent input noise source $S_{n,\text{eq,in}}(f)$. Both are schematically depicted in figure 4.18. The most important observation,

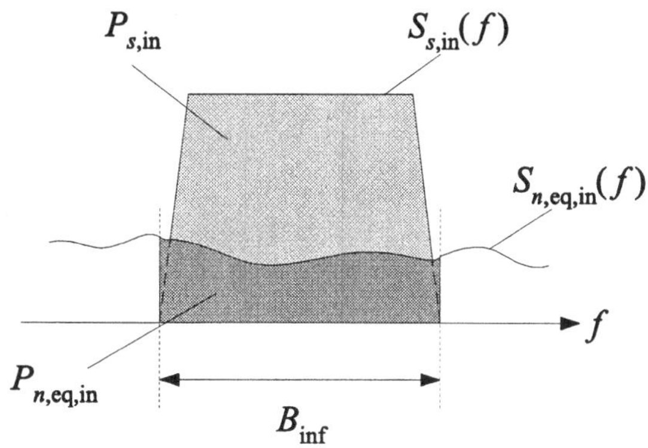


*Figure 4.18.* Schematic representation of the noise- and signal power spectral densities.

which is generally true, to be made from this figure is that the noise is spread over a much wider frequency range than the information signal, which is bound to the *information bandwidth* $B_{inf}$ as shown.

From an information theoretical point of view, only the noise that cannot be distinguished from the information signal fundamentally limits the information handling capacity. In figure 4.18, this is the noise located inside the same frequency band $B_{inf}$ as the information signal. Any noise located outside this bandwidth can in principle be removed by means of frequency filtering. Consequently, the noise power located in the information bandwidth is a lower bound, and the corresponding SNR an upper bound on the SNR that can exist within the amplifier.

For this reason, we define the equivalent input noise power $P_{n,eq,in}$ to be part of the power produced by the equivalent input noise source that resides in the same frequency range as the information signal. Power equals the area under the power density spectrum, such that:

$$P_{n,\text{eq,in}} \triangleq \int_{B_{inf}} S_{n,\text{eq,in}}(f)df. \tag{4.20}$$

In the next section, we discuss how the power spectral density $S_{n,eq,in}(f)$ can be determined.

### 4.3.2 Power Spectral Density

As we have seen, the power spectral density is the key towards the calculation of the equivalent input noise power, and the equivalent (available) input SNR. In this section, we discuss in which way the spectral density corresponding to the equivalent input noise source can be determined.

Formally, according to its definition, the power spectral density should be obtained through Fourier transformation of the so called *autocorrelation function*, denoted by $R(\tau)$:

$$S(f) \triangleq \int_{-\infty}^{\infty} R(\tau) \exp\left(-j2\pi f\tau\right) d\tau. \tag{4.21}$$

The autocorrelation function expresses 'how fast a stochastic process fluctuates in time on average' (a formal definition is postponed to section 4.3.3), such that $S(f)$ describes the frequencies associated with these fluctuations. Since for physical stochastic processes $R(\tau)$ is a symmetrical real-valued function, $S(f)$ is also real valued (and symmetrical around $f = 0$).

In the calculation of the equivalent amplifier input noise power, however, we can circumvent the elaborate calculation of the autocorrelation function, and subsequent Fourier transformation. This is due to the fact that the power density spectra of the various *uncorrelated* circuit noise processes are generally known. Therefore, we only have to express the power spectral density of the equivalent input noise in terms of the spectral densities of the various noise processes; the frequency-domain equivalent of equation (4.13). For this purpose, we can use the Wiener-Kintchine theorem, which (basically) states that when a noise process $e_{n,y}(t)$ equals the convolution of a linear-time invariant impulse-response $h(\tau)$ and a noise process $e_{n,x}(t)$, i.e.:

$$e_{n,y}(t) = h * e_{n,x}(t), \tag{4.22}$$

then its power spectral density, $S_{n,y}(f)$, is related to the power spectral density $S_{n,x}(f)$ of $e_{n,x}(t)$ through:

$$S_{n,y}(f) = \left|H(j2\pi f)\right|^2 S_{n,x}(f), \tag{4.23}$$

where the transfer function $H(j2\pi f)$ is the Fourier transform of $h(\tau)$. By application of this theorem to equation (4.13), we find that the power spectral density of the equivalent input noise can be expressed as:

$$S_{n,\text{eq,in}}(f) = \sum_{i=1}^{n} |H_{2i}(j2\pi f)|^2 \, S_{v_{n,2i}}(f) + \sum_{k=0}^{m} |H_{2k+1}(j2\pi f)|^2 \, S_{i_{n,2k+1}}(f), \tag{4.24}$$

where it is assumed that all involved noise processes are uncorrelated (see section 4.3.3).

Throughout the literature, the power spectral density of an electrical noise process is often given in terms of the mean-square value of the noise associated to a frequency band $[f, f+\Delta f]$. This value is nothing else than the power contained in a frequency band $\Delta f$ around frequency $f$. Consequently if $\overline{v_n^2}(f, \Delta f)$ denotes the mean-square value of a noise voltage $v_n(t)$ in a frequency band $\Delta f$ around $f$, and $S_{v_n}(f)$ denotes the associated power spectral density, then the two are related through:

$$S_{v_n}(f) = \lim_{\Delta f \to 0} \frac{\overline{v_n^2}(f, \Delta f)}{\Delta f}. \tag{4.25}$$

Examples of such representations are given in section 4.4.

### 4.3.3 Correlation and Autocorrelation Function

The discussions in foregoing sections sometimes referred to properties or assumptions regarding the correlation between various noise processes. For example, the general expression for the power density spectrum of the equivalent input noise, equation (4.24), assumed that all noise processes $v_{n,2i}$ and $i_{n,2k+1}$ are uncorrelated. In section 4.2, however, we saw that noise transformations can result in *correlated* noise sources. Thus, apparently, the assumption leading to equation (4.24) is not always true. Therefore, in order to use this equation, we have to check whether the assumption is true or not. Fortunately, as we will see in this section, we can always make this assumption true by rearranging our formulation of the equivalent input noise. Further, section 4.3.2 mentioned that the power spectral density is related to a time-domain function; the autocorrelation function.

In this section, we briefly review the notions of correlation and the autocorrelation function to the extend required to apply equation (4.24) in the correct way. For a more advanced treatment, the reader is referred to texts on probability theory and stochastic processes.

#### Correlation

In essence, correlation is a measure for the extend to which two noise processes are related to each other; they have something in common. Generally, the existence of a relation between two processes indicates that they somehow have the same origin. An obvious example of this is given by the noise transforms discussed in section 4.2. Reconsider the V-shift transform as illustrated by figure 4.5. The two noise sources $v_{n,1}$ and $v_{n,2}$ are correlated, due to the fact that they are related through $v_{n,1} = v_{n,2}$. This relation is caused by the fact that $v_{n,1}$ and $v_{n,2}$ share the same origin: the source $v_n$.

An alternative way to express that two noise sources are correlated is that each of them contains information about the other one. Returning to the V-shift transform example, we see that once we know the value of $v_{n,1}$, we can predict the value of $v_{n,2}$ with 100% accuracy (or, in stochastic terminology, with probability 1). Thus, $v_{n,1}$ contains all information required to predict the value of $v_{n,2}$ completely. These two noise processes are therefore *fully correlated.* The opposite situation occurs when two noise processes do not contain any information about each other; in that case, knowledge of one of them is of no use in an attempt to predict the value of the other one. Such processes are called *uncorrelated.* In general, when two noise processes $e_{n,1}(t)$ and $e_{n,2}(t)$ are correlated, $e_{n,2}$ contains some information about $e_{n,1}$, but not all information required to predict its value completely. This is the case when $e_{n,2}$ is a mixture of several (uncorrelated/unrelated) noise processes. For example, let $e_{n,1}$ be given by:

$$e_{n,1} = ae_{n,2} + be_{n,3}, \tag{4.26}$$

where $a$ and $b$ are arbitrary constants. Then knowledge of $e_{n,2}$ reduces the uncertainty about the value of $e_{n,1}$, but cannot eliminate it completely; for this, we also need to know $e_{n,3}$. For this reason, $e_{n,1}$ and $e_{n,2}$ are correlated, but not fully correlated.

The significance of correlation for our noise calculations is that it affects the mean square value (also the mean square value associated to a frequency band $\Delta f$) and the power spectral density of a process. This can be demonstrated by calculating the mean square value of the process $e_1$, given by equation. (4.26). When E(...) denotes the expectation operator (i.e. a kind of weighted average), then we can write:

$$\begin{aligned} \mathrm{E}\left(e_{n,1}^2\right) &= \mathrm{E}\left[\left(ae_{n,2} + be_{n,3}\right)^2\right] \\ &= a^2\mathrm{E}\left(e_{n,2}^2\right) + b^2\mathrm{E}\left(e_{n,3}^2\right) + 2ab\mathrm{E}\left(e_{n,2}e_{n,3}\right). \end{aligned} \tag{4.27}$$

Likewise, the variance associated to a frequency band $\Delta f$ can be written as:

$$\overline{e_{n,1}^2}(f, \Delta f) = a^2\overline{e_{n,2}^2}(f, \Delta f) + b^2\overline{e_{n,3}^2}(f, \Delta f) + 2ab\Re\left\{\overline{e_{n,2}e_{n,3}}(f, \Delta f)\right\}, \tag{4.28}$$

where $\Re(\ldots)$ denotes the real part of its argument; the cross-term in equation (4.28) is generally complex valued. Notice that equation (4.28) contains cross-terms between $e_{n,2}$ and $e_{n,3}$, whereas equation (4.24) contains no cross-terms at all. Apparently, the cross-terms vanish for uncorrelated processes. This can be seen as follows. When two processes $e_{n,2}$ and $e_{n,3}$ are uncorrelated, then the following equation hold:

$$\mathrm{E}\left(e_{n,2}e_{n,3}\right) = \mathrm{E}\left(e_{n,2}\right)\mathrm{E}\left(e_{n,3}\right). \tag{4.29}$$

The quantities at the right-hand side of this equation equal the mean (i.e. average) values of $e_{n,2}$ and $e_{n,3}$, respectively. It is known that the mean value of all electrical noise processes of interest in this book equals zero. Therefore, the right-hand side of equation (4.29) vanishes. Consequently, for zero-mean, uncorrelated processes $e_{n,2}$ and $e_{n,3}$ we can write:

$$\overline{e_{n,1}^2}(f,\Delta f) = a^2\overline{e_{n,2}^2}(f,\Delta f) + b^2\overline{e_{n,3}^2}(f,\Delta f). \tag{4.30}$$

This is the result used in equation (4.24).

### Autocorrelation Function

So far, we have assumed that the power spectral density, or mean square value for a frequency band $\Delta f$ of the various noise processes in equation (4.24) are given. In the rare cases that this is not the case, it has to be derived from the autocorrelation function through Fourier transformation, as explained in section 4.3.2. Here, we will give a qualitative explanation of the meaning of the autocorrelation function, and its definition formulas.

Our discussion on correlation was implicitly restricted to values of different processes evaluated at the same time instant. In contrast, the autocorrelation function considers the relation between two values of the same noise process at different time instances. This is illustrated by figure 4.19, which depicts the noisy signal $e_n(t)$ as a function of time. The autocorrelation of $e_n$ is a

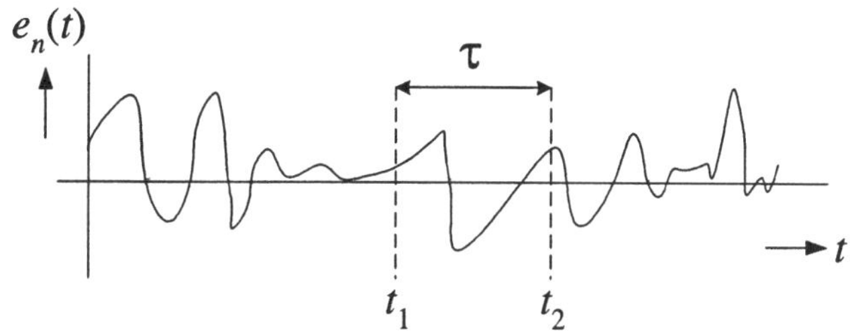


*Figure 4.19.* A noisy signal as a function of time.

measure for how good the value of $e_n$ at time $t = t_1$ predicts the value of $e_n$ at time $t = t_2 = t_1 + \tau$. It is not difficult to see that this is tightly related to the bandwidth of $e_n(t)$. When this bandwidth is very small, it is unable to change its value in time very quickly, while when it is large, $e_n(t)$ can change rapidly. In other words, when the bandwidth of $e_n(t)$ is small, $e_n(t_1)$ will quite accurately predict the value of $e_n(t_2)$ up to relatively large time intervals $\tau$. On the other hand, when the bandwidth is much larger, the prediction will be accurate only for very small $\tau$. Of course, for extremely large $\tau$, the prediction, i.e. the correlation between $e_n(t_1)$ and $e_n(t_2)$ will become very inaccurate in both cases; the values become uncorrelated. Thus if the bandwidth of $e_n(t)$ is relatively small (corresponding to a narrow power density spectrum), the autocorrelation will be a relatively wide 'peak', whereas it is a narrow 'peak' for large bandwidths.

The formal definition of the autocorrelation function of $e_n(t)$ is:

$$R_n(t,\tau) \triangleq \mathrm{E}\left\{e_n(t)e_n(t+\tau)\right\}. \tag{4.31}$$

Thus, in principle, it is a function of both the absolute time $t$ and the time-interval $\tau$. For a large class of noise processes, however, the autocorrelation is only a function of the time interval $\tau$, and not of $t$. Such processes are called (wide-sense) *stationary* [3]. All important electrical noise processes are wide-sense stationary, such that we can write:

$$R_n(\tau) \triangleq \mathrm{E}\left\{e_n(t)e_n(t+\tau)\right\}. \tag{4.32}$$

Expression (4.32) is not very suited to determine the autocorrelation from measurement results, since it requires the joint probability density function of $e_n(t)$ and $e_n(t+\tau)$ to be known for all time-differences $\tau$. For such purposes, one often uses an alternative definition:

$$R_n(\tau) = \lim_{T\to\infty} \frac{1}{T}\int_{-\frac{T}{2}}^{\frac{T}{2}} e_n(t)e_n(t+\tau)dt. \tag{4.33}$$

Although both formulations are quite different they yield the same function $R_n(\tau)$ for a large class of stochastic processes. Such processes are called (weakly) *ergodic*. Again, all major electrical noise processes are (assumed) weakly ergodic.

### 4.3.4 Example

As an example of an equivalent input noise power calculation, we determine the equivalent input noise of the amplifier of figure 4.9. The equivalent input noise voltage of this amplifier was determined already in section 4.2.6, and is given by equation (4.19).

[3]One usually distinguishes wide-sense stationary and strictly stationary processes. Although significant in probability theory, this distinction is not important in our present discussion.

**Step 1: Detect Correlation between Noise Sources**

The first step we have to take is determine whether the various noise sources$v_{n,2i}$ and the current source $i_{n,1}$ appearing in this expression are uncorrelated. This depends on whether these sources have the same origin, i.e. originate from the same electrical noise mechanism, or not. Although we have not discussed the noise phenomena present in electronic circuit components yet, we will assume that all noise source have zero mean and are uncorrelated, except the sources $i_{n,1}$ and $v_{n,2}$. As explained in later sections, these two sources represent the equivalent input noise of the nullor implementation, originating from the first transistor stage of the circuit implementing the nullor.

We will further assume that $v_{n,2}$ and $i_{n,1}$ can be expressed as a linear combination of three zero-mean, uncorrelated noise processes $v_{n,12}$, $i_{n,3}$ and $i_{n,5}$ as:

$$v_{n,2} = v_{n,12} + Bi_{n,3} \tag{4.34}$$
$$i_{n,1} = i_{n,5} + D * i_{n,3}, \tag{4.35}$$

The three independent noise sources are usually the main noise sources of the input transistors of which the details will be given in section 4.4.4 and section 4.4.5. The parameters that are used to transform the shot noise of the collector or drain current to the input of the transistor are$B$ and $D$. Parameter $B$ has the dimension of an impedance and$D$ is the dimensionless current-gain factor. For this example, $B$ is chosen to be constant and $D$ to be frequency dependent. So, for $D$ the impulse response $d(\tau)$ is used (a "typical FET-case").

**Step 2: Rearrange Expression for the Equivalent Noise**

Before we can apply equation (4.24) to determine the power spectral density, expression (4.19) for $v_{n,eq,in}$ has to be rearranged in such a way that it consist of a sum of mutually uncorrelated terms only. For this purpose, we substitute equation (4.34) and equation (4.35) for $v_{n,2}$ and $i_{n,1}$ into equation (4.19), and collect the terms for the uncorrelated processes$i_{n,3}$, $i_{n,5}$, and $v_{n,12}$. This yields:

$$\begin{aligned} v_{n,eq,in} = {} & \left[\left(R_s + \frac{R_1R_2}{R_1+R_2}\right) D(s) - B\right] i_{n,3} - \left(\frac{R_2}{R_1+R_2}\right) v_{n,4} \\ & + \left(R_s + \frac{R_1R_2}{R_1+R_2}\right) i_{n,5} + \left(\frac{R_1}{R_1+R_2}\right) v_{n,6} + v_{n,10} - v_{n,12}, \end{aligned} \tag{4.36}$$

where $d(\tau)$ has been represented by its Laplace transform$D(s)$ for convenience. Al terms in this expression are uncorrelated, such that equation (4.24) can be applied to it.

**Step 3: Equivalent Input Power Density Spectrum**

In order to calculate the power density spectrum of $v_{n,eq,in}$, we apply equation (4.24) to equation (4.36). Subsequently, the expressions for the power density spectra of the noise current- and voltage sources appearing in the right-hand side of equation (4.36) and the Fourier transform $D(f)$ of $d(\tau)$ have to be substituted. Usually, all these spectra are known in advance, such that the corresponding autocorrelation function doesn't have to be determined first.

We will assume that the power density spectra of all noise voltage sources are so called *white* spectra, i.e. frequency independent. This means that e.q. $S_{v_{n,3}}(f) \equiv S_{v_{n,3}}$. For the noise current sources we will assume:

$$S_{i_{n,3}}(f) = S_3\left(1 + \frac{f_{l,3}}{f}\right), \tag{4.37}$$

$$S_{i_{n,5}}(f) = S_5\left(1 + \frac{f_{l,5}}{f}\right), \tag{4.38}$$

and for $D(f)$ we take:

$$D(f) = D_0 + j2\pi f D_1. \tag{4.39}$$

Substitution of all these expressions finally yields:

$$\begin{aligned} S_{v_{n,eq,in}}(f) &= \\ &\left[\left(R_s + \frac{R_1R_2}{R_1+R_2} + D_0 - B\right)^2 + \left(R_s + \frac{R_1R_2}{R_1+R_2}\right)^2 D_1^2(2\pi f)^2\right] \\ &\times S_3\left(1 + \frac{f_{l,3}}{f}\right) \\ &+ \left(\frac{R_2}{R_1+R_2}\right)^2 S_{v_{n,4}} + \left(R_s + \frac{R_1R_2}{R_1+R_2}\right)^2 S_5\left(1 + \frac{f_{l,5}}{f}\right) \\ &+ \left(\frac{R_1}{R_1+R_2}\right)^2 S_{v_{n,6}} + S_{v_{n,10}} + S_{v_{n,12}}. \end{aligned} \tag{4.40}$$

Although this is a formidable expression, we can get a good impression of its behavior by looking at the various frequency asymptotes contained in it. The following asymptotes can be identified:

- an asymptote inversely proportional with frequency ($f^{-1}$);
- an asymptote independent of frequency ($f^0$);
- an asymptote proportional with frequency ($f^1$);
- an asymptote proportional to the square of frequency ($f^2$).

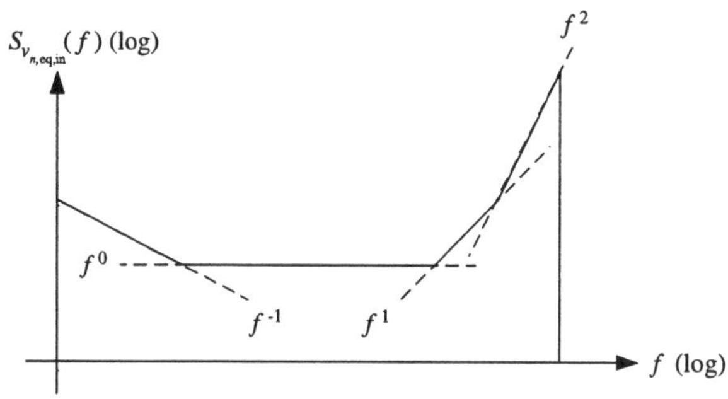


*Figure 4.20.* Asymptotic representation of $S_{v_{n,eq,in}}(f)$.

These asymptotes are sketched in figure 4.20 on a double-logarithmic scale (Bode Plot). As observed from this figure, the information signals located in the frequency region where the noise spectrum is white (flat) experience the smallest contamination by noise. In effect, to obtain an as low as possible equivalent input noise power, it is preferable to locate all signal information in this frequency range.

**Step 4: Equivalent Input Noise Power**

Finally, the equivalent input noise power is obtained by integration of the equivalent input noise power spectral density over the information bandwidth. This can be done separately for each of the four identified asymptotes.

Generally, the message information is located in the white part of the spectrum, such that the equivalent input noise power increases linearly with the information bandwidth; it simply equals the product of the power spectral density in that region and the information bandwidth. We will not give the general result for the noise power that can be obtained from equation (4.20), since it doesn't contribute much to the understanding of the noise behavior.

### 4.3.5 Summary

This section discussed the techniques required to determine the equivalent input noise power from the equivalent input noise source.

This power was defined to be the part of the noise power produced by the equivalent input noise source that is located in the same frequency range as the information signal, the so called information bandwidth. This part of the noise power fundamentally limits the Signal-to-Noise Ratio that can exist within the amplifier. All noise power produced by the equivalent source located outside this region can, in principle, be eliminated.

The distribution over frequency of the noise power produced by the equivalent source is described by its *power spectral density*. This power spectral density can be derived from the power spectral densities of the various contributing noise processes, which are generally known in advance, and the transfers of these sources to the equivalent input source. A constraint on this way of calculation is that all contributing noise sources have a zero mean value, and are uncorrelated.

Correlation is a measure for the extend that one noise process contains information about the value of another process. Two processes are called *fully correlated* when the value of one of them can be used to predict the value of the other one completely. When two processes are uncorrelated, they do not contain any information about each other. Correlated noise processes generally have a common origin. In the amplifier noise analysis, correlated noise sources can always be expressed in terms of linear combinations of uncorrelated noise processes.

The autocorrelation function is a measure for the extend that the present value of a noise process is able to predict future values. The Fourier transform of this function equals the power spectral density of the noise process. In amplifier noise analysis, however, it is not necessary to determine the power spectral density in this way.

## 4.4 Noise Models for Electronic Circuit Components

Throughout the foregoing sections, the various noise analysis techniques without reference to the origin of the noise sources have been discussed. For most of the discussions this was simply not necessary. An exception is the step were the presence of correlation between noise sources has to be detected, as illustrated by the example calculation in section 4.3.4. At such a point, the (common) origin of noise sources has to be identified.

In noise analysis of electronic circuits, however, the noise sources will not be given in advance, but have to be obtained from the noise models of the various circuit components. In addition, these models reveal the origin of the noise processes, and their mutual correlations.

In this section, we will discuss the noise models of the most frequently used electronic circuit components, the mutual correlations that exist and the associated noise power spectral densities. Together with the techniques discussed in section 4.2 and section 4.3, this provides all information necessary for the noise analysis of electronic amplifiers.

The various electronic noise processes have very similar characteristics, that will be considered in section 4.4.1. Sections 4.4.2 through 4.4.5 discuss the noise models of the various circuit components. Section 4.4.6 illustrates the use of this model for the voltage amplifier used in foregoing sections. Finally, section 4.4.7 summarizes the results.

### 4.4.1 General Characteristics of Electronic Circuit Noise

Although several different types of electronic circuit noise, with slightly different causes can be identified, all of them are fundamentally due to the quantized nature of electric charge on a microscopic scale. As a result of this quantized nature, electric current is not a continuous flow of charge, as it is modelled in circuit theory, but a stream of charged particles moving on average in one direction. Due to several (microscopic) mechanisms, their actual path of motion can be very irregular and very different from this average direction. As a consequence, the instantaneous current flowing through electronic components will slightly fluctuate in a random manner: it contains noise.

Due to this common cause, the statistical characteristics of electronic noise processes are very similar. All these processes are due to random movements of large numbers of equally charged particles. From stochastic theory is known that the probability density function of such processes approaches a normal/Gaussian density function. For this reason, all electrical noise processes considered in this book possess an (approximately) Gaussian probability function $p(e_n)$:

$$p(e_n) = \frac{\exp\left[-\frac{(e_n-\mu_n)^2}{2\sigma_n^2}\right]}{\sigma_n\sqrt{2\pi}}. \tag{4.41}$$

The parameters $\mu_n$ and $\sigma_n$ in this expression represent the average (expected) value and the variance of the process respectively. Since electrical noise is generally defined as the fluctuations of currents and voltages with respect to their average value, the average value of electrical noise processes equals zero:

$$\mu_n \equiv 0. \tag{4.42}$$

The value of the variance differs among the types of noise processes, and is determined by circuit and component parameters. In the sequel, the variance for each noise process will be specified in terms of the mean square value associated to a frequency band $\Delta f$, which is related to the power spectral density through equation (4.25).

### 4.4.2 Resistor Noise Model

Resistors produce a type of circuit noise that is called *thermal noise.* It is caused by (thermal) kinetic energy gained by free charge carriers when the temperature rises (thermal agitation). If the temperature is the same everywhere in the resistance material, this thermal motion of the charge carriers has no preference for any direction. As a result, thermal energy causes carriers to move randomly, uniformly distributed among all directions, which generates a random current/voltage.

In addition to the thermal agitation, an external applied source gives the motion of the charge carriers a drift component, in the same (or opposite) direction of the external current. The total external noticeable current/voltage equals the superposition of both effects, and therefore contains a random component: thermal noise.

Nyquist has shown that the voltage fluctuations (voltage noise) observed at the terminals of a resistor with value $R$ due to thermal agitation possesses a mean-square value associated to a frequency band $\Delta f$ equal to:

$$\overline{v_n^2}(f, \Delta f) = 4kTR\Delta f, f \geq 0, \tag{4.43}$$

where $k = 1.38 \cdot 10^{-23}$ (J/K) denotes Boltzmanns constant, and $T$ the absolute temperature in Kelvin. Thus, resistors produce white noise[4]. Figure 4.21 depicts the Thevenin and Norton equivalents of the resistor with thermal noise voltage source $v_n$ and noise current source $i_n$, respectively. Using the Thevenin-

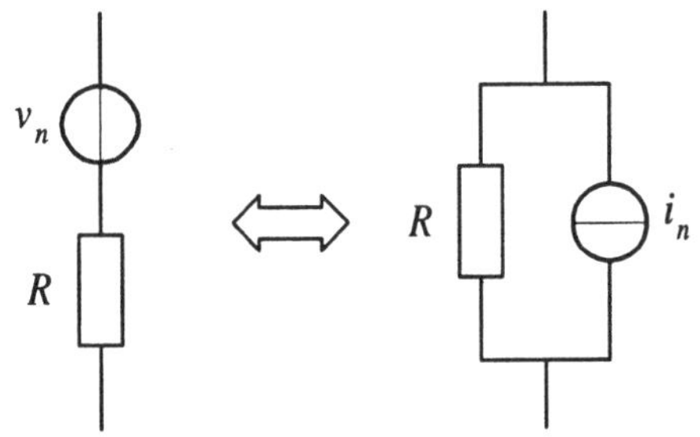


*Figure 4.21.* Noise model of a resistor.

Norton transform, equation (4.24) and equation (4.43), we find that the mean square value of the noise current in the Norton equivalent equals:

$$\overline{i_n^2}(f, \Delta f) = \frac{4kT\Delta f}{R}, f \geq 0. \tag{4.44}$$

In resistors made of materials with a very irregular internal structure, another type of noise, so called $1/f$ noise or flicker noise, may yield a non-negligible contribution to the resistor noise production. This is sometimes the case for resistors made of compressed coal powder. The variance of this flicker noise can be written as:

$$\overline{v_n^2}(f, \Delta f) = 4kTR\frac{f_l}{f}\Delta f, f \geq 0, \tag{4.45}$$

where $f_l$ denotes the frequency above which the white thermal noise dominates. For frequencies below $f_l$, the flicker noise dominates. The flicker noise is uncorrelated with the thermal white noise.

[4]In reality, the thermal noise of the resistor has an extremely large, but finite bandwidth. Its power spectral density, however, remains white up to frequencies far beyond the range of interest to electronic circuit design.

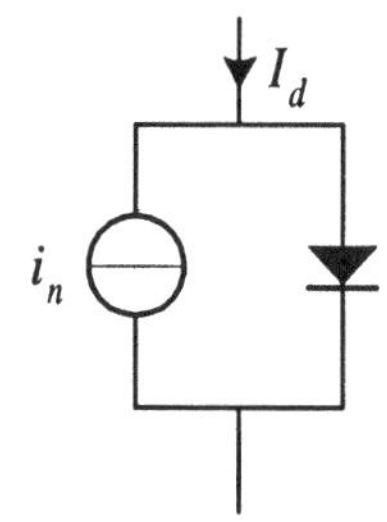


*Figure 4.22.* Noise model of a diode.

### 4.4.3 Diode Noise Model

Diodes, either pn-junction or Schottky, produce a type of noise that is called *shot noise.* As is known from semiconductor physics, the conduction in diodes is due to the diffusion mechanism, that injects minority carriers in the n-region and p-regions across the junction. The crossing of the individual carriers across the junction is a random process, that causes slight fluctuations, i.e. noise, in the external diode current.

The crossing events of the individual carriers can be assumed to occur independently from each other. The probability of such events is known to possess a Poisson distribution, which approaches the Gaussian distribution if the number of crossings becomes large. Figure 4.22 depicts the noise model of a diode, consisting of the diode and the shot noise current source connected in parallel. The mean square value of the noise current is proportional to the average external diode current $I_d$, i.e. the diode bias current:

$$\overline{i_n^2}(f, \Delta f) = 2q(I_d + 2I_s)\Delta f \approx 2qI_d\Delta f, f > 0, \qquad (4.46)$$

where $I_s$ denotes the diode saturation current. Thus, also the diode shot noise has a white power spectral density.

### 4.4.4 Bipolar Transistor Noise Model

Bipolar transistors produce a combination of shot noise, thermal noise and flicker noise. If forward biased in the normal operating region, the one of interest to amplifier design, this noise is determined by four uncorrelated noise processes.

Since the device essentially consists of two pn-junctions, shot noise is a principal cause of noise in bipolar transistors. In normal operation the base-emitter junction is forward biased, while the base-collector junction is reverse biased. The diffusion currents across the latter one are therefore extremely small, such that their noise contribution is negligible (see equation. (4.46)). The two diffusion currents crossing the forward biased base-emitter junction are not negligible, and therefore produce non-negligible shot noise. The diffusion current injected from the emitter into the base, constituting the collector current $I_c$, is transported out of the base through the reverse biased base-collector junction. This current therefore results in a shot noise source connected between the collector and emitter. The other diffusion current is injected from the base into the emitter, constituting the base current $I_b$, and therefore results in a shot noise source connected between the base and the emitter. Both sources, denoted by $i_c$ and $i_b$ respectively, are depicted in figure 4.23. Their mean square values

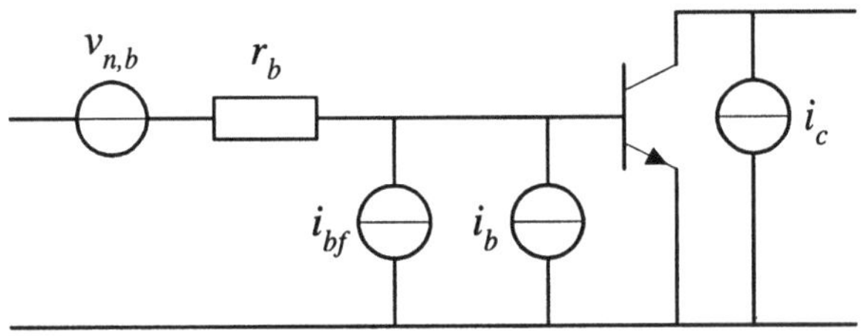


*Figure 4.23.* Noise model of a bipolar transistor.

are given by:

$$\overline{i_c^2}(f, \Delta f) = 2qI_c\Delta f, f \geq 0 \tag{4.47}$$

$$\overline{i_b^2}(f, \Delta f) = 2qI_b\Delta f, f \geq 0, \tag{4.48}$$

where $I_c$ and $I_b$ denote the DC collector and base currents, respectively.

Flicker noise, due to a variety of imperfections, yields another contribution to the noise current between the base and the emitter. In figure 4.23, it is represented by a current source $i_{bf}$. Its mean square value can be expressed as:

$$\overline{i_{bf}^2}(f, \Delta f) = 2qI_b\frac{f_l}{f}\Delta f. \tag{4.49}$$

For modern bipolar transistors, the frequency $f_l$ is usually very small (up to a few Hz), such that $1/f$ noise is mostly negligible.

The resistance of the bulk material, connected to the intrinsic base-emitter and base-collector junction produces thermal noise. The (small-signal) bulk resistance of the base, denoted by $r_b$, is generally much larger than the collector and emitter bulk resistances, and is usually the only one that produces non-negligible noise. The mean square value of the noise voltage $v_{nb}$ (see figure 4.23) produced by it equals:

$$\overline{v_{nb}^2}(f, \Delta f) = 4kTr_b\Delta f, f \geq 0. \tag{4.50}$$

### 4.4.5 Field-Effect Transistor Noise Model

The operation of Field-Effect Transistors (FETS), like Junction FETs and MOSFETs, is based on modulation of the resistance (or conductance) between source and drain by a third terminal, the gate. The noise produced by these devices is therefore mainly of thermal origin. Generally, this noise is determined by five different noise processes. Two or three of them, however, can often be neglected.

The principle cause of noise in FETs is the thermal noise produced by the drain-source channel resistance. The value of this resistance, and also its noise production, is generally dependent on the gate-source and gate-drain voltages. The lowest resistance is observed in the triode region, where the channel contains free charge carriers over its entire length between source and drain. In amplifiers, FETs are used in the saturation region, i.e. above pinch-off, where part of the channel is depleted, resulting in a higher resistance value. The thermal noise in the channel is generally represented by a noise current source $i_d$, connected between drain and source, as depicted in figure 4.24. Although the

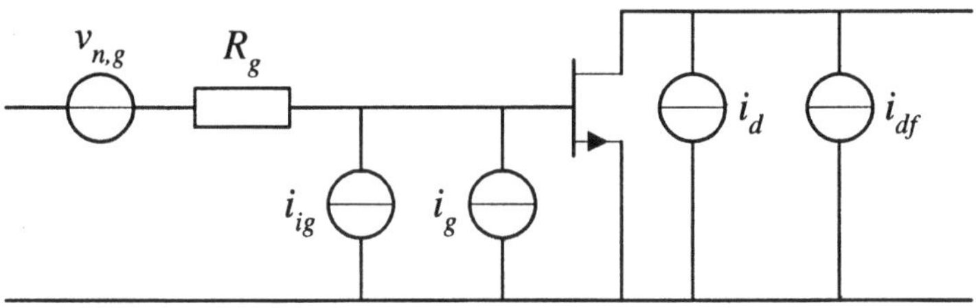


*Figure 4.24.* Noise model of a JFET and a MOSFET.

mean square value of $i_d$ is determined by the channel conductance between source and drain, it is customary to express it in terms of the *transconductance* $g_m$, relating the small-signal drain current to the gate-source voltage, and a proportionality constant $c$ (the ratio of the channel conductance and the transconductance) [3, 23, 24]:

$$\overline{i_d^2}(f, \Delta f) = 4kTcg_m\Delta f. \tag{4.51}$$

Although $c$ depends on the bias voltages of the device, one usually adopts a single value for it. From theoretical considerations, it follows that $c = 2/3$ for JFETs in saturation. For MOSFETs, theory predicts values between $c = 2/3$ and $c = 4/3$. However, practical values can range up to about $c = 2$.

Besides a thermal component, the channel noise generally also contains a flicker noise, $1/f$ noise component. Its influence is considerably larger than in bipolar transistors, due to the fact that the channel is located close to the surface, and therefore contains many defects and traps. The $1/f$ noise is represented by a current source $i_{df}$ between source and drain, that possesses a mean-square value equal to:

$$\overline{i_{df}^2}(f, \Delta f) = 4kTcg_m\frac{f_l}{f}\Delta f. \tag{4.52}$$

For JFETs, the frequency $f_l$ can be up to several tens of kHz, while for MOS-FETs, it can even be several hundreds of kHz.

Another, often omitted thermal noise contribution is generated by the resistance $R_g$ in series with the gate. Its value is usually small, since the gate is generally made of a highly doped semiconductor, or a metal. In some cases, however, the gate is made of a material with a relatively high-ohmic resistance, resulting in a non-negligible noise contribution. The mean-square value of the associated noise voltage $u_{n,g}$ equals:

$$\overline{v_{n,g}^2}(f, \Delta f) = 4kTR_g\Delta f. \tag{4.53}$$

In JFETs, a small shot noise contribution is generated by the saturation current $I_g$ flowing through the reverse biased gate-channel (source and drain) pn-junction. This noise current, represented by $i_g$ in figure 4.24, possesses a mean-square value equal to:

$$\overline{i_g^2}(f, \Delta f) = 2qI_g\Delta f. \tag{4.54}$$

At very high frequencies, around the transistor transit frequency, so called induced gate noise becomes noticeable. It is caused by the capacitive coupling that exists between the channel and the gate terminal. Charged particles travelling through the channel induce a small current through the gate terminal, i.e. the other plate of the gate-channel capacitance. In figure 4.24, the induced gate noise is represented by the current source $i_{ig}$ connected between the gate and the source. Its mean-square value equals [3, 23, 24]:

$$\overline{i_{ig}^2}(f, \Delta f) = 4kT\frac{(2\pi f)^2 C_{gs}^2}{3g_m}\Delta f. \tag{4.55}$$

The noise current $i_{ig}$ and the thermal drain noise current $i_d$ are correlated, due to capacitive gate-channel coupling. However, since their cross-correlation is complex valued with a zero real part, its does not contribute to the noise power spectral density (see section 4.3). For this reason, $i_{ig}$ and $i_d$ can be considered to be uncorrelated in calculation of the equivalent amplifier input noise power.

### 4.4.6 Example

In this section we illustrate the use of the noise models for electronic circuit components in the calculation of the equivalent amplifier input noise. For this purpose, we continue the voltage amplifier example, and implement the first stage of it by a JFET in common-source configuration, as illustrated in figure 4.25. The noise of the JFET, i.e. the equivalent input noise of the nullor implementation, is represented by the noise sources $i_{n,1}$ and $v_{n,2}$. An expression for them has to be derived from the JFET noise model, depicted in figure 4.24. The channel noise sources $i_d$ and $i_{df}$ have to be transformed to the JFET input

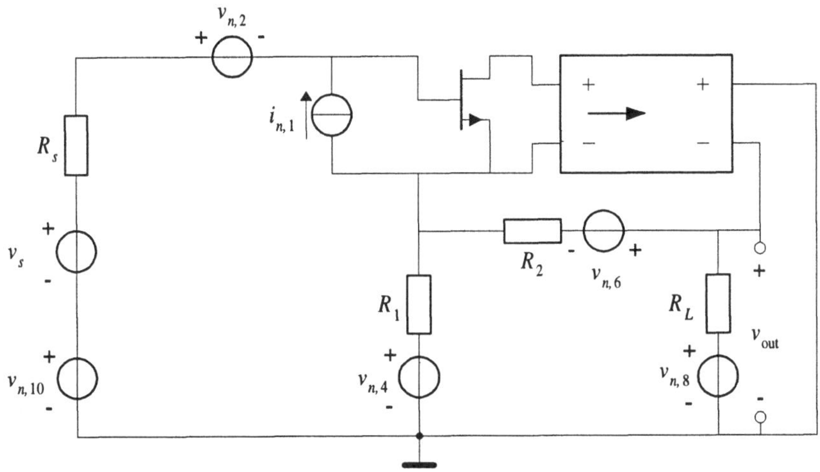


*Figure 4.25.* Voltage amplifier with a JFET input stage.

using the two-port transform. The chain parameters of the JFET required for this, $B$ and $D$, equal:

$$B \approx -\frac{1}{g_m}, \tag{4.56}$$

$$D(f) \approx -\frac{j2\pi f\,(C_{gs} + C_{gd})}{g_m}. \tag{4.57}$$

The result after transformation is depicted in figure 4.26. Subsequently, the

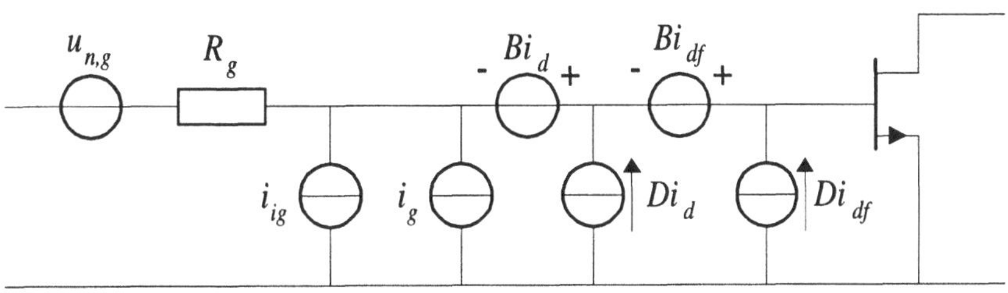


*Figure 4.26.* JFET noise model with channel noise transformed to the input.

noise current sources have to be shifted through the gate resistance, to the input port of the model. This is achieved by combined application of the I-shift and the Norton-Thevenin transform. The result is depicted in figure 4.27. This

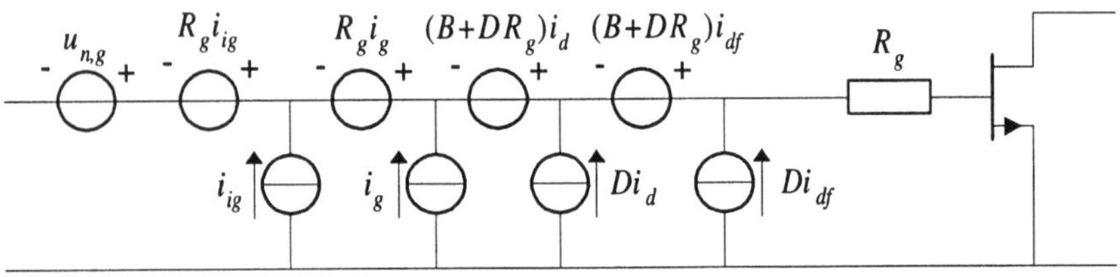


*Figure 4.27.* Equivalent JFET input noise.

yields the following result for $i_{n,1}$ and $v_{n,2}$:

$$i_{n,1} = i_g + i_{ig} - \frac{j2\pi f\,(C_{gs} + C_{dg})}{g_m}\,(i_d + i_{df}) \tag{4.58}$$

$$v_{n,2} = -u_{n,g} - R_g i_g - R_g i_{ig} + \left[\frac{1}{g_m} + \frac{j2\pi f R_g\,(C_{gs} + C_{gd})}{g_m}\right](i_d + i_{df})\,. \tag{4.59}$$

Obviously, the two sources are correlated, since they contain several contributions with the same origin.

The expression for the equivalent input noise voltage $v_{n,eq,in}$ is obtained by substitution of these expressions into equation (4.19). The resulting expression has to be re-grouped into uncorrelated terms before the equivalent input noise power spectral density can be determined; all contributions that originate from the same source have to be collected. The result obtained for $S_{v_{n,eq,in}}(f)$ is:

$$\begin{aligned} S_{v_{n,eq,in}}(f) &= \left(R_s + R_g + \frac{R_1 R_2}{R_1 + R_2}\right)^2 \left[S_{i_g}(f) + S_{i_{ig}}(f)\right] \\ &+ \left[\left(R_s + R_g + \frac{R_1 R_2}{R_1 + R_2}\right)^2 \frac{(2\pi f)^2\,(C_{gs} + C_{gd})^2}{g_m^2} + \frac{1}{g_m^2}\right] \\ &\times \left[S_{i_d}(f) + S_{i_{df}}(f)\right] \\ &+ \left(\frac{R_2}{R_1 + R_2}\right)^2 S_{v_{R_1}}(f) + \left(\frac{R_1}{R_1 + R_2}\right)^2 S_{v_{R_2}}(f) + S_{v_{R_s}}(f), \end{aligned} \tag{4.60}$$

where it has been assumed that $v_{n,4}$ represents the (thermal) noise of resistor $R_1$, $v_{n,6}$ the noise of $R_2$. Further, it has been assumed that the noise produced by the source equals the (thermal) noise of the source impedance $R_s$, modelled by $v_{n,10}$. In general, however, the noise produced by the source can be very different from the thermal noise of the source impedance. Expressions for the various component noise power spectral densities in equation (4.60) are obtained from the component noise models. Substitution of them finally yields:

$$
\begin{aligned}
S_{v_{n,eq,in}}(f) &= 4kT\left(R_s + R_g + \frac{R_1R_2}{R_1+R_2}\right) \\
&+ \left(R_s + R_g + \frac{R_1R_2}{R_1+R_2}\right)^2 \left[2qI_g + 4kT\frac{(2\pi f)^2 C_{gs}^2}{3g_m}\right] \\
&+ 4kT\left[\left(R_s + R_g + \frac{R_1R_2}{R_1+R_2}\right)^2 \frac{(2\pi f)^2 \left(C_{gs}+C_{gd}\right)^2}{g_m^2} + \frac{1}{g_m^2}\right] \\
&\times cg_m\left(1+\frac{f_l}{f}\right).
\end{aligned}
\tag{4.61}
$$

Usually, the contributions of $i_g$ and $i_{ig}$ are negligible, such that the second term in this expression can be omitted.

An important conclusion that can be drawn from equation (4.61) is that the feedback resistors $R_1$ and $R_2$ affect the amplifier noise behavior in exactly the same way as a resistance connected in series with the source, with a resistance value equal to the parallel connection of $R_1$ and $R_2$. It appears that analogous behavior is observed in the other three single-loop amplifier configurations. This can be used to simplify the noise analysis considerably, as will be demonstrated in the remaining sections of this chapter.

### 4.4.7 Summary

This section considered the final step in the amplifier noise analysis; the identification and modelling of the electronic circuit noise processes. Electronic noise is due to the quantized nature of electric current at a microscopic scale. Each noise mechanism involves large numbers of independently behaving charge carriers, which causes the probability density of virtually all electronic noise processes to be (approximately) Gaussian.

The three most important types of noise encountered in electronic components are:

- thermal noise, due to the thermal energy of charge carriers;
- shot noise, due to barrier crossings;
- flicker noise, $1/f$-noise, due to surface defects/traps.

Expressions for the power spectral densities of all relevant noise sources have been given that can readily be used in calculations of the equivalent input noise of a circuit. It appears that most electronic noise processes have a white noise power spectral density, at least as far as the frequency range covered by electronic circuit design is concerned.

In the example analysis, performed for a voltage amplifier with a JFET input stage, it was shown that the (impedance) feedback network affects the noise in the same way as an impedance connected in series with the source, with a value equal to the parallel connection of the feedback network impedances. Similar result can be obtained for the other three single-loop amplifier types. This result can be used to simplify the noise analysis considerably, as will be demonstrated in subsequent sections.

## 4.5 Feedback network noise optimization

The noise analysis techniques discussed in the foregoing sections are important, but not sufficient to attain an amplifier with an optimal noise performance. In addition, we need design techniques to minimize/optimize the noise contribution of both the feedback network and the nullor implementation.

This section shows in which way the feedback network should be designed, in order to arrive at an amplifier with an optimal noise performance.

The foregoing sections already demonstrated that the feedback network can affect the amplifier noise performance in two different ways:

- it can produce noise by itself;
- it can enlarge the noise contribution of the nullor implementation.

For example, the expression for the equivalent input noise power spectral density in section 4.4.6, equation (4.61), shows that the feedback resistors $R_1$ and $R_2$ do not only contribute through their own thermal noise (first term), but also by enlarging the noise contribution of the JFET (second and third term). These effects can be minimized, starting from a generic model for the feedback network. This finally leads to the feedback networks that are optimum with respect to noise.

### 4.5.1 Generic Model for the Feedback Network

In order to find all possibilities to design a feedback network with an optimal noise performance, we should start from a very general description that initially does not contain any information about its construction. Such a model can be obtained from the chain matrix, that was used already to describe the ideal feedback amplifier configurations, and the nullor.

In the same way as we did for the nullor and the entire amplifier, we can describe the feedback network transfer through its chain matrix $K_{fb}$, given by:

$$K_{fb} = \begin{pmatrix} A_{fb} & B_{fb} \\ C_{fb} & D_{fb} \end{pmatrix}. \tag{4.62}$$

The input of the feedback network is connected to the output of the amplifier, while the output is connected to the input of the amplifier. The transform described by the chain matrix $K_{fb}$ is thus directed from amplifier input to amplifier output.

From this chain matrix, an especially useful circuit representation can be derived, consisting of a controlled source, an input impedance and an output impedance. The controlled source represents the idealized transfer of the feedback network. For example, the circuit model of the feedback network of a transimpedance amplifier contains a voltage-controlled current source (VCCS). It is not difficult to show that this type of representation also holds for impedance feedback networks. The input and output impedance of the feedback network, $Z_{fb,in}$ and $Z_{fb,out}$, can both be expressed in terms of the chain parameters

$$Z_{in,fb} = \frac{A_{fb}Z_{L,fb} + B_{fb}}{C_{fb}Z_{L,fb} + D_{fb}}, \tag{4.63}$$

$$Z_{out,fb} = \frac{B_{fb} + D_{fb}Z_{s,fb}}{A_{fb} + C_{fb}Z_{s,fb}}, \tag{4.64}$$

where $Z_{L,fb}$ denotes the load of the feedback network (combination of the nullor input and the signal source), and $Z_{s,fb}$ the impedance of the 'source' driving the feedback network, i.e. the amplifier output impedance in combination with the amplifier load impedance. Expression (4.63) and (4.64) are the key to obtain the optimal feedback networks.

### 4.5.2 Magnification of Noise by the Feedback Network

As stated before, the feedback network can increase the amplifier noise production through magnification of the noise of the nullor implementation, even when it does not produce noise by itself.

In this section, we explain the origin of this magnification effect, and derive conditions on the feedback network that should be met to prevent it.

We derive the results for an amplifier with a voltage input (input information signal is a voltage). Figure 4.28 represents the input part of an amplifier with a voltage input, using the previously discussed generalized model for the feedback network. The voltage source $v_{fb}$ represents the signal fed back from the amplifier output to the input, while the impedance $Z_{out,fb}$ is given by equation (4.64). The sources $i_{n,1}$ and $v_{n,2}$ represent the equivalent input noise of the nullor implementation. Possible noise production by the feedback network is currently disregarded.

The magnification effect becomes clearly visible by calculation of the equivalent input noise voltage for this configuration. The first observation to be made is that the output impedance $Z_{out,fb}$ is connected in series with the source impedance. Therefore, the noise current source $i_{n,1}$ transforms into a noise voltage source by application of the Norton-Thevenin transform to the combination

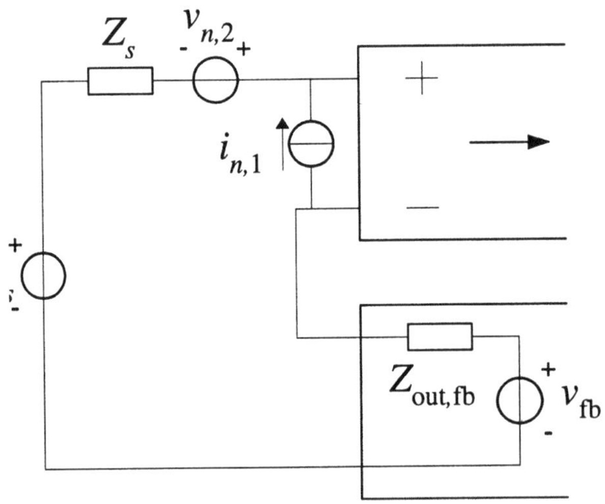


*Figure 4.28.* Model of an amplifier with a voltage input in the presence of noise.

$Z_s + Z_{out,fb}$. This yields:

$$v_{n,eq} = (Z_s + Z_{out,fb})\, i_{n,1} + v_{n,2}. \tag{4.65}$$

This expression shows that the output impedance of the feedback network magnifies the contribution of the input noise current of the nullor implementation by increasing the 'source impedance' seen by the current noise source. Similarly, in the case of a current input, the output impedance of the feedback network magnifies the contribution of the input voltage noise of the nullor implementation by decreasing the source impedance seen by the noise voltage source.

Consequently, to prevent magnification, we should require:

- $Z_{out,fb} = 0$ for a voltage input;
- $Z_{out,fb} \to \infty$ for a current input.

These conditions can be translated into conditions on the chain parameters via equation (4.64). The impedance $Z_{s,fb}$ in this expression equals the amplifier output impedance. When the amplifier has a voltage output, then $Z_{s,out} = 0$, while $Z_{s,out} \to \infty$ in case of a current output. Combination of this information then yields the requirements on the chain parameters listed in table 4.1.

Notice that the chain parameters required to be nonzero are exactly those that determine the transfer of the amplifier; they equal the (ideal) amplifier gain $A_{t\infty}$. Furthermore, only two of the four chain parameters are subjected to the conditions. The other two can, as far as noise magnification is concerned, be chosen arbitrarily.

| | | Amplifier Output V ($Z_{s,fb} = 0$) | I ($Z_{s,fb} \rightarrow \infty$) |
|---|---|---|---|
| Amplifier Input | V ($Z_{out,fb} = 0$) | $A_{fb} \neq 0, B_{fb} = 0$ | $C_{fb} \neq 0, D_{fb} = 0$ |
| | I ($Z_{out,fb} \rightarrow \infty$) | $A_{fb} = 0, B_{fb} \neq 0$ | $C_{fb} = 0, D_{fb} \neq 0$ |

*Table 4.1.* Conditions on the chain parameters of the feedback network to prevent noise magnification.

### 4.5.3 Noise Production by the Feedback Network

The second constraint on the feedback network for an optimum noise performance is that the feedback network itself should not generate noise. Using the ideal lossless network components, the transformer and the gyrator satisfy this requirement. Obviously, components like transistors, diodes and resistors do not satisfy this constraint.

A more general conclusion that can be drawn is that if the feedback network does not produce noise, its power gain cannot be larger than unity. As far as known, physical mechanisms that provide power amplification also cause the production of noise. A noise-optimal feedback network therefore cannot provide power amplification. A noise-optimum feedback network must therefore be *passive*. This matches very well to the conclusion in section 2.6.1, which also states that the feedback network should merely consist of passive components. But even losses in the feedback network should if possible be avoided. Power losses, like the power dissipation of resistors, always coincide with noise generation. So, the most favorable network is loss-free, i.e. its power gain equals unity. An other reason why this is desirable, is given in the next section.
It can be concluded that both power amplification and power dissipation (power attenuation) are processes that are generally contaminated by noise. So, inevitably, the circuit implementing the nullor will produce noise, but for the feedback network it is possible to avoid this.

### 4.5.4 Feedback Networks with Optimum Noise Performance

The previous sections already revealed the requirements that have to be satisfied by the feedback network to assure an as low as possible noise production. In order to arrive at an optimal feedback network, however, one additional

| | | Amplifier Output | |
|---|---|---|---|
| | | V ($Z_{in,fb} \to \infty$) | I ($Z_{in,fb} = 0$) |
| Amplifier Input | V ($Z_{L,fb} \to \infty$) | $A_{fb} \neq 0, C_{fb} = 0$ | $A_{fb} = 0, C_{fb} \neq 0$ |
| | I ($Z_{L,fb} = 0$) | $B_{fb} \neq 0, D_{fb} = 0$ | $B_{fb} = 0, D_{fb} \neq 0$ |

*Table 4.2.* Requirements on the chain parameters of a feedback network that does not load the nullor.

requirement has to be fulfilled. By adding this requirement, we derive the noise-optimal feedback network configurations in this section.

As discussed in chapter 1, orthogonality should be established in order to allow separate optimization of the design requirements (noise, distortion and bandwidth). Optimization of the noise behavior of the feedback network should therefore not result in a degeneration of the distortion behavior or bandwidth (LP-product), that cannot be repaired in later stages of the design procedure.

The orthogonality between amplifier noise and amplifier distortion is directly affected by the power gain/attenuation of the feedback network. The amplifier distortion is minimal when the output stage of the nullor implementation has to deliver an as low as possible amount of power to the load and the feedback network. The amount to be delivered to the load cannot be influenced by the designer. The amount of power required to drive the feedback network, however, can be influenced by proper design. The best possible situation is attained when all power provided by the nullor is used in the load, i.e. when nothing is spoilt to/lost in the feedback network. To achieve this, the feedback network should be loss-free.

The orthogonality constraint can also be translated into requirements on the input impedance of the feedback network. The network should not load the nullor. This means that if the amplifier has a voltage output (i.e. if it is a voltage or transimpedance amplifier) the input impedance of the feedback network should be infinite. When the amplifier has a current output (current and transconducance amplifier), the input impedance of the feedback network should be zero. Through equation (4.63), these requirements can be translated to requirements on the chain parameters, see table 4.2. Note that the requirement are not contradictory with the respective requirements listed in table 4.1.

Combining all requirements on the feedback network it is found that a noise-optimum feedback network should:

- not cause magnification of the noise of the nullor implementation;
- be passive and lossless to prevent noise production;
- not load the nullor to maximize orthogonality between noise and distortion.

Looking to the requirements as shown in table 4.1 and table 4.2, and considering the chain matrices of the ideal transformer and gyrator, it can be concluded that the optimal feedback network for a voltage amplifier and a current amplifier is an (ideal) transformer, while the optimal feedback network for a transconductance and transimpedance amplifier is a gyrator. Unfortunately, both the ideal transformer and the gyrator are circuit theoretical abstractions, that cannot be realized as an electronic circuit component. So, during the noise design, also more practical feedback networks, as shown in figure 2.8 constructed from impedances, will have to be examined. These networks will enlarge the amplifier noise and/or distortion level, but still an optimization is possible and must therefore be carried out, knowing that the absolute minimum that could be reached with a transformer or a gyrator is beyond reach. Only the feedback networks of current- and voltage buffers, consisting of just "wires" (zero impedance), are practical noise-optimal feedback networks.

### 4.5.5 Impedance Feedback Networks

When the noise-optimal feedback networks with transformers are not realizable, lacking the proper transformer for the application or because a gyrator is needed that cannot be realized noiseless in practice, one usually has to resort to impedance networks. Since these networks are not noise-optimal, they inevitably somehow affect the amplifier noise and distortion. In this section it will be shown how to minimize the influence of the impedance feedback network on the amplifier noise and distortion. This will also yield a very useful "short cut" for the evaluation of the equivalent input noise.

#### Noise Magnification

As discussed in section 4.5.2, magnification of the noise of the nullor implementation by the feedback network is the result of a non-zero or finite feedback network output impedance, for a voltage or current input, respectively. The output impedance of impedance feedback networks is always non-zero and finite, with a possible exception for a few discrete frequency points, like $f = 0$ in case of a capacitance/inductance network. Consequently, impedance feedback networks always enlarge the noise of the nullor implementation.

If the amplifier has a voltage input the output impedance of the feedback network is connected in series with the source impedance, as depicted in figure 4.29(a), and enlarges the effect of the input current noise of the nullor implementation. As far as magnification of the noise of the nullor implementation is concerned, the configuration of figure 4.29(a) therefore behaves identical to the configuration of figure 4.29(b), in which the source impedance has been increased from $Z_s$ to $Z_s + Z_{out,fb}$. This circuit is preferable in noise calcula-

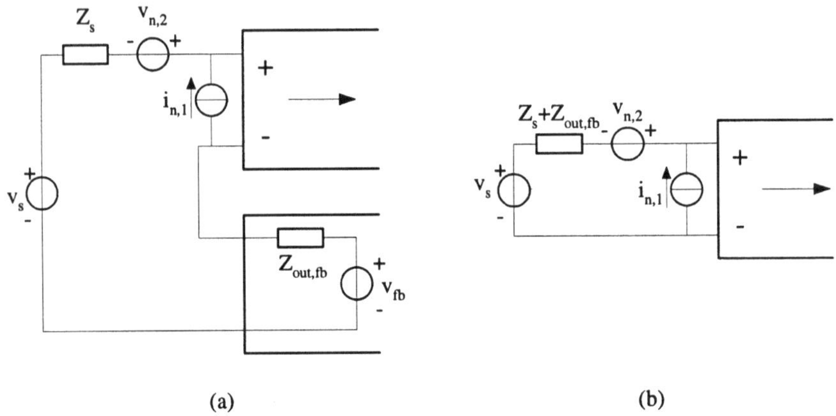


*Figure 4.29.* Amplifier with a voltage input (a), equivalent circuit w.r.t. magnification of the noise of the nullor implementation (b).

tions, since it reduces the number of noise transformations required to find the equivalent input noise voltage. Furthermore, this circuit shows that to prevent noise magnification, $Z_{out,fb}$ should be small, but also that it does not make sense to make it much smaller than $Z_s$.

If the amplifier has a current input, the output impedance of the feedback network is connected in parallel with the source impedance, as depicted in figure 4.30(a), and enlarges the effect of the input voltage noise of the nullor implementation. The configuration depicted in figure 4.30(b), with a source impedance equal to the parallel connection of $Z_s$ and $Z_{out,fb}$, yields the same noise magnification as the circuit in figure 4.30(a). Also this circuit simplifies the evaluation of the equivalent input noise source. In this case, the best noise performance is obtained when the output impedance is large> But it is also clear from figure 4.30(b) that it does not make sense to make it much larger than $Z_s$.

### Noise Production

When resistors are used, the feedback network does not only magnify the noise of the nullor implementation, but also produces noise itself. We already obtained the noise-equivalent circuits of figure 4.29(b) and figure 4.30(b), that provide a short-cut to the evaluation of the noise magnification by the feedback network. The result derived below is that these equivalent circuits also correctly represent the noise production of the feedback network itself in the case of resistors. Thus, when the feedback network consists of resistors only,

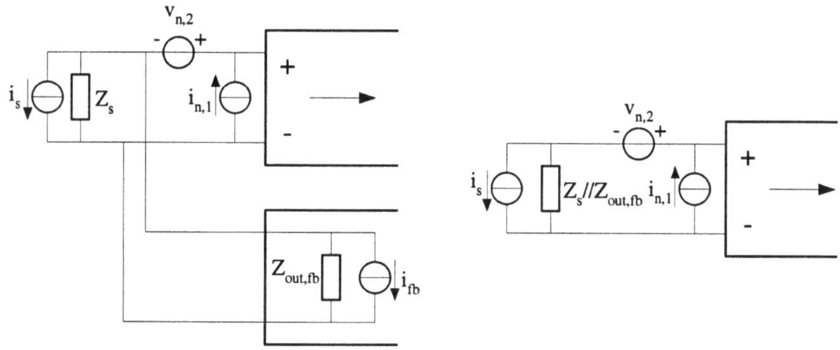


(a) (b)

*Figure 4.30.* Amplifier with a current input (a), equivalent circuit w.r.t. magnification of the noise of the nullor implementation (b).

its contribution to the equivalent input noise is equal to the noise of a resistor with a value $R_{out,fb}$ connected in series/parallel with the source.

The output impedance of a feedback network is, in the case of resistors only, determined by parallel and series connections of these resistors. It can easily be shown that for a series connection of two resistors $R_1$ and $R_2$ the power spectral density of the equivalent voltage noise is described by:

$$S_v = 4kT(R_1 + R_2) \tag{4.66}$$

and analogous for a parallel connection of two conductances ($G_1$ and $G_2$) the power spectral density of the current noise is given by:

$$S_i = 4kT(G_1 + G_2) \tag{4.67}$$

Subsequently, via induction follows for an one-port consisting of an arbitrary connection of resistors, the power spectral density of the equivalent voltage noise at the terminals is given by:

$$S_v = 4kTR_{port} \tag{4.68}$$

in which $R_{port}$ is the equivalent port resistance.

In section 4.5.2 the feedback network is modelled as a controlled source and an output impedance. From the previous discussion it follows that for the case of a passive resistive feedback network the associated noise at the output of the feedback network can be described by:

$$S_v = 4kTR_{out,fb} \quad \text{or} \quad S_i = \frac{4kT}{R_{out,fb}} \tag{4.69}$$

| | | Amplifier Output V | I |
|---|---|---|---|
| Amplifier Input | V | $S_v = 4kT(R_1//R_2)$ | $S_v = 4kTR$ |
| | I | $S_i = \frac{4kT}{R}$ | $S_i = \frac{4kT}{R_1+R_2}$ |

*Table 4.3.* Noise production for the feedback networks of figure 2.8 on page 39 for resistive feedback elements.

For the 4 configurations depicted in figure 2.8 on page 39 the power spectral densities for the corresponding noise sources are given in table 4.3.

For general passive networks, thus including capacitors and inductors, the noise production can not be be taken into account via this short-cut. It is easily illustrated by the following reasoning.

Assume that the power spectral density of the equivalent voltage noise of the feedback network is modelled correctly by

$$S_{v,eq} = 4kT\Re\{Z_{out,fb}\} \tag{4.70}$$

Analogous, for the power spectral density of the equivalent current noise should hold then:

$$S_{i,eq} = \frac{4kT}{\Re\{Z_{out,fb}\}} \tag{4.71}$$

These two power spectral densities relate to each other as:

$$S_{v,eq} = S_{i,eq}|Z_{out,fb}|^2 \tag{4.72}$$

Substituting the expressions for $S_{v,eq}$ and $S_{i,eq}$ yields:

$$4kT\Re\{Z_{out,fb}\} = \frac{4kT}{\Re\{Z_{out,fb}\}}|Z_{out,fb}|^2 \tag{4.73}$$

This is only true if and only if

$$\Re\{Z_{out,fb}\} = |Z_{out,fb}| \tag{4.74}$$

which is only true for real (resistive) networks.

The correct way for complex networks is to connect explicitly the complete feedback network in series/parallel, instead of a single element representing $Z_{out,fb}$. For instance, for a current amplifier the series connection of $Z_1$ and $Z_2$ should be connected in parallel with the input of the active part.

| | | Amplifier Output V | I |
|---|---|---|---|
| Amplifier Input | V | $A_{fb} = 1 + \frac{Z_2}{Z_1}$ $(\neq 0)$<br>$C_{fb} = \frac{1}{Z_1}$ $(0)$ | $A_{fb} = 1$ $(0)$<br>$C_{fb} = Y$ $(\neq 0)$ |
| | I | $B_{fb} = Z$ $(\neq 0)$<br>$D_{fb} = 1$ $(0)$ | $B_{fb} = Z_2$ $(0)$<br>$D_{fb} = 1 + \frac{Z_2}{Z_1}$ $(\neq 0)$ |

*Table 4.4.* The chain parameters of impedance feedback networks (figure 2.8) compared with the constraints for non-loading the nullor.

**Orthogonality between Noise and Distortion**

To have maximum orthogonality between noise and distortion, it was argued in section 4.5 that the feedback network should not load the nullor such that no increase of distortion results.

In table 4.4 the chain parameters of impedance feedback networks (figure 2.8) are compared with the constraints for not loading the nullor (see table 4.2).

For the voltage amplifier, the feedback network is not loading the nullor for the situation that $Z_2 = 0$ and $Z_1 = \infty$. This yields a voltage follower, i.e. the feedback network is just a wire. For other amplification factors the ratio of the two impedances is determined by the specified gain and the impedance level of $Z_1$ and $Z_2$ is free to choose. For the situation that the impedance level can be chosen such that $Z_1 \ll Z_s$ and $Z_1 + Z_2 \gg Z_L$, orthogonality exists between noise and distortion. When in this case for $Z_1$ and $Z_2$ capacitors or inductors are used, the feedback network does not produce noise and it is close to noise optimal.

Analogous conclusions hold for the current amplifier. Only the current follower ($Z_2 = 0$ and $Z_1 = \infty$) is perfectly non-loading the nullor. For other amplification factors loading may arise. However, with the freedom of the impedance level, this can be optimized.

For the transimpedance and transconductance amplifier no freedom exists. This is easily explained by the fact that the feedback impedance sets the transfer and it is the same impedance loading the output.

### 4.5.6 Example

In this section, again the equivalent voltage noise source is determined for the voltage amplifier of figure 4.9, but now the introduced short-cut for impedance feedback networks is used.

**Step 1: $v_{n,8}$**

This step is identical to step 2 on page 98. The result is that$v_{n,8}$ can be ignored. This is shown in figure 4.31.

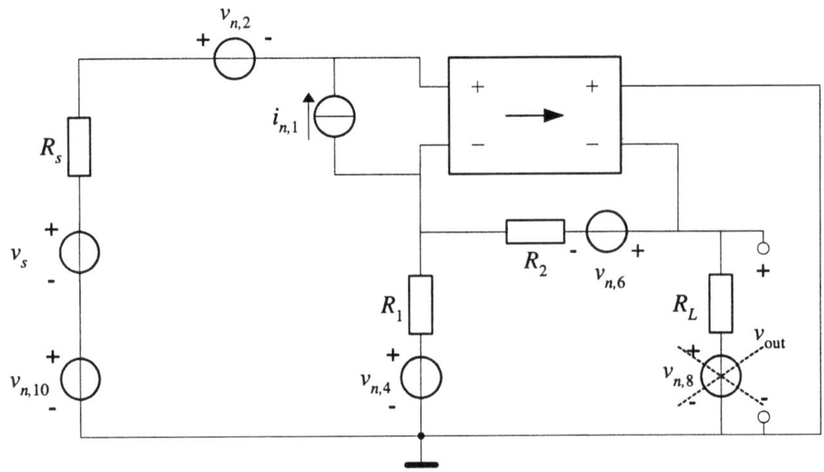


*Figure 4.31.* Contribution of $v_{n,8}$ to the equivalent noise source.

**Step 2: Short-cut**

Subsequently, the short-cut as introduced in section 4.5.5 can be used. This results in the noise-equivalent diagram of figure 4.32. The effect of the feed-

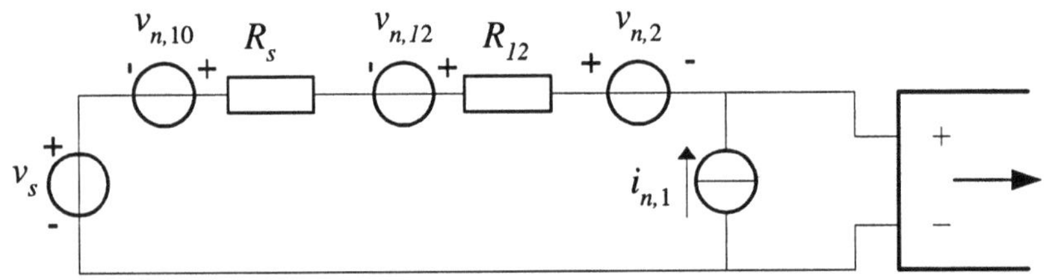


*Figure 4.32.* Noise-equivalent diagram when short-cut of section 4.5.5 is used.

back network on the noise magnification is equivalent to a resistor $R_{12}$, being the parallel connection of $R_1$ and $R_2$, in series with the signal source. The contribution to the noise of the feedback network, $v_{n,12}$, is equal to the noise of the parallel connection of $R_1$ and $R_2$. The power spectral density of this source equals $S_{vn,12} = 4kTR_{12}$.

**Step 3: $i_{n,1}$**

The last step in finding the equivalent source is transforming $i_{n,1}$ via the Norton-Thevenin transform. This results in the situation as depicted in figure 4.33.

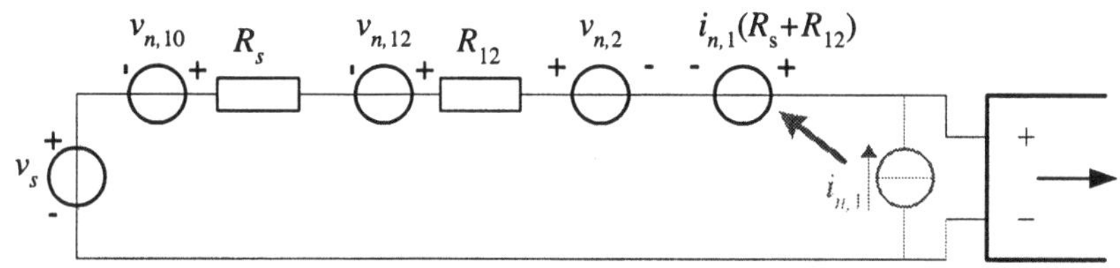


*Figure 4.33.* Transforming $i_{n,1}$ to the input.

Finally, the equivalent noise source is found to be:

$$v_{n,eq,in} = (R_s + R_{12})i_{n,1} - v_{n,2} + v_{n,10} + v_{n,12} \tag{4.75}$$

When compared with equation 4.19 on page 102, the difference is found in the way the noise contribution of the feedback network is represented. In expression 4.19 the contributions of the resistors $R_1$ and $R_2$ are indicated separately, i.e. $v_{n,4}$ and $v_{n,6}$, respectively, whereas in the expression found here, the noise of the total feedback network arises as a single term, i.e. $v_{n,12}$. When the power spectral densities are derived, for identical results are obtained for the two expressions.

Clearly, using this short-cut, a considerable reduction in the number of transforms is obtained, which is useful in the design process.

## 4.6 Design of the nullor input stage

The nullor implementation has to comply simultaneously with the requirements with respect to noise, distortion and bandwidth. To achieve this, we orthogonalize the requirements by confining them to different stages inside the nullor implementation in the following way. Since small signals are more vulnerable to noise than large ones, it is likely that the amplifier input has a dominant influence on the noise performance; the power contents of the information signal is minimum here. Therefore, it makes sense to design the nullor input stage for minimum noise, and assure that the noise contribution of other stages is negligible. In this section, we concentrate on the design of the input stage for low noise.

The input stage of the nullor in a low-noise amplifier has to comply with two requirements:

- it has to assure orthogonality between noise and the other requirements
- its own noise production should be minimal.

To assure orthogonality, the gain of the input stage should be made as large as possible. This was noticed already by Friis in 1944 [22] in conjunction with repeaters for telegraph lines. It also directly follows from the two-port transformation discussed in section 4.2, as illustrated by figure 4.34.

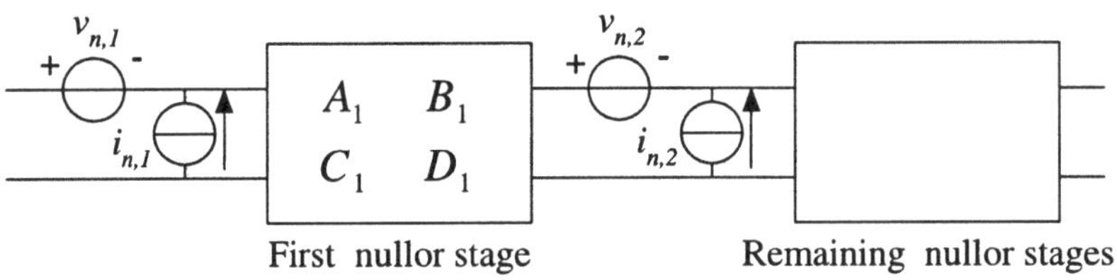


*Figure 4.34.* Assuring orthogonality with respect to noise by maximizing the gain of the first stage.

The two-port at the left of this figure represents the nullor input stage, while the one at the right represents the following stages. The equivalent input noise sources of these stages, represented by $v_{n,2}$ and $i_{n,2}$, transform to the amplifier input through the chain parameters of the input stage. According to figure 4.8 on page 96, their contribution to the equivalent amplifier input noise vanishes when all chain parameters of the input stage equal zero, i.e. when the gain of the input stage is infinite. So, it is important to use a first stage that has a gain as high as possible. If done so, the noise contribution of all other stages becomes negligible and the noise optimization of the amplifier can be concentrated on just this first stage.

The transistor stage that best approximates this behavior is the CS stage (common source) for MOSFETS/JFETS, or the CE stage (common emitter) for bipolar transistors. All parameters of the chain matrix are the smallest compared to the CD and CG, and the CC and CB stage, respectively. CS/CE-stages are the only stages that do not contain local feedback, like the source/emitter follower where $A = 1$ or the current follower where $D = 1$ (see exercise 3). Of course, also shunt- or series stages at the input are not preferred since they have a reduced gain and on top of that the local feedback network produces noise or at least increases the noise contributions of the transistor.

In the case that a CE or a CS stage does not suffice, (see section 2.6 for the correct reasons), an voltage/current follower is the next choice since the noise *production* of these stages is almost identical to the noise production of the CE/CS stage. Clearly, since for the voltage/current follower the chain parameter A/D equals one, the noise contribution of a second stage may become significant.

## 4.7 Noise Optimizations

After selection of the feedback network and the nullor input stage, there are still some degrees of freedom left to optimize the amplifier noise performance. These optimizations will be discussed below. In general, one or more of the following three types of optimizations can be applied:

- noise matching to the source via a transformer
- optimization of the bias current of the first stage
- connecting several input stages in series/parallel.

All of them can essentially be viewed as variants of the first type, as shown below.

### 4.7.1 Noise matching to the source via a transformer

The principle of noise matching to the signal source can be explained with the aid of figure 4.35. The sources $v_n$ and $i_n$ represent the equivalent amplifier input

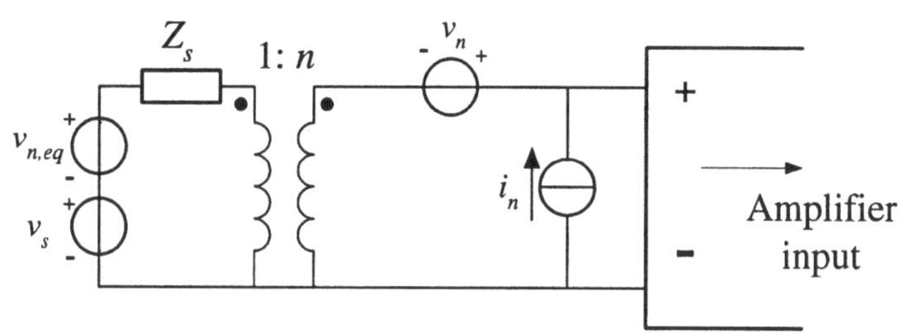


*Figure 4.35.* Principle of noise matching to the source.

voltage- and current noise, respectively. As shown in section 4.2, an equivalent noise voltage source $v_{n,eq}$ can be calculated. Via the (ideal) transformer, the ratio between the contribution of $v_n$ and $i_n$ to $v_{n,eq}$ can be adjusted. This can be done in such a way that the available power of the equivalent noise source $v_{n,eq}$ becomes minimum. Since it leaves the equivalent power of the signal source $v_s$ unchanged, this approach maximizes the equivalent amplifier input Signal-to-Noise ratio (SNR). The details are as follows. As known from circuit theory, the available power of $v_s$ and $v_{n,eq}$ equals:

$$P_{av,v_s} = \frac{\overline{v_s^2}}{4\Re\{Z_s\}} \tag{4.76}$$

$$P_{av,v_{n,eq}} = \frac{\overline{v_{n,eq}^2}}{4\Re\{Z_s\}} \tag{4.77}$$

The equivalent amplifier input SNR equals the ratio of these two. The transformer has no effect on equation (4.76), but it does affect equation (4.77), via the relation between $v_{n,eq}$ and $v_n$, in:

$$v_{n,eq} = \frac{v_n}{n} + nZ_s i_n \tag{4.78}$$

where $n$ is the transformation ratio. Consequently, the amplifier input noise has an effect as if the source impedance were equal to $n^2 Z_s$. The transformation ratio $n$ is chosen such that equation (4.77) is minimized, while equation (4.76) remains unchanged. For simplicity, assume that $Z_s = R_s$ is real, and $v_n$ and $i_n$ are uncorrelated (which is not always true). Then the optimal ratio $n_{opt}$ satisfies:

$$n_{opt} = \frac{1}{R_s}\sqrt{\frac{\overline{v_n^2}}{\overline{i_n^2}}} \tag{4.79}$$

A similar result is found when the signal source is a current source. For the optimal ratio, the contributions of $v_n$ and $i_n$ are equal. The advantage of this approach is that all noise sources in the amplifier are included in the optimization. The main disadvantage is that a transformer is not available in many cases or, as in microwave designs, realizes a match over a limited frequency range only.

### 4.7.2 Optimization of the bias current

This type of optimization relies on the fact that the intensities of several transistor noise sources, the collector- and base shot noise in a bipolar transistor and the channel noise in a MOS/JFET, depend on the transistor bias current. In addition, the chain parameters are also bias dependent.

The principle is illustrated by figure 4.36, which depicts the input stage with all noise sources. The source $i_{n,1}$ represents the bias dependent output noise current source (collector shot noise/channel noise), present for both a bipolar and MOS/JFET input stage. As depicted, it transforms to the input through the chain parameters $B$ and $D$. The bias-dependent input noise source $i_{n,2}$ (shot noise of the base current) is present only in case of a bipolar input stage. The power spectral density of both $i_{n,1}$ and $i_{n,2}$ is proportional to the bias current $I_{c,d}$. Two different situations can be distinguished.

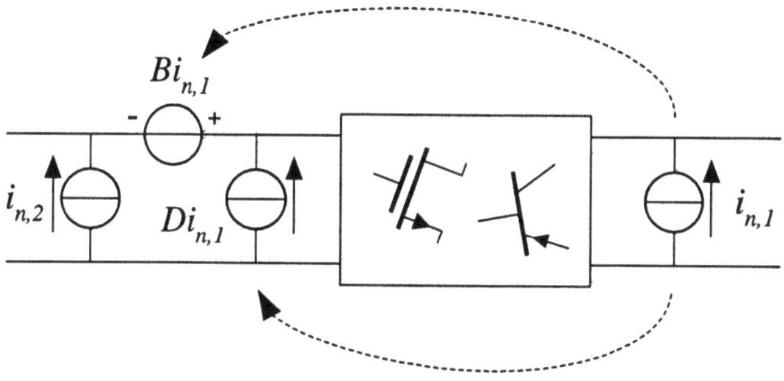


*Figure 4.36.* The bias dependent noise sources in a first stage.

- For a **FET** input stage, both $B$ and $D$ are inversely proportional to the drain bias current $\left(\frac{1}{I_d}\right)$ and consequently, the *power* contents of all equivalent input noise sources due to $i_{n,1}$ is proportional to $\frac{I_d}{I_d^2} = \frac{1}{I_d}$. In this case, no global noise optimum exists. To minimize the noise, the bias current, $I_d$, should be chosen as large as possible.

- For a **bipolar** input stage, a global noise optimum does exist. The chain parameter $D$ is independent of the collector bias current $I_c$ in this case. The power contents of the input current noise is proportional to $I_c$. The power contents of the equivalent input voltage noise is, similar to a FET input stage, proportional to $\frac{1}{I_c}$. So, there is a possibility to optimize the noise contribution of a bipolar transistor stage by choosing the optimal collector bias current. In figure 4.37 the transformations are indicated. The equivalent

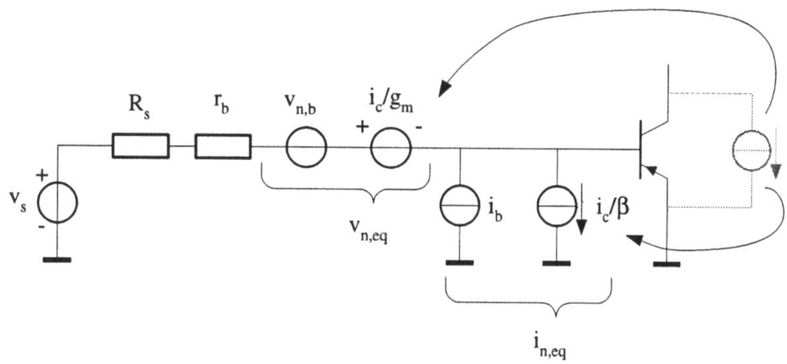


*Figure 4.37.* Equivalent noise sources of bipolar transistor in the case of a voltage input ($R_s$ models the source resistance and the feedback network).

input noise is given by:

$$v_{n,eq} = v_{n,b} + \frac{i_c}{g_m} \tag{4.80}$$

$$i_{n,eq} = i_b + i_c/\beta \tag{4.81}$$

Transforming $i_{n,eq}$ to a voltage source in series with $v_{n,eq}$ yields:

$$v_{n,bip} = v_{n,eq} + (R_s + r_b)i_{n,eq} \tag{4.82}$$

The power spectral density is readily found to be (see section 4.4.4)

$$S_{v,bip} = 4kTr_b + \frac{2qI_c}{g_m^2} + \left(2qI_b + \frac{2qI_c}{\beta^2}\right)(R_s + r_b)^2 \tag{4.83}$$

Rearranging and assuming $\beta \gg 1$ yields:

$$S_{v,bip} = 4kT\left(r_b + \frac{1}{2}r_e\right) + 2qI_b\,(R_s + r_b)^2 \tag{4.84}$$

in which $r_e$ is $\frac{kT}{qI_c}$. By differentiating this expression with respect to $i_c$ and subsequently equating the result to zero yields the optimum collector current:

$$I_{c,opt} = \frac{V_T\sqrt{\beta}}{R_s + r_b} \tag{4.85}$$

in which $V_T$ is $kT/q$. The optimum is mostly rather weak, such that biasing is not very critical.

The strength of bias current optimization is that it is virtually always applicable. No additional circuitry is required. Its main disadvantage is the limited scope. It can only optimize the contribution of bias dependent sources such as $i_c$ and $i_b$. The contribution of other sources cannot be influenced by it, whereas this could be done with a transformer.

### 4.7.3 Connecting stages in-series/in-parallel

A third optimization method is connecting several input stages in series or in parallel, that is scaling of the input stage.

When two identical stages are placed in parallel, their identical but uncorrelated input current sources add, resulting in a total noise current of $\sqrt{2}$ times larger than the noise current of a single stage. It can be shown that at the same time, the equivalent input noise voltage source of the two stages is reduced by a factor $\sqrt{2}$.

In case of two series connected input stages, exactly the opposite situation occurs. The voltage noise increases by factor $\sqrt{2}$, while the current noise decreases by a factor $\sqrt{2}$.

When $n$ stages are placed in series or in parallel, the same magnification/reduction is observed, but now with a factor $\sqrt{n}$.

The optimal scaling factor of the input stage can again be found from equation (4.79). In full-custom IC's, $n$ can be chosen arbitrarily (within certain boundaries), but in semi-custom IC's and discrete circuits, only integer values can be realized. But again, as the optimum is relatively flat, choosing only integer values leads to a slightly smaller SNR than the maximum one.

## 4.8 Exercises

### Exercise 4.1

Given the differential pair and its noise sources, as depicted in figure 4.38.

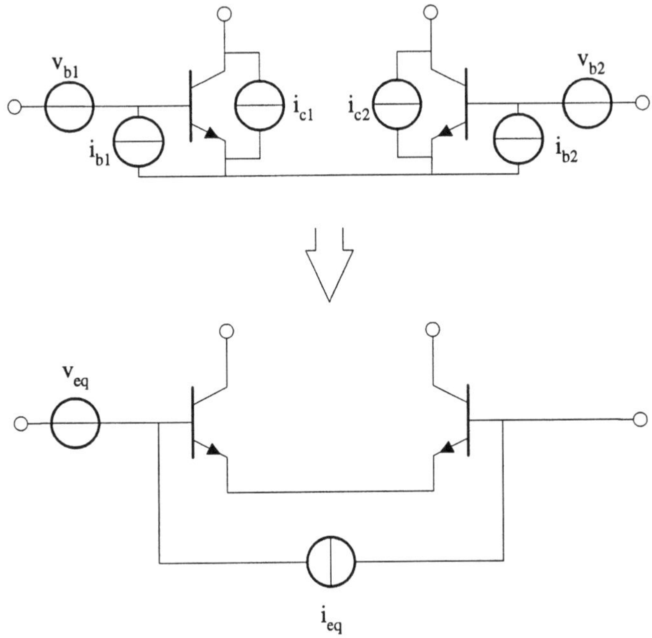


*Figure 4.38.* A differential pair and its (equivalent) noise sources.

1. Calculate the equivalent noise sources at the input of the differential pair.

2. How do these noise sources compare to the equivalent noise sources of a CE stage?

3. Calculate the equivalent noise sources at the input of two parallel connected CE stages .

4. How do these sources compare to the equivalent noise sources of a single CE stage?

## Exercise 4.2

A transadmittance amplifier is designed of which the gain is specified to be $\frac{1}{R}$. When compared to the noise specifications, it appears that the resistor which is required to set the transfer function of the trans-admittance amplifier, as depicted in figure 4.39, contributes too much noise. The only solution seems to be a reduction of the resistance. Consequently, the transfer function becomes too high. This can be corrected by a current attenuator cascading the amplifier (also depicted below).

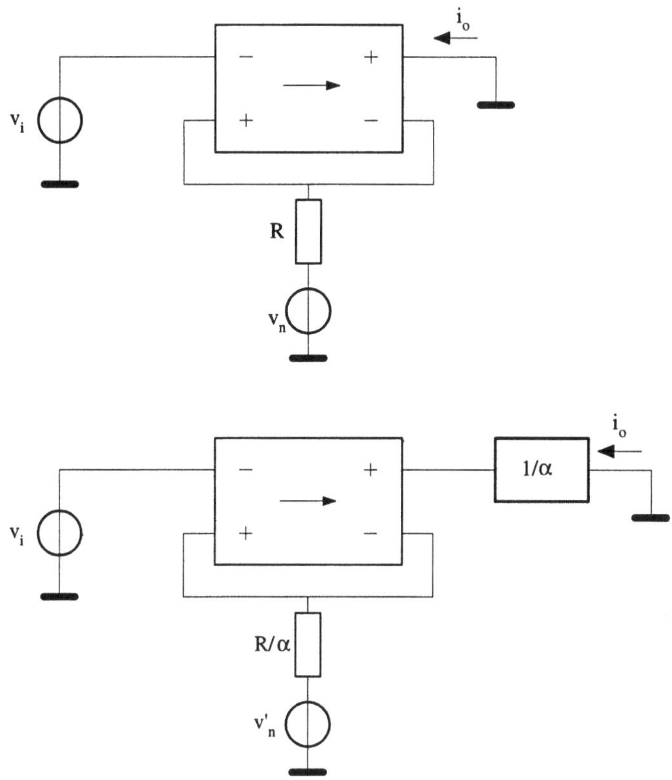


*Figure 4.39.* Two transconductance amplifiers

1. Is it possible to reduce the noise level of the amplifier in this way?
2. Are there any drawbacks?
3. To reduce the transfer to the original value, is it a good choice to use a voltage divider at the input of the amplifier instead of a current attenuator at the output?

## Exercise 4.3

Given the voltage amplifier built with an LF356 opamp as depicted in figure 4.40.

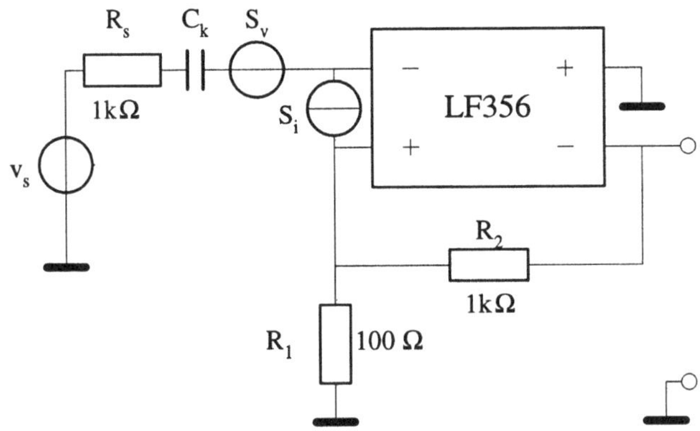


*Figure 4.40.* A voltage amplifier built with an LF356 opamp.

For the LF356 the following specifications apply:

- $S_v = 1.44 \cdot 10^{-16}$ V$^2$/Hz (=12 nV/$\sqrt{\text{Hz}}$)
- $S_i = 10^{-28}$ A$^2$/Hz (=0.01 pA/$\sqrt{\text{Hz}}$)

For the total amplifier holds:

- $R_i = 10^{12}$ Ω

For the signal holds:

- $B_{signal} = 50$ Hz .. 1 MHz

1. Determine $C_K$ on bandwidth constraints.
2. Identify all the noise sources.
3. Determine the equivalent noise source at the input (give the analytical expression).
4. What is the influence of $C_K$ on the noise behavior, what do you conclude ?

## Exercise 4.4

Given the voltage amplifier as depicted in figure 4.41.

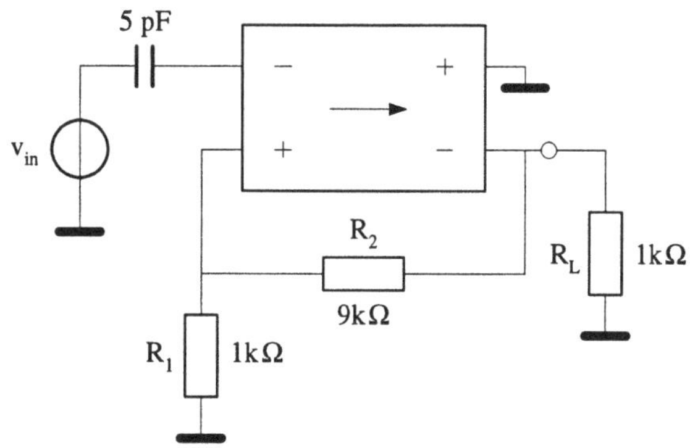


*Figure 4.41.* A voltage amplifier.

The source is a transducer with a capacitive impedance. The source voltage is the information-carrying quantity. The information supplied by the transducer is concentrated in a relatively small band centered at 1 kHz.

For the input transistor of the nullor implementation the following devices are available:

- BC549C : NPN, $\beta = 500, r_b = 500\ \Omega$
- J108 : NJFET, $I_{dss} = 150$ mA, $g_{m,max} = 70$ mA/V

1. Which transducer can be modelled as described above?
2. Design the first stage of the amplifier. The specification for the equivalent noise voltage at the input is:

$$v_{neq} \leq 25nV/\sqrt{Hz} \quad (4.86)$$

   Calculate the optimal bias current for the bipolar transistor and for the JFET.

3. Explain/interpret the differences.

## Exercise 4.5

Given the signal diagram of an amplifier (figure 4.42).

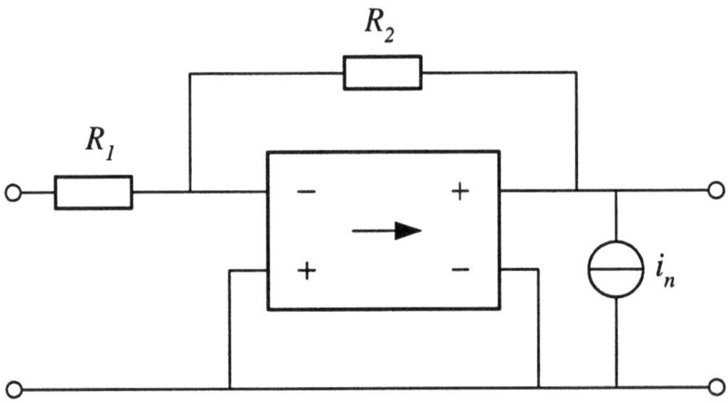


*Figure 4.42.* An amplifier.

1. Calculate the chain matrix of this amplifier assuming the active part to be a nullor.
2. When a noise source $i_n$ is connected at the output, calculate the equivalent sources at the input of the amplifier.
3. When the nullor is replaced by a single CE-stage, calculate again the equivalent sources at the input of the amplifier.

## Exercise 4.6

A capacitive source as depicted in figure 4.43, sources a current$i_s$. A noise voltage source $v_n$ represents the equivalent noise source of an amplifier that is connected to the source.

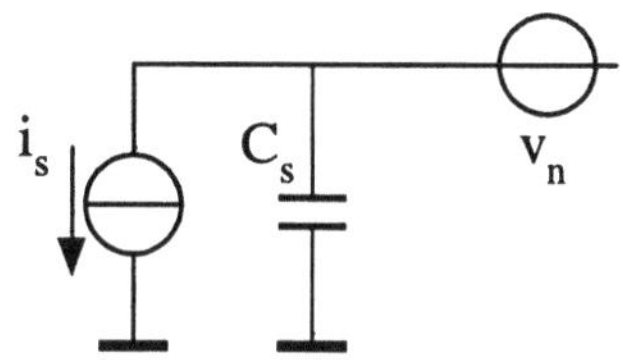


*Figure 4.43.* A capacitive current source

1. Is more information necessary about the amplifier to evaluate the noise behavior or is everything defined via the equivalent noise source? Why (not)?
2. The amplifier is implemented with a nullor as an active element. Calculate the equivalent noise source that can be used to evaluate the signal-to-noise ratio.
3. In a practical situation, the source will be connected to the amplifier via a coax cable, that introduces a parasitic capacitance of about$3C_s$. What will be the consequences for the noise behavior?

## Exercise 4.7

Figure 4.44 depicts a transimpedance amplifier. It comprises a nullor and a feedback resistor with a value of 1 MΩ.

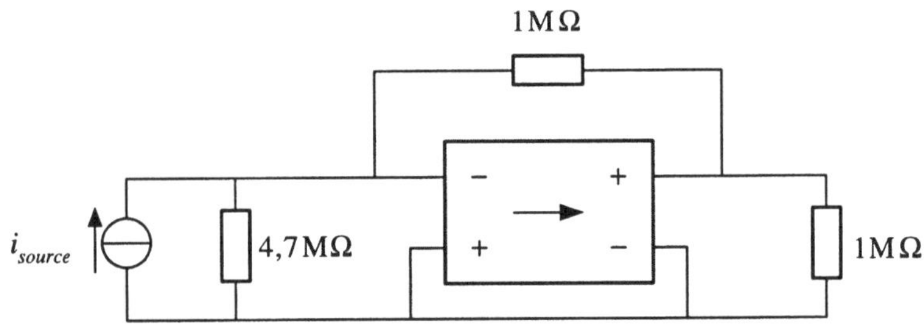


*Figure 4.44.* A transimpedance amplifier.

1. Calculate the value of the power spectral density of the equivalent noise source at the input.
2. Calculate the value of the power spectral density of the equivalent noise source at the output.

   With a "true-rms voltmeter" the power spectral density of the noise at both the input and the output of the amplifier is measured.

3. What is the measured value of the power spectral density of the the noise at the input.
4. What is the measured value of the power spectral density of the the noise at the output.
5. What is the influence of the load resistor on the*measurements* ?

## Exercise 4.8

Below, a number of signal sources has been depicted. Each of the sources is connected to an amplifier that has a bipolar transistor as first stage. For each

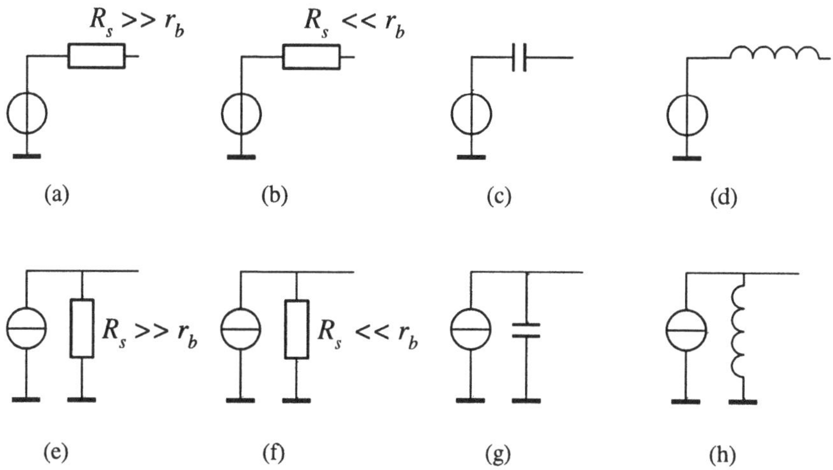


*Figure 4.45.* Various sources.

source, indicate what the effect is on the noise for the following situations / components:

1. The base resistance $r_b$.
2. The base noise current $i_b$.
3. For the frequency-dependent source impedances, evaluate both, the situation in which the amplifier is used to amplify very low frequencies only, and in which the amplifier is used to amplify very high frequencies only.
4. Indicate for which sources it might be better to use a FET as first device in the amplifier.

## Exercise 4.9

For the circuit depicted in figure 4.46, sketch the power density spectrum in figure 4.47, for the equivalent noise source at $v_s$, due to $i_n$ . The resonance peak of the source impedance is exactly in the middle of the information band.

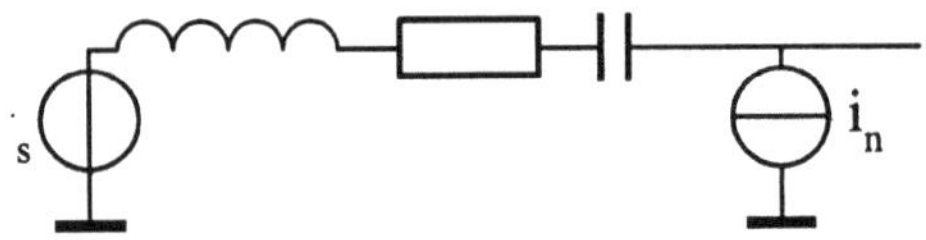


*Figure 4.46.* Source with a resonant source impedance.

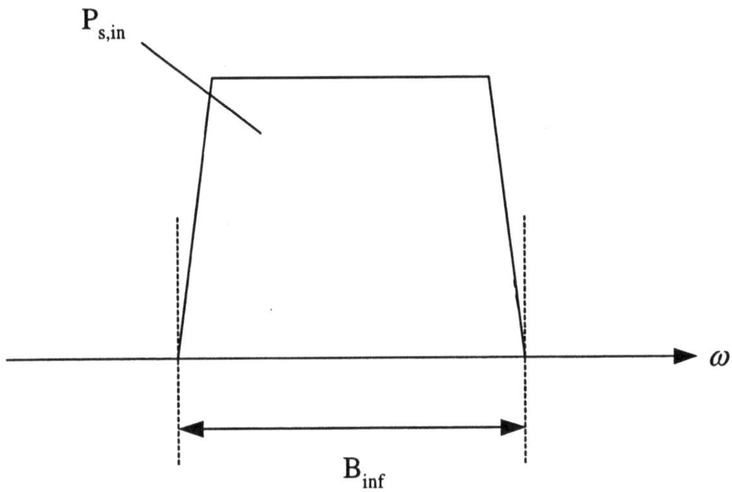


*Figure 4.47.* The signal power.

# 5

# NONLINEAR DISTORTION

## 5.1 Introduction

During the design of the amplifier circuit, linear models are used which are in the end approximated by real life nonlinear active devices. The behavior of linear models is independent of the applied signals. The nonlinear active devices can approximate this linear behavior up to a large extent. They deviate, however, in three distinct ways from the linear models, they have:

- input and output offsets
- a limited output range
- a curved input-output relation

The fact that the input and output offsets are not zero means that signals can not be applied to the nonlinear device as they are applied to the linear models. To the signals a level shift needs to be added in order to compensate for the offset. The addition of these offsets is called biasing. How to do this systematically is extensively described in chapter 8, Biasing.

This chapter is dedicated to the fact that the nonlinear device is not linear (not affine) in the sense that the output range is limited and that the input-output relation is curved, see figure 5.1. The limited output range of the active device results in the so-called *clipping distortion.* As the output signal cannot become larger beyond a certain level, i.e. the clipping level, any increase of input signal does not result in a change in the output signal. This is a rather aggressive type of distortion.

Due to curved relation, the output signal deviates (slightly) from the expected signal (based on a linear transfer). These deviations are called *weak distortion.* It results in additional harmonics in the case of a single-tone input signal or in intermodulation when a multi-tone signal is applied.

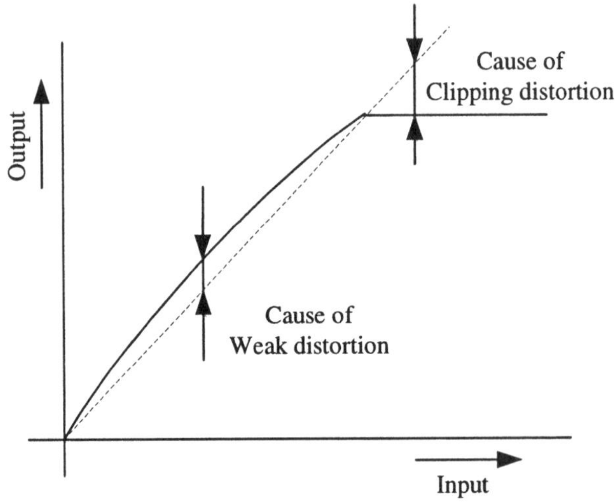


*Figure 5.1.* The deviations of a nonlinear active device with respect to a linear model.

Distortion is in general a nonlinear dynamic effect. So, to describe distortion with a high-level of accuracy, nonlinear dynamic models need to be used. However, these models tend to result in a high complexity, which make them difficult to use for design purposes. Therefor, in this chapter the focus is mainly on instantaneous distortion, i.e. the models are nonlinear and static.

This chapter is organized as follows. First, section 5.2 focusses in more detail on the origin of clipping and weak distortion. Subsequently, section 5.3 treats the most commonly used ways to give a measure for the level of distortion. For high-performance amplifiers the distortion should be low. Section 5.4 describes the design steps for preventing clipping distortion and minimizing weak distortion. At the end of this chapter, section 5.5 describes recent developed models derived with Volterra series. These models can predict relatively easily the generation of nonlinear-dynamic distortion of single amplifying stages.

## 5.2 The origin of distortion

Most of the design of the amplifier is done using linear models. Since at the system level a linear behavior is expected from an amplifier, this is the first choice[1]. The linear controlled sources in these models do not exist in practice,

[1]Linearity has already been the main paradigm for electronic design for many decades. Although the world around us is not linear at all, the linear models help us very much to synthesize circuits with a predictable behavior. Deviations from the linear behavior are called distortion and are considered to be undesirable. The fact that these deviations can still be rather predictable is usually ignored in the sense that they are not used to improve circuit behavior. Circuits that truly exploit the dynamic non-linear capabilities of the electronic components are rarely designed. Probably the design theory for circuits like that is not mature enough.

but using them in the first phases of the design makes designing relatively easy. The linear controlled sources that are commonly used in electronic design, are depicted in figure 5.2 and figure 5.3. They are the so-called "small-signal

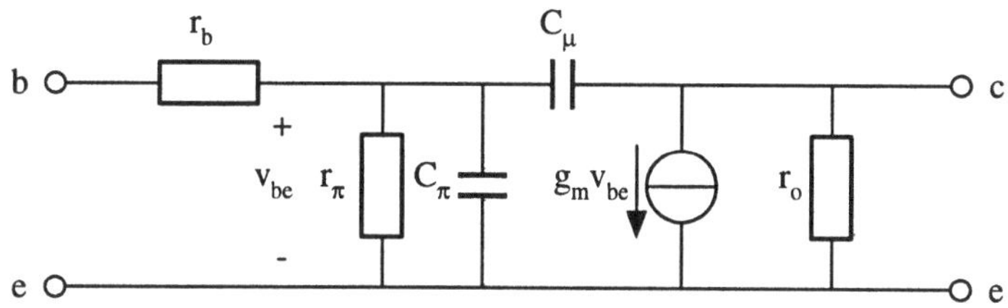


*Figure 5.2.* The small-signal model for the bipolar transistor comprising a *linear* voltage-controlled current source.

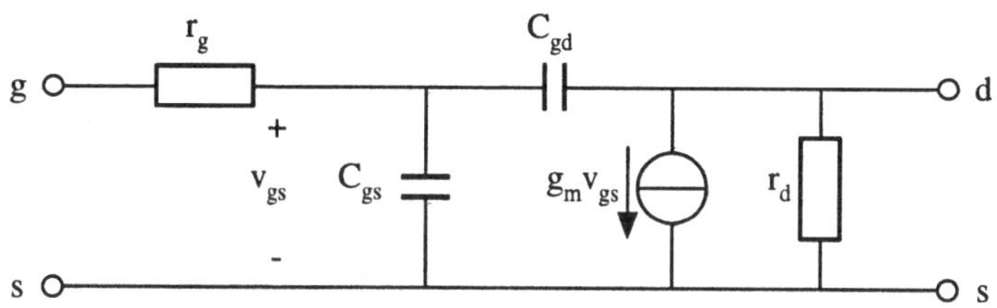


*Figure 5.3.* The *linear* small-signal model for the field effect transistor comprising a *linear* voltage-controlled current source.

models" of the bipolar transistor and the field-effect transistor. *This merely means that when these models are used in the first design stages, it is likely that later on for a limited signal range their behavior can also be realized using a bipolar transistor or a field-effect transistor.* Merely this, because more elements than just a transistor are needed to really implement something that may behave like the small-signal model. Inspecting figure 5.2 and figure 5.3, it can be seen that both models contain a source. A transistor does not contain a source. The sources in the models can generate power, a transistor can not. But a *biased* transistor can. Being a *non-linear* element, a transistor can convert power from a DC-source (a bias source, a battery) to any other frequency. So the combination of a transistor and a DC-source can generate the necessary power at the right frequency and thus perform the role of the controlled source that is present in the small-signal model. It is important to understand what the small-signal model really stands for, since this will provide the proper insight in the origin of distortion.

The source-transistor combination has a non-linear transfer. This combination can only approach the linear signal behavior in a limited signal range. The characteristic of the combination has been sketched in figure 5.4. The characteristic of the small-signal model would be a straight line through the origin. The first thing to do is to find a tangent to the non-linear function with the

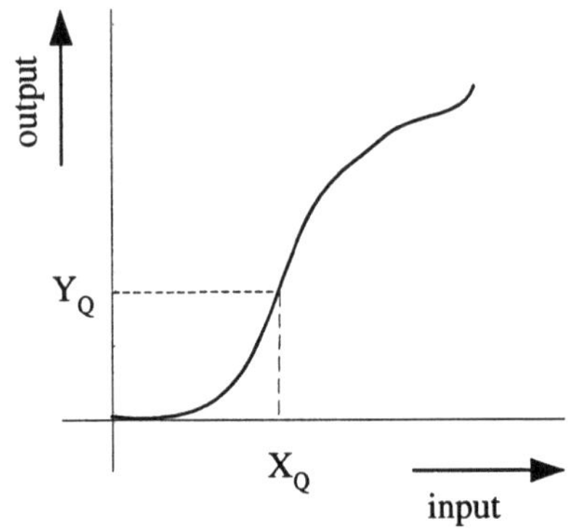


*Figure 5.4.* A non-linear transfer function.

same slope as the characteristic of the linear transfer. When a point is found on the non-linear characteristic for which this is true, this is chosen to be the operating point $(X_Q, Y_Q)$. Then the parameters of the small-signal model are the derivatives in this chosen operating point.

At the desired operating point a Taylor expansion of the transfer can be made:

$$\begin{aligned} y = Y_Q + y_s &= \tilde{f}(X_Q) \\ &+ A_t(x - X_Q) \\ &+ \frac{\tilde{f}''(X_Q)}{2!}(x - X_Q)^2 \\ &+ \frac{\tilde{f}'''(X_Q)}{3!}(x - X_Q)^3 \\ &+ \ldots \end{aligned} \tag{5.1}$$

in which $x$ and $y$ are the total input and output quantities, $X_Q$ and $Y_Q$ the input and output bias and $y_s$ the output signal, respectively. Further

$$A_t = \tilde{f}'(X_Q) \tag{5.2}$$

is the parameter that expresses the desired transfer function and therefore is found in the small-signal model. In equation (5.1) the desired expression:

$$y_s = A_t x_s \tag{5.3}$$

is "visible" ($x_s = x - X_Q$). All other terms in equation (5.1) are distortion terms that make the transfer non-linear. The methods to deal with the higher-order terms and with the DC terms ($X_Q$ and $Y_Q$) differ. The DC terms (offsets) are removed by adding bias signals to the information carrying signals. The higher-order terms can be reduced via the various methods described in this chapter. Main issue is to keep the signal small compared with the biasing. In that case the higher-order terms can be made negligible.

To have a linear transfer, the characteristic should at least be centered around the origin, see section 1.5.2. To have an overall linear transfer in this respect, first the input signal is translated to the operating point by adding $X_Q$ to it and after passing the device the signal is translated back to the origin by subtracting $Y_Q$ from it. In this way a linear (small-signal) transfer with the value $A$ for the information is achieved.

So, when a device with a non-linear characteristic $\tilde{f}(.)$ is used to implement a small-signal model with a transfer $A_t$, the operating point has to become the new origin. This is achieved by translating the signal both at the input and the output by adding bias signals. In figure 5.5 the principles of operation are shown. First an offset $X_Q$ is added to the input signal $x_s$. Then, the total signal

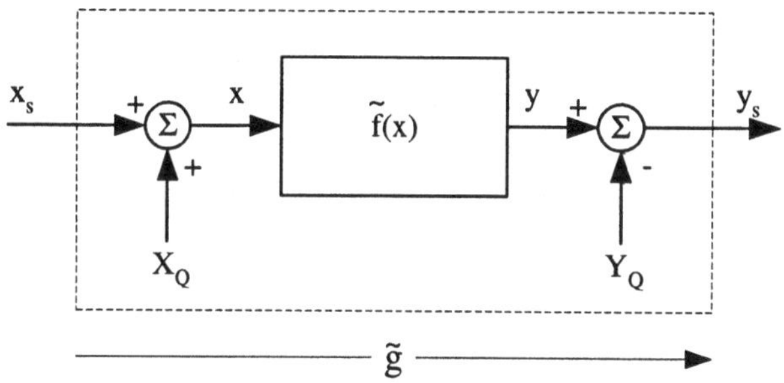


*Figure 5.5.* The principle of shifting the origin with bias signals.

$x$ is transferred via the non-linear device to the output. At the output the offset $Y_Q$ is subtracted and the output signal $y_s$ remains. When $x_s$ is relatively small, it is amplified with a "constant" factor $A$ to $y_s$. The addition and subtraction of the offsets (bias) $X_Q$ and $Y_Q$ have produced the required centering of the transfer around the origin, as depicted in figure 5.6.

What remains is a new non-linear input-output relation $y = \tilde{g}(x)$ in which only the distortion caused by the higher-order terms remains present. How to reduce this distortion, will be discussed in the next sections. The translations will be dealt with in more detail in chapter 8.

In conclusion, the non-linear elements in the amplifier are used around an operating point. This operating point is made the new origin of the transfer by adding and subtracting the proper bias signals as depicted in fig.5.6. From the fact the the device is nonlinear and biased, two types of distortion van be distinguished:

- Clipping distortion
- Weak distortion

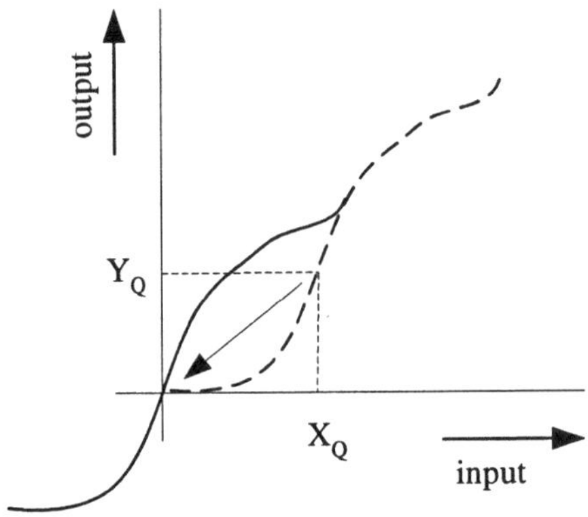


*Figure 5.6.* The translation of the operating point to the origin.

### 5.2.1 Clipping distortion

Clipping distortion is a direct consequence of the fact that a nonlinear device is biased. The bias level my appear to be insufficient, resulting in clipping distortion. The origin of clipping distortion is explained in more detail, referring to figure 5.7. For correct functioning of the nonlinear device with the input-

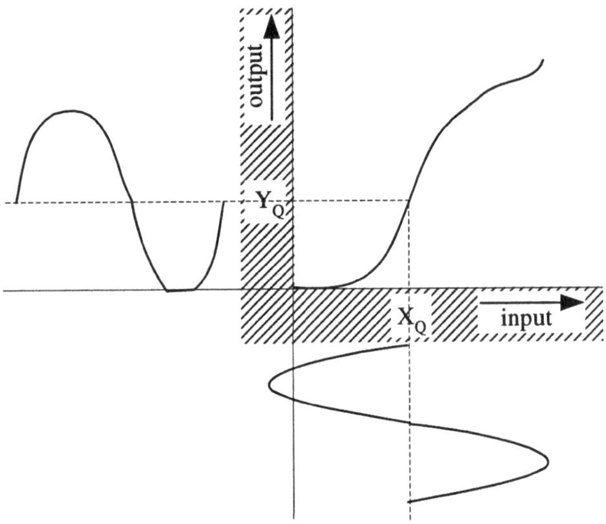


*Figure 5.7.* Origin of clipping distortion.

output relation as depicted in figure 5.7, the total applied signal, small signal *plus* biasing, should result in a device operation in the first quadrant *only*. Pushing the device in one of the other regions causes the device to switch off. Clearly, no signal processing is possible then. This is illustrated in the figure. On top of the bias, $X_Q$, a sinusoidal is superimposed. The amplitude of the sinusoidal is too large as it causes for a part of the period, the device to operate in the third quadrant. In this quadrant the device is switched off, and thus the output signal clips to the line $y = 0$.

For transistors analogous regions of operation can be identified. Bipolar transistors as well as FETs should operate in the forward active region. A transistor is a three terminal device[2], with terminal currents and port voltages, so numerous graphs can be depicted in which the correct operating region can be identified. However, as the transistors are assumed to supply gain, it is only necessary to consider the output characteristics. For a FET device a typical output characteristic is depicted in figure 5.8. For proper operation of the FET

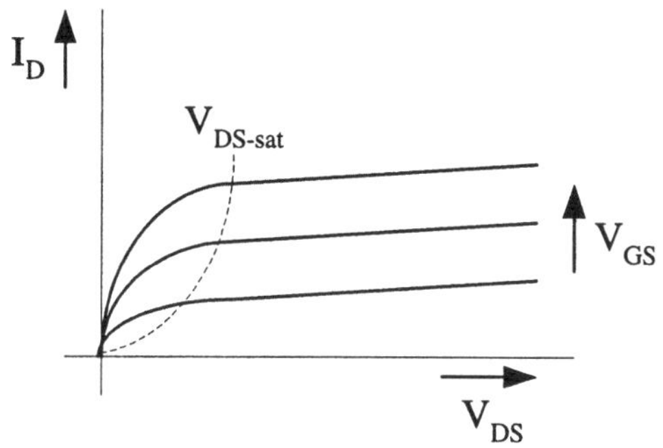


*Figure 5.8.* A typical output characteristic of a FET (N-type).

(N-type), i.e. for applying it as a gain stage, it should operate in a region with boundaries:

$$v_{DS} > V_{DS-sat} \tag{5.4}$$
$$i_D > 0 \tag{5.5}$$

The first boundary states that the FET should be in the saturation region. For drain-source voltages below $V_{DS-sat}$ the transistor is in the linear region. In that region the output resistance of the FET is relatively low, resulting in a reduced gain. The second boundary states that the transistor always should have a positive drain-source current.

For the bipolar transistor (N-type) same type of boundaries can be identified:

$$v_{CE} > V_{CE-sat} \tag{5.6}$$
$$i_C > 0 \tag{5.7}$$

[2]It is assumed that the substrate or well terminal is connected to the appropriate supply line

The collector-emitter voltage should always be larger than the saturation voltage. For lower voltages, the bipolar transistor starts saturating and, consequently, gain is reduced. On top of that, for proper functioning, the collector should be positive.

For P-type devices the boundaries are analogous, it is only a matter of polarities and changing ">" into "<".

Note that in figure 5.7 also the effect of weak distortion is visible. The positive top at the output is distorted by the nonlinear input-output relation.

In OPAMP design "Slewing" is a commonly used term. Slewing can be considered as a specific type of clipping distortion. This term is used for the effect that an amplifying stage clips in the current domain because of a relatively large capacitive loading. In the case of relatively high frequencies, the current required to charge/discharge the load capacitor becomes too large such that all the bias current of the driving stage is used for charging/discharging the capacitor. Consequently, the driving transistor is switched off and the feedback loop is broken. Of course, this situation should never occur in a good design. As the design measures for preventing slewing and clipping distortion are identical, slewing is not treated separately in the chapter.

### 5.2.2 Weak distortion

Weak distortion is caused by the weak nonlinearity of the input-output relation of the device. When considering weak distortion it is assumed that the proper bias signals are added, i.e. no clipping distortion present and the required small-signal behavior is obtained. The corresponding input-output relation $y = \tilde{g}(x)$ in which the required quiescent point is shifted to the origin, can be described by a Taylor approximation (see also equation (5.1) ):

$$y_s = A_t x_s + \frac{\tilde{g}''(0)}{2!} x_s^2 + \frac{\tilde{g}'''(0)}{3!} x_s^3 + \cdots \tag{5.8}$$

with:

$$A_t = \tilde{g}'(0) \tag{5.9}$$

The first term, $A_t$, of this Taylor approximation is the desired linear term.

Clearly, as long as $x_s$ is relatively small, indeed the nonlinear transfer is rather well approximated by:

$$y_s = A_t x_s \tag{5.10}$$

However, the higher-order terms are not zero as long as the input signal is not zero. These always present higher-order terms in the output signal are due to the *weak distortion.* They are designated as "weak" as these higher-order terms are for high-performance amplifiers relatively small; they can be orders of magnitude smaller than the intended output signal. Therefor, the terms with

| Amplifying stage | $\kappa_2$ | $\kappa_3$ |
|---|---|---|
| CE stage | $\frac{r_\pi}{2(r_\pi+r_s)}$ | $\frac{r_\pi(r_\pi-2r_s)}{6(r_\pi+r_s)^2}$ |
| Diff. CE stage | $\frac{\delta_{I_Q} r_\pi}{2r_\pi+r_s}$ | $\frac{-2r_\pi[(2-2\delta)r_\pi+r_s]}{3(2r_\pi+r_s)^2}$ |
| CS stage | 1/4 | 0 |
| Diff. CS stage | $\frac{\delta_{I_Q}}{4}$ | $\frac{\delta_{I_Q}^2}{8}$ |

*Table 5.1.* The normalized output related distortion coefficients for a CE and CS stage and the differential version.

order higher than three are ignored, yielding:

$$y_s = A_t x_s + A_{t2} x_s^2 + A_{t3} x_s^3 \tag{5.11}$$

For obtaining a compact equation in terms of bias-related and transistor-related parameters, expression (5.11) is rewritten in terms of a load-referred distortion by substituting $x_s = y_{s1}/A_t$, [25], [26] and [27]:

$$y_s = y_{s1} + A_{l2} y_{s1}^2 + A_{l3} y_{s1}^3 \tag{5.12}$$

in which $A_{ln} = A_{tn}/A_t^n$. When deriving this expression for the bipolar transistor and MOS devices, it appears that a key parameter is the relative current swing, $m$. It is defined as:

$$m = \frac{\hat{i}_o}{I_Q} \tag{5.13}$$

in which $\hat{i}_o$ is the total output current and $I_Q$ is the bias current. In terms of the intended output current, i.e. the linear part, the relative current swing equals:

$$m_1 = \frac{\hat{i}_{o1}}{I_Q} \tag{5.14}$$

in which $\hat{i}_{o1}$ is the linear portion of the output current. Substituting the expressions for the relative current swing in expression 5.12, yields [26] and [25]:

$$m = m_1 + \kappa_2 m_1^2 + \kappa_3 m_1^3 \tag{5.15}$$

So, an expression is obtained in terms of bias-related parameters, $m$, and transistor-related parameters, $\kappa$. Calculating the weak distortion for a single bipolar and MOS transistor stage, yields the $\kappa's$ as summarized in table 5.1 [25]]]] In the expressions for $\kappa_i$, $r_s$ is the resistance of the driving source and $\delta_{I_Q}$ is the matching error of the quiescent current of the two transistors in the differential stage. A difference between the bipolar transistor and FET is the dependency on the driving-source resistance. For the FET, the distortion is independent of $r_s$. This is because in the instantaneous situation the input resistance of the FET is infinite. For the bipolar transistor the input resistance is $r_\pi$, yielding the factor $r_\pi/(r_s + r_\pi)$ in the expressions. This dependency gives rise to the definition of two types of weak distortion:

- $\beta$-distortion
- $g_m$-distortion

The first type of distortion arises in the bipolar transistor when it is current driven. Looking to table 5.1 reveals that in that case the bipolar transistor does not introduce any distortion. In this text this is a consequence of the assumption that the current-gain factor of the bipolar transistor is independent of the collector current, i.e. a constant. Would the current-gain factor be modelled as collector-current dependent, then for an infinite driving resistance, solely this effect would be responsible for the distortion.

The second type of weak distortion is the $g_m$-distortion. This type of distortion is caused by the collector-current (drain-current) dependency of the transconductance of the transistors. Clearly, when the driving resistance is zero, i.e. an ideal voltage drive, the $g_m$ distortion dominates in the transistor. As FET devices have a infinite DC current-gain factor, and all the non-zero chain parameters are a function of $g_m$, the FET can only exhibit $g_m$-distortion.

For the dynamic situation, i.e. for frequencies beyond the pole of the transistor, one might expect that also the bipolar transistor my dominantly introduce $g_m$-distortion. The relatively low impedance of the transistors input capacitance causes the intrinsic base-emitter junction to be voltage driven, yielding $g_m$-distortion. This will be seen in more detail in section 5.5 where the dynamic distortion is considered.

Using the normalized distortion coefficients of table 5.1, relatively simple expressions can be obtained for the weak distortion of negative-feedback amplifiers. For a nullor implemented by means of three amplifying stages, the weak distortion of the corresponding amplifier obtained at the output, is described by:

$$A_{t2} = \frac{1}{F\nu}\left(\frac{\kappa_{2a}}{\alpha_b\alpha_c I_{Qa}} + \frac{\kappa_{2b}}{\alpha_c I_{Qb}} + \frac{\kappa_{2c}}{I_{Qc}}\right) \tag{5.16}$$

$$\begin{aligned} A_{t3} &= \frac{1}{F\nu^2}\left(\frac{\kappa'_{3a}}{\alpha_b^2\alpha_c^2 I_{Qa}^2} + \frac{\kappa'_{3b}}{\alpha_c^2 I_{Qb}^2} + \frac{\kappa'_{3c}}{I_{Qc}^2}\right. \\ &\quad \left. - \frac{2\kappa_{2a}\kappa_{2b}}{\alpha_b\alpha_c^2 I_{Qa} I_{Qb}} - \frac{2\kappa_{2a}\kappa_{2c}}{\alpha_b\alpha_c I_{Qa} I_{Qc}} - \frac{2\kappa_{2b}\kappa_{2c}}{\alpha_c I_{Qb} I_{Qc}}\right) \end{aligned} \tag{5.17}$$

in which $F = 1 - L$ and $\nu$ the output loading factor of the asymptotic-gain model. Further, $\kappa'_3 = \kappa_3 - 2\kappa_2^2$, which is listed in table 5.2 for the bipolar and

| Amplifying stage | $\kappa_3'$ |
|---|---|
| CE stage | $\frac{-r_\pi}{3(r_\pi+r_s)}$ |
| Diff. CE stage | $\frac{-2r_\pi}{3(2r_\pi+r_s)}$ |
| CS stage | -1/8 |
| Diff. CS stage | 0 |

*Table 5.2.* The normalized output related distortion coefficients,$\kappa_3' = \kappa_3 - 2\kappa_2^2$, for a CE and CS stage and the differential version.

FET basic stages. Further, $\alpha_b$ and $\alpha_c$ are the current gain of the middle stage and output stage, respectively (stages are labelled from input to output with$a$ to $c$, respectively). In the derivation of this expression a relatively high loop gain was assumed.

The equation clearly shows some general aspects of distortion generation in feedback amplifiers:

- The larger the return difference, $F$, or approximately, the loop gain, the lower the distortion. This is because the higher the loop gain, the better the nullor is approached by the active part. In the case of infinite loop gain, i.e. a nullor, no distortion is obtained at all.

- The larger the bias current, the lower the distortion, which is also in line with common intuition. A larger biasing current means a smaller relative current variation ($m$), which is preferable for low distortion, see equations (5.14) and (5.15).

- The type of amplifying stage (CE/CS or differential CE/CS) and what type of weak distortion is dominant for the stage ($\beta$ or $g_m$) is described by the $\kappa s$.

- The specific location in the loop, i.e. at what stage the distortion originates, has two consequences.

  Firstly, the closer to the input of the amplifier the distortion originates, the less overall distortion is obtained. This is because of the fact that for equal biasing currents and equal $\kappa s$ for the stages, the relative current swing reduces when going from output stage to input stage, which is indicated in the equation by the $\alpha$s in the denominators of the corresponding terms. In that case the distortion is dominated by the output stage.

  Secondly, from the equation follows that having additional gain after a stage with distortion is more advantageous for the overall distortion than having additional gain before the stage with distortion. This is most easily seen when looking to the distortion due to the middle stage, i.e. stage b. In the denominators of the corresponding terms for the second and third-order distortion, only $\alpha_c$ and $\alpha_c^2$ are found, respectively, no terms relating to stage are present. Reminding that the gain of all the stages can appear linearly in the expression for the return difference, leads to the following conclusions. Increasing the gain of the input stage, stage a, yields in the best case a proportional reduction of the second and third-order distortion due to stage b. Whereas increasing the gain of the output stage, stage c, may yield in the best case a quadratic and cubic reduction of the second and third-order distortion due to the middle stage, respectively. This difference is explained as follows. Assume that the input-output relation of the middle stage (ignoring the bias term) is given by:

$$e_{ob} = b_1 e_{ib} + b_2 e_{ib}^2 + b_3 e_{ib}^3 + \cdots \tag{5.18}$$

in which $e_{ib}$ and $e_{ob}$ are the input signal and output signal, respectively, and $b_i$ are the Taylor coefficients. In the case that the gain of the output stage is increased by $A_c$, the input signal of the middle stage is reduced, because of the feedback, to $e_{ib}/A_c$, yielding:

$$e_{ob} = b_1 \left(\frac{e_{ib}}{A_c}\right) + b_2 \left(\frac{e_{ib}}{A_c}\right)^2 + b_3 \left(\frac{e_{ib}}{A_c}\right)^3 + \cdots \tag{5.19}$$

Clearly, second-order distortion reduced quadratically with the increase of gain, and the third-order distortion reduces cubically with the increase of gain. In the case that the gain of the input stage is increased by $A_a$, the output signal of the middle stage decrease by the same factor, as the output signal of the feedback amplifier remains the same. Thus:

$$\frac{e_{ob}}{A_a} = b_1 \frac{e_{ib}}{A_a} + b_2 \frac{e_{ib}^2}{A_a} + b_3 \frac{e_{ib}^3}{A_a} + \cdots \tag{5.20}$$

A proportional reduction is obtained for all orders.

Equations (5.19) and (5.20) express the best case. For a specific amplifier the loop gain does not need to depend linearly on the gain of all the stages; it might be sub-linear. Consequently, some lower reduction is obtained in both cases. Conclusion from these equations is that the stages with a higher nonlinearity should be placed as close to the input as possible whereas the stages with the higher gain should be placed closer to the output.

In order to illustrate the use of the distortion expressions, the distortion of three-stage amplifiers, for bipolar and FET technology, is derived by using expressions (5.16) and (5.17).

**Example of a three-stage bipolar amplifier.** Figure 5.9 shows an example of a three-stage bipolar transconductance amplifier. For finding the distortion

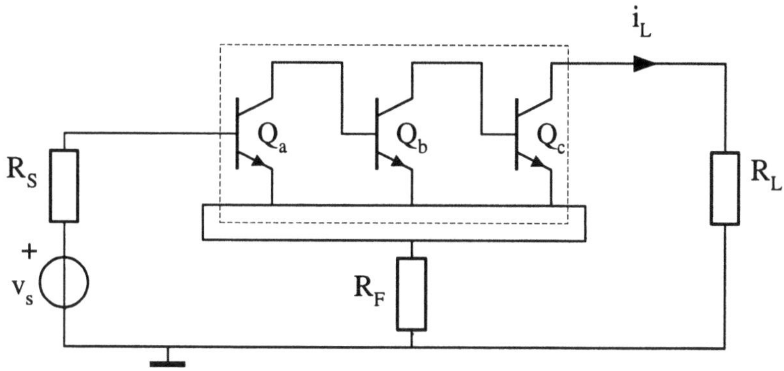


*Figure 5.9.* A three-stage bipolar transconductance amplifier.

of this amplifier the $\alpha s$, $F$ and $\nu$ must be calculated. The $\kappa s$ can be found from tables 5.1 and 5.2. The $\alpha s$, being the current-gain of the stages equal $-\beta_f$. $F$ can be approximated by minus the loop gain:

$$F \approx -L = \frac{R_F \beta_{fa} \beta_{fb} \beta_{fc}}{r_{\pi a} + R_S + R_F} \tag{5.21}$$

The output loading factor is given by (according to the asymptotic-gain model):

$$\nu = \left. \frac{e_c}{i_L} \right|_{e_s = 0} \tag{5.22}$$

For this amplifier holds:

$$\nu = -1 \tag{5.23}$$

The normalized output related distortion coefficients follow from the tables 5.1 and 5.2. In the simplest models it is assumed that the output conductance of the transistors are zero. In that case the $\kappa s$ become:

$$\kappa_{2a} = \frac{r_{\pi a}}{2(r_{\pi a} + R_S)} \tag{5.24}$$

$$\kappa_{2b,c} = 0 \tag{5.25}$$

$$\kappa'_{3a} = \frac{-r_{\pi a}}{3(r_{\pi a} + R_S)} \tag{5.26}$$

$$\kappa'_{3b,c} = 0 \tag{5.27}$$

The $\kappa s$ for stages $b$ and $c$ are zero as those stages are current driven and thus only exhibit $\beta - distortion$, which is assumed to be zero. Subsequently, the resulting expression for the second and third-order distortion is given by:

$$
\begin{aligned}
A_{t2} &= -\frac{r_{\pi a}+R_S+R_F}{R_F\beta_{fa}\beta_{fb}\beta_{fc}}\left[\frac{r_{\pi a}}{2(r_{\pi a}+R_S)\beta_{fb}\beta_{fc}I_{Qa}}\right] \\
&\approx -\frac{V_T}{2\beta_{fb}^2\beta_{fc}^2 R_F I_{Qa}^2}
\end{aligned}
\qquad (5.28)
$$

$$
\begin{aligned}
A_{t3} &= \frac{r_{\pi a}+R_S+R_F}{R_F\beta_{fa}\beta_{fb}\beta_{fc}}\left[\frac{r_{\pi a}}{3(r_{\pi a}+R_S)\beta_{fb}^2\beta_{fc}^2 I_{Qa}^2}\right] \\
&\approx \frac{V_T}{3\beta_{fb}^3\beta_{fc}^3 R_F I_{Qa}^3}
\end{aligned}
\qquad (5.29)
$$

The last approximations are valid when $r_{\pi_a} \gg R_s, R_F$. The expressions show indeed that the second and last stage do not add any distortion. In contrast, the first stage is not ideally current driven and thus introduces $g_m$-distortion. So, for this amplifier, in the instantaneous case, the input stage dominates the distortion!

**Example of a three-stage FET amplifier.** As indicated before, FET devices always introduce $g_m$-distortion, independent whether the device is voltage or current driven. Therefor, it might be expected when considering a three-stage FET implementation of the transconductance, that all the stages contribute to the distortion. The FET amplifier is depicted in figure 5.10. When considering

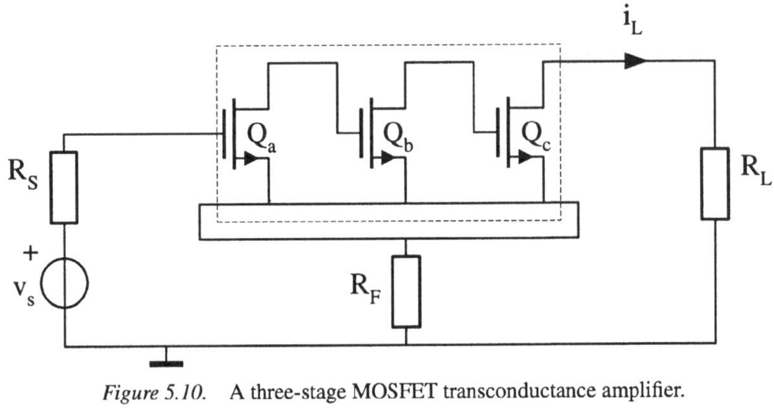


*Figure 5.10.* A three-stage MOSFET transconductance amplifier.

this amplifier, a finite output resistance needs to be assumed. Otherwise, the DC gain of the FETs is infinite and no distortion is obtained at all! Assuming that the output resistance of the FET, $Q_c$, is much larger than $R_F$ and $R_L$, and the transistors have equal voltage gain and channel length modulation, the following expressions can be obtained:

$$A_{t2} = -\frac{V_A}{4\mu_a\mu_b\mu_c R_F I_{Qc}^2}\left[\frac{1}{\mu_b\mu_c} - \frac{1}{\mu_c} + 1\right] \approx -\frac{V_A}{4\mu_a\mu_b\mu_c R_F I_{Qc}^2} \quad (5.30)$$

$$A_{t3} = -\frac{V_A}{8\mu_a\mu_b\mu_c R_F I_{Qc}^3}\left[1 - \frac{1}{\mu_c} + \frac{1}{\mu_c^2} + \frac{1}{\mu_b\mu_c} - \frac{1}{\mu_b\mu_c^2} + \frac{1}{\mu_b^2\mu_c^2}\right] \approx -\frac{V_A}{8\mu_a\mu_b\mu_c R_F I_{Qc}^3} \quad (5.31)$$

For calculating these expressions the following relations were assumed:

$$F = \mu_a\mu_b\mu_c\frac{R_F}{r_{dc}} \quad (5.32)$$

$$\nu = -1 \quad (5.33)$$

$$\kappa_2 = 1/4 \quad (5.34)$$

$$\kappa_3' = -1/8 \quad (5.35)$$

$$r_{di} = \frac{V_A}{I_{Qi}} \quad (5.36)$$

$$\alpha_i = -r_{d_{i-1}}g_{m_i} \quad (5.37)$$

**Comparing the bipolar and FET amplifier.** Comparing the distortion expressions for the bipolar amplifier with the corresponding expressions for the FET amplifier, a high similarity between the two sets of equations can be seen. However, some specific differences can be identified:

- In the case of the bipolar amplifier, the current-gain factors of the transistors are key parameters, whereas for the FET amplifier, the voltage-gain factors of the comprising FETs are key parameters. The voltage-gain factor of the FET can easily be increased by choosing a longer channel. In the case of the bipolar transistor the current-gain factor cannot be increased by the designer.
- For this specific example, the distortion of the bipolar amplifier is dominated by the input stage because of the zero $\beta$-distortion of the second and third stage, whereas for the FET amplifier, in the case of large voltage-gain factors of the FETs, the output stage dominates the weak distortion.
- For the bipolar amplifier the second and third-order distortion reduces with, $\beta_{fb}^2\beta_{fc}^2$ and $\beta_{fb}^3\beta_{fc}^3$, respectively. For the FET amplifier, both, the second and third-order distortion reduces proportional to $\mu_a\mu_b\mu_c$, which corresponds to the increase of loop gain. The difference is due to specific location at which the distortion originates as explained before.

## 5.3 Measures of distortion

As discussed in the previous section, the output signal of an amplifier is distorted due to the nonlinearity of the comprising devices. For electronics holds: the lower the distortion the better it is and the more difficult to reach. However, it is not straightforward to say, for instance, for two arbitrary amplifiers that one is better than the other. At least a quality measure should be available to rank the amplifiers. Subsequently, when one amplifier is higher in the ranking than the other, the one is better.

Nonlinearity of a device can have several different effects on a signal and thus, several different ways are available to qualify a device in the context of distortion. Which quality measure needs to be chosen depends on the application of the device.

The measures of quality are based on exciting the amplifier with a reference signal, often a single tone, and comparing the required output with the obtained output. This comparison can be done in different ways. On top of that, the outcome can be presented in different ways, i.e. obtained distortion for a certain input or the maximum input for obtaining a maximum level of distortion.

In this section several measures of distortion are treated. To illustrate, a fictitious amplifier is assumed that has an instantaneous signal behavior with distortion up to the third order. Thus, its signal behavior can be described by equation (5.11), which is repeated here for convenience:

$$y_s = A_t x_s + A_{t2} x_s^2 + A_{t3} x_s^3 \tag{5.38}$$

Assume the amplifiers input signal is a single tone, like:

$$y_s = a\cos(\omega t) \tag{5.39}$$

Substituting this expression for the input signal in equation (5.38), yields an expression for the output signal in terms of several harmonic component:

$$y_s = y_0 + y_1\cos(\omega t) + y_2\cos(2\omega t) + y_3\cos(3\omega t) \tag{5.40}$$

in which:

$$y_0 = \frac{1}{2}a^2 A_{t2} \tag{5.41}$$

$$y_1 = aA_t + \frac{3}{4}a^3 A_{t3} \approx aA_t \tag{5.42}$$

$$y_2 = \frac{1}{2}a^2 A_{t2} \tag{5.43}$$

$$y_3 = \frac{1}{4}a^3 A_{t3} \tag{5.44}$$

The approximation in the expression for $y_1$ is allowed for the case of relatively low-distortion amplifiers. Note that the ignored term is three times the third-order component, $y_3$. Besides the three harmonics also a DC term is obtained.

This term is due to rectification which results from the second-order nonlinearity; it is often called DC-shift. The components $y_1$, $y_2$ and $y_3$ are depicted in figure 5.11. In the figure, the output signal at the fundamental frequency

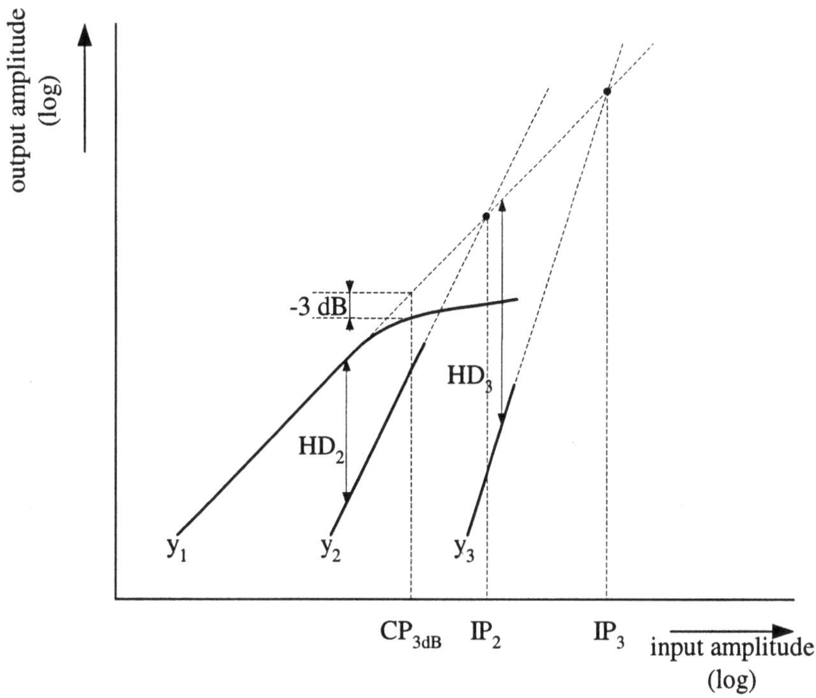


*Figure 5.11.* The amplitude of the fundamental output signal ($y_1$), the second-order output component ($y_2$) and the third-order output component ($y_3$) as a function of the input amplitude (CP: Compression Point, IP: Intercept Point).

increases linearly with the input amplitude, whereas, the output signal at the second and third-order component increases quadratically and cubically with the input amplitude, respectively. Further, in the figure the distortion measures based on a single tone, to be discussed subsequently, are indicated.

## 5.3.1 Harmonic Distortion

The Harmonic Distortion (HD) figure of merit is obtained by comparing the amplitude at the n-th order harmonic component, $y_n$, with the amplitude of the fundamental harmonic, $y_1$, in the output signal:

$$\mathrm{HD}_n \triangleq \frac{y_n}{y_1} \tag{5.45}$$

Using the expressions (5.42)-(5.44), yields for the second-order (HD2) and third-order (HD3) harmonic distortion:

$$\mathrm{HD}_2 = \frac{1}{2}a\left|\frac{A_{t2}}{A_t}\right| \tag{5.46}$$

$$\mathrm{HD}_3 = \frac{1}{4}a^2\left|\frac{A_{t3}}{A_t}\right| \tag{5.47}$$

When specifying a harmonic distortion, also the input amplitude needs to be specified. This is because the fundamental component and the higher harmonic components depend differently on the input amplitude. Inspection of figure 5.11 clearly shows that indeed the harmonic distortion depends on the amplitude of the input signal.

Related to Harmonic Distortion is the Total Harmonic Distortion (THD). This measure is used when for the distortion more than one harmonic at a time needs to be taken into account. It is defined as:

$$\mathrm{THD} = \sqrt{\sum_{n=2}^{m} |\mathrm{HD}_n|^2} = \frac{\sqrt{\sum_{n=2}^{m} |y_n|^2}}{|y_1|} \tag{5.48}$$

Application of this measure can be, for instance, in the case of audio amplifiers. Assume a test tone of 1 kHz with a certain amplitude, than because of the distortion several harmonics are in the audio band (< 20 kHz). By specifying the THD, a measure of the total power of all the harmonics is given.

### 5.3.2 Intercept points

A different way to present the information of harmonic distortion is by using the intercept voltages. From equations (5.41)-(5.44) and figure 5.11 clearly follows the different dependencies on the input amplitude for $y_1$, $y_2$ and $y_3$. The second-order component increases quadratically with the input amplitude, whereas the fundamental component increases only linearly. So, at a certain input amplitude $y_2$, intercepts $y_1$. The corresponding input amplitude at which this occurs, is called $IP_2$, the intercept amplitude for the second harmonic. Likewise is $IP_3$ defined. These two points are indicated in figure 5.11. The intercept point can be considered to be the input amplitude at which the corresponding harmonic distortion is 0 dB.

The intercept points can be calculated from:

$$IP_2 = 2\left|\frac{A_t}{A_{t2}}\right| \tag{5.49}$$

$$IP_3 = 2\sqrt{\left|\frac{A_t}{A_{t3}}\right|} \tag{5.50}$$

### 5.3.3 Compression points

inxxdistortion, compression points In contrast to intercept points and harmonic distortion, the compression point is used to indicate the error in the *fundamental* harmonic due to distortion. To find a compression point, equation (5.42) without the approximation, is considered (and repeated here):

$$y_1 = aA_t + \frac{3}{4}a^3 A_{t3} \tag{5.51}$$

This expression shows that due to third-order distortion an additional term at the fundamental frequency is obtained. In most of the cases, the polarity of this additional term is opposite to the polarity of the required fundamental term. Consequently, the gain of the amplifier is smaller than expected. A 3 dB compression point ($CP_{3\ dB}$) is defined as the input amplitude at which the gain of the amplifier is reduced by 3 dB. Of course, this reduction should be because of the nonlinear distortion and not because of dynamic effects. In the same way also a 1 dB compression point is used, $CP_{1\ dB}$

The 3 dB compression point corresponding to equation (5.51), equals:

$$CP_{3\ dB} = \sqrt{\frac{4}{3}\frac{A_t}{A_{t3}}\left(\frac{1}{2}\sqrt{2} - 1\right)} \approx \frac{5}{8}\sqrt{\left|\frac{A_t}{A_{t3}}\right|} \tag{5.52}$$

### 5.3.4 Intermodulation

inxxdistortion, intermodulation Last measure of distortion to be discussed is the intermodulation (IM). When two or more tones are applied to a nonlinear amplifier, mixing products arise. Especially in communication systems a bad intermodulation performance is severe. In that case the information from one channel can be mixed into an other channel.

Commonly used intermodulation tests are based on two and three tone measurements. Here, the two tone intermodulation is considered. Assume that the input signal of the amplifier with input-output relation (5.38), is given by:

$$x = a\cos(\omega_1 t) + a\cos(\omega_2 t) \tag{5.53}$$

For convenience, the amplitudes of both spectral components is assumed to be equal. The resulting spectral components and the corresponding relative amplitudes are depicted in figure 5.12. In this figure, both harmonic distortion (HD) and intermodulation (IM) products are depicted. Whereas harmonic distortion is due to nonlinearity and a single tone, intermodulation (IM) products depend on both tones. For instance, the $IM_3$ products are the components at:

- $2\omega_1 - \omega_2$
- $2\omega_2 - \omega_1$

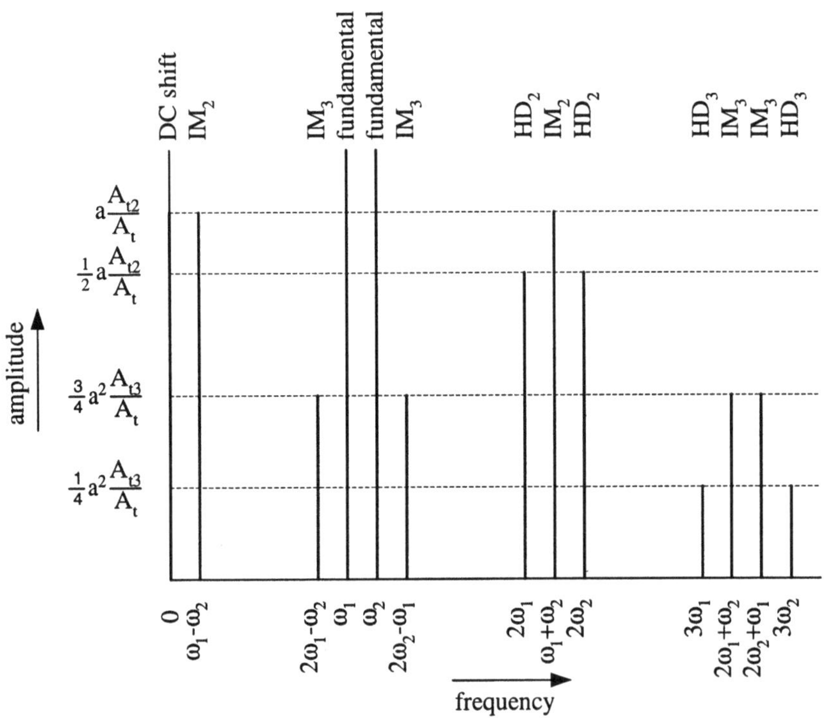


*Figure 5.12.* The resulting spectral components and their relative amplitudes for a two tone input with equal amplitudes.

- $2\omega_1 + \omega_2$
- $2\omega_2 + \omega_1$

The frequency of each of these components is a result of mixing three components with frequency $\omega_1$ and $\omega_2$. Note that the amplitude of the corresponding components of $IM_3$, $IM_2$, $HD_3$ and $HD_2$ are equal.

For different amplitudes and different relative frequencies of the two tones, the plot may appear differently.

## 5.4 Design for low distortion

In the previous sections two basic different types of distortions were discussed. The main differences are found in the origin of the distortion and in the corresponding effect. Clipping distortion is an aggressive type of distortion resulting from a limited output range of a device. When clipping distortion occurs, the feedback loop is broken resulting in a severe reaction of the amplifier. Weak distortion is much milder. It originates from the weak nonlinearities in the devices input-output relation. The consequences are harmonic and intermodulation distortion. In contrast with clipping distortion, it has no significant effect on the negative-feedback loop.

As the origins and consequences of the two types of distortion are very different, it might be expected that also the design measures to be taken are different.

In the next sections the design measures for preventing clipping distortion and reducing weak distortion are treated.

## 5.4.1 Clipping distortion

Main cause of the clipping distortion is that the bias of a device is too small. A signal may cause the transistor to clip and thus to switch off. Consequently, the feedback loop is broken and is not able to counteract the clipping. This should, of course, never occur. So, the design measure to be taken for clipping distortion should be a measure for*preventing* the amplifier from clipping.

To be able to prevent clipping distortion of a device, one should know what the maximum expected output signal is for the device. Then, the bias can be chosen such that clipping is prevented.

When considering a negative-feedback amplifier, for which the active part is implemented by a cascade of amplifying stages, it is likely that the output stage is the hot spot for clipping. Of course, in general, all stages are vulnerable for clipping, this will, however, be treated at the end of this section. For now, it is assumed that the output stage is the suspected stage concerning clipping.

### 5.4.1.1 Clipping in the output stage

Determining the maximum expected signal means that the maximum expected voltage *and* current need to be determined, i.e. the peak values. Remind that the bias of a nonlinear device is both, in the current domain and the voltage domain. To determine the peak value of the voltage and current at the output of the amplifier, consider the amplifier output as depicted in figure 5.13. In general, both output terminals of the amplifier are loaded. For instance, for a current amplifier one terminal is connected to the load and the other terminal is loaded by the parallel connection of the two feedback impedances (in the case of a feedback network based on impedances). By specification, either the peak voltage ($\hat{V}_o$) or the peak current ($\hat{I}_o$) of the amplifier is given somehow. Which of the two is specified depends on the type of output quantity of the amplifier.

**Voltage output.** For a voltage amplifier and a transimpedance amplifier, the peak voltage is specified in one way or another. For these types of amplifiers one terminal is connected to the load and feedback network and the other terminal is grounded, i.e. either $Z_1$ or $Z_2$ is zero. Thus, the peak current is determined by the parallel connection of the load impedance and the input impedance of the feedback-network. Clearly, to have minimum contribution of the feedback

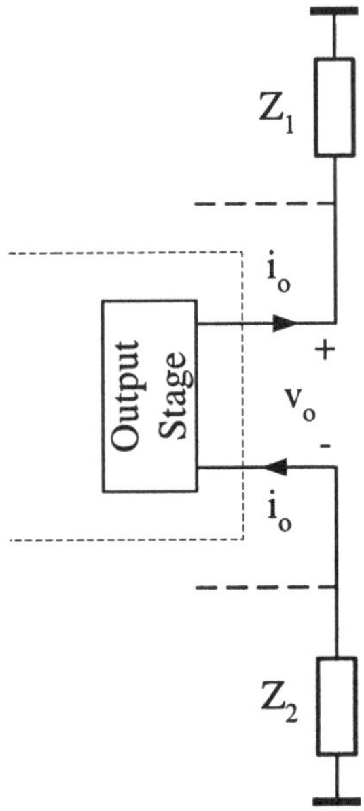


*Figure 5.13.* Model of a general amplifier output for determining the peak voltage and peak current at the output.

network to the total loading, the input impedance of the feedback network should be negligibly high with respect to the load impedance.

**Current output.** In the case of a current amplifier and a transconductance amplifier, the peak current is somehow specified. Consequently, the peak voltage needs to be determined. For these amplifier configurations one output terminal is connected to the load and the other terminal is connected to the feedback network, i.e. $Z_1$ and $Z_2$ are non-zero. As the output current at both terminals have opposite sign, the peak voltage is determined by the series connection of the load impedance and the input impedance of the feedback network. Thus, for minimum peak voltage, given the peak current, the input impedance of the feedback network should be low compared with the load impedance.

#### 5.4.1.2 Clipping in a stage other than the output stage

For the input and intermediate stage, also clipping distortion needs to be prevented. As at those stages the signal swing is relatively small, the chance on clipping is also small. But in general, it needs to be checked. The intermediate stage, is after the output stage, the stage with highest risk on clipping distortion. Consider the cascade of two amplifying stages, which are a part of a nullor implementation, as depicted in figure 5.14. Transistor $Q_x$ needs to be able

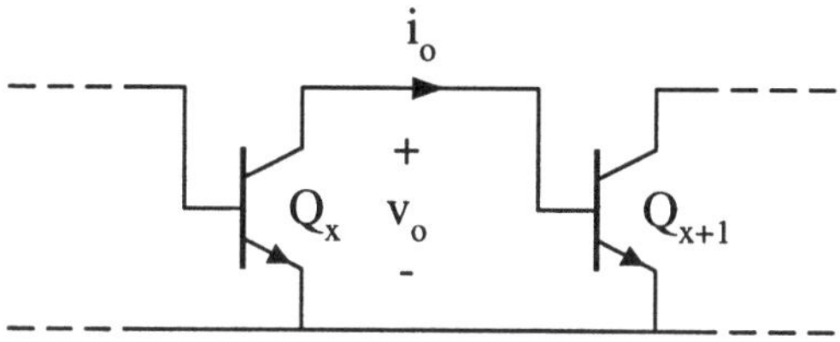


*Figure 5.14.* A cascade of two amplifying stages for determining the clipping level of stage $Q_x$.

to drive stage $Q_{x+1}$. The input of that stage can be considered as a parallel connection of a resistor and a capacitor ($r_\pi$ and $c_\pi$). For a stage, other than the output stage, it is not always straightforward to determine the peak voltage and peak current. By using the product of the chain matrices of the stages between $Q_x$ and the output it is possible to relate the peak voltage and peak current at the amplifier output with the peak voltage and peak current at the output of stage $Q_x$. A completely different way to determine these peak values is to use a small-signal analysis at the worst case condition, i.e. often maximum output signal and maximum frequency. As in a small-signal analysis clipping cannot occur, the peak current and peak voltage are the current and voltage obtained for $Q_x$ in this worst case situation. For SPICE-like simulators an AC analysis would give the required information. From these peak values, the required bias can be derived to prevent clipping.

#### 5.4.1.3 Design guidelines

When both the peak current and the peak voltage are determined, the biasing of the output stage can be determined, taking the constraints of equations (5.4) - (5.7) into account. Of course, some margin to the clipping level should be incorporated. For instance, when the peak current is 1 mA, the bias current should be reasonably higher than 1 mA. Would the bias current be chosen slightly beyond 1 mA, the output stage would be close to clipping at the peaks, a small unexpected increase in the peak value would lead to clipping again. On top of that, considering weak distortion, the$m$ factor, i.e. the ratio between the peak current and the bias current is close to one. Further, when considering the dynamic performance of the transistor, the ratio between the peak collector current and the minimum current collector is very large. Taking into account that the transit frequency of the transistor depends on the collector current, leads to the conclusion that also the transit frequency is changing a lot as a function of the amplitude. Generally speaking, the small-signal analysis may cease to be valid.

#### 5.4.1.4 Power bandwidth

Peak voltage and peak currents for an amplifier are frequency dependent. Of course, for preventing clipping distortion, the worst case situation needs to be considered.

In several applications it may appear that, besides the signal bandwidth, i.e. the band in which the transfer for *small-signals* should correspond to the specified transfer, also the band is specified in which maximum signal swing may arise. This is the so called *power bandwidth*. Consequently, the worst case situation for determining peak current and peak voltage, is at maximum output signal at the power bandwidth. Especially in those applications in which the power bandwidth is smaller than the signal bandwidth, the bias conditions for the corresponding stage are relieved.

### 5.4.2 Weak distortion

Weak distortion arises from the weak nonlinearity of the comprising devices. In section 5.2.2 expressions were derived indicating the level of second and third-order distortion as a function of amplifier and device characteristics. From these expressions, equations (5.16) and (5.17), design guidelines can be derived. Three different approaches can be used for lowering the weak distortion:

- decreasing the $\kappa_i$;
- increasing the bias current $I_{Qi}$;
- increasing the return difference $F$.

Often also local feedback is considered as an option to improve distortion performance. At the end of this section, this statement is shown to be incorrect.

#### 5.4.2.1 Decreasing $\kappa$

The level of distortion depends linearly on the normalized nonlinearity coefficients, $\kappa$, of the comprising stage. Thus, when $\kappa$ of a stage is reduced the distortion lowers proportionally. $\kappa$ can be lowered in two ways. Firstly, choose an other stage with a lower $\kappa$. For instance, changing all the unbalanced stages into balanced stages, nullifies, theoretically, the second-order distortion. Secondly, when looking to the $\kappa$s of the bipolar transistor, it appears that a current drive is advantageous. In the ideal case the $\kappa$s become zero. For instance, when the output resistance of a stage is considerably lower than the input resistance of the next stage, a current buffer (CB or CG stage) may improve the distortion performance by increasing the driving resistance.

#### 5.4.2.2 Increasing the bias currents

Straightforward method is to increase the bias current of one or more stages. This helps for lowering the distortion as the bias currents appear in the denominator of the distortion terms, see equations (5.16) and (5.17). On top of that, by increasing the bias current(s) the return difference may increase as well. This is, of course, dependent on the specific topology.

#### 5.4.2.3 Increasing the return difference

The distortion is inversely proportional to the return difference:

$$F = 1 - L \approx L \tag{5.54}$$

in which $L$ is the loop gain of the amplifier. Thus, when the loop gain is increased, the distortion lowers proportionally. Depending on how the loop gain is increased a larger reduction can be obtained. For instance, as discussed before, when the bias current of a stage is increased, besides an increase in loop gain also an reduction in the relative current swing of the corresponding amplifying stage is obtained.

A straightforward way to increase the loop gain is by adding a stage. Assuming that one stage dominantly determines the distortion level, for instance the output stage, the minimally required return difference can be determined and from that the minimum number of amplifying stages.

When the output determines the weak-distortion level, the minimum required return difference to obtain a certain level of distortion can be derived from equations (5.16) and (5.17). It is given by:

$$F = \max\left(\frac{\kappa_{2c}}{\nu A_{t2} I_{Qc}}, \frac{\kappa_{3c}}{\nu^2 A_{t3} I_{Qc}^2}\right) \tag{5.55}$$

#### 5.4.2.4 The effect of local feedback

Often, local feedback is considered to be a measure for reducing the overall distortion level. However, when calculating the distortion of an amplifier comprising local feedback, it appears that the distortion increases and in the best case does not change at all. This is verified by means of calculating again $A_{t2}$ and $A_{t3}$ for a negative-feedback amplifier comprising local feedback.

Assume, again, a negative-feedback amplifier implemented by means of three amplifying stages. The stages are numbered from input to output as $a$, $b$ and $c$. To each of these stages local feedback is applied. The level of feedback is indicated by the corresponding return differences of the local loops as: $F_a$, $F_b$ and $F_c$ for the input, middle and output stage, respectively. Further, $F_{max}$ is the maximum return difference of the overall loop, i.e. in the case that no local feedback is present, $F_a = F_b = F_c = 1$. Then, for the second and third-order distortion the following expressions are found [25]:

$$A_{t2} = \frac{1}{F_{max}\nu}\left(\frac{\kappa_{2a}F_b^2F_c^2}{\alpha_b\alpha_c I_{Qa}} + \frac{\kappa_{2b}F_aF_c^2}{\alpha_c I_{Qb}} + \frac{\kappa_{2c}F_aF_b}{I_{Qc}}\right) \tag{5.56}$$

$$\begin{aligned} A_{t3} &= \frac{1}{F_{max}\nu^2}\left(\frac{\kappa'_{3a}F_b^3F_c^3}{\alpha_b^2\alpha_c^2 I_{Qa}^2} + \frac{\kappa'_{3b}F_aF_c^3}{\alpha_c^2 I_{Qb}^2} + \frac{\kappa'_{3c}F_aF_b}{I_{Qc}^2}\right. \\ &\quad \left. - \frac{2\kappa_{2a}\kappa_{2b}F_bF_c^3}{\alpha_b\alpha_c^2 I_{Qa}I_{Qb}} - \frac{2\kappa_{2a}\kappa_{2c}F_b^2F_c}{\alpha_b\alpha_c I_{Qa}I_{Qc}} - \frac{2\kappa_{2b}\kappa_{2c}F_aF_c}{\alpha_c I_{Qb}I_{Qc}}\right) \end{aligned} \tag{5.57}$$

From these expression follows that applying local feedback at a stage does not change its contribution to the overall distortion. What happens is that the distortion of the other stages become more important as a result of the reduced overall loop gain. On top of that, because of the local feedback the driving impedance for a stage might become different resulting in an increased$\kappa$. So, local feedback in the best case has no effect on the distortion. Usually, the output stage is considered to be the most likely stage to apply this local feedback to, for reducing the distortion. However, the second and third-order distortion resulting from the other stages increase quadratically and cubically, respectively! This is because the gain preceding the distorting stage is lowered, cf. equations (5.19) and (5.20). Therefor, when local feedback is applied in amplifiers it should have a different reason then reducing distortion. For instance, local feedback can be used for designing the dynamic behavior of the overall loop. But in any case, it should be reminded that the distortion performance can be seriously affected when applying local feedback, especially in the output stage.

## 5.5 Dynamic Nonlinear Distortion

The presented description of the nonlinear behavior of a negative-feedback amplifier is valid in the instantaneous situation only. It gives yet, however, a rather good insight in what key aspects are in obtaining low distortion. In this section, the description of the distortion of a single nonlinear*dynamic* amplifying stage is treated. It will be seen that, again, relatively simple expression can be obtained for the second and third-order distortion. The extension to a description of a cascade of amplifying stages including overall feedback is currently still an issue of research. The modelling presented in this section is based on the Volterra series.

### 5.5.1 Volterra Series

The input-output relation of a linear time-invariant system is described by its impulse response, $h(t)$, according to the familiar convolution integral:

$$y(t) = \int_{-\infty}^{\infty} h(\tau)u(t-\tau)d\tau \tag{5.58}$$

in which $u(t)$ is the input signal and $y(t)$ is the output signal. For causality should hold that $h(\tau)$ for $\tau < 0$ equals zero.

This convolution integral can be generalized to describe the input-output relation of an $n$-th order nonlinearity. This yields an $n$-dimensional convolution integral [28], [29]:

$$\begin{aligned} y(t) &= \mathbf{H_n}[u(t)] \\ &= \int_{-\infty}^{\infty} \cdots \int_{-\infty}^{\infty} h_n(\tau_1, \cdots, \tau_n) u(t-\tau_1) \cdots u(t-\tau_n) d\tau_1 \cdots d\tau_n \end{aligned} \tag{5.59}$$

In this expression, $\mathbf{H_n}[u(t)]$ is called the $n$-th order Volterra operator, and the terms $h_n(\tau_1, \cdots, \tau_n)$, are called the $n$-th order Volterra Kernels. Note that in the expression the input signal is present with order $n$. Thus the $n$-th order Volterra operator expresses the convolution of the input signal with the $n$-th order Volterra kernel. Therefor, this Volterra kernel is also called the $n$-dimensional impulse response of the system.

Clearly, for a general system with nonlinearities up to infinite order, the output is described by:

$$y(t) = \mathbf{H}_1[u(t)] + \mathbf{H}_2[u(t)] + \mathbf{H}_3[u(t)] + \cdots \tag{5.60}$$

This expression is called a Volterra series.

Usually, in circuit design, a frequency domain representation is used in stead of a time domain representation. To switch for linear systems to the frequency domain, the Laplace transform is used. The corresponding transfer function for a linear time-invariant circuit is the Laplace transformation of the convolution in equation (5.58), yielding the well-known transfer function $H(s)$:

$$Y(s) = H(s)U(s) \tag{5.61}$$

In an analogous way, an $n$-dimensional impulse response or $n$-th order Volterra kernel, can be transformed to a frequency domain transfer by using an $n$-th order Laplace transform, [28]:

$$Y_n = H_n(s_1, s_2, \cdots, s_n)U(s_1)U(s_2)\cdots U(s_n) \tag{5.62}$$

Thus, for a nonlinear system comprising nonlinearities up to order 3, the input-output relation is described by a sum of three transfer functions:

$$\begin{aligned} Y(s_1, s_2, s_3) &= H_1(s_1)U(s_1) \\ &+ H_2(s_1, s_2)U(s_1, s_2) \\ &+ H_3(s_1, s_2, s_3)U(s_1, s_2, s_3) \end{aligned} \tag{5.63}$$

This equation is schematically depicted in figure 5.15. When describing weak

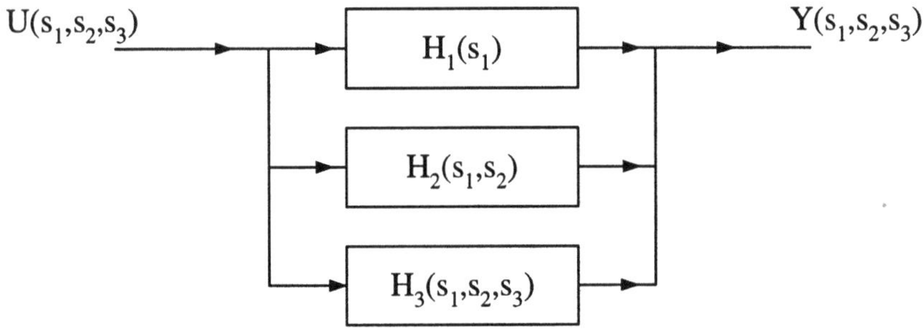


*Figure 5.15.* Input-output relation of a system comprising nonlinearities up to the 3-rd order.

distortion by means of Volterra kernels, the higher-order kernels decrease rapidly. In the remainder of this section, it is assumed that the stages are neatly described by the first three Volterra kernels.

To derive expressions for the harmonic distortion, the Volterra kernels need to be determined at single frequency. Thus:

$$Y(s,s,s) = H_1(s)U(s) + H_2(s,s)U(s,s) + H_3(s,s,s)U(s,s,s) \tag{5.64}$$

Subsequently, the $n$-th order harmonic distortion is given by, [29]:

$$HD_n(|U(s)|,s) = \left(\frac{|U(s)|}{2}\right)^{n-1} \left|\frac{H_n(s,\cdots,s)}{H_1(s)}\right| \tag{5.65}$$

### 5.5.2 Behavior of CE stage

To determine the nonlinear dynamic distortion of a single CE stage, the nonlinear signal diagram of figure 5.16 is used. It is assumed that the CE stage

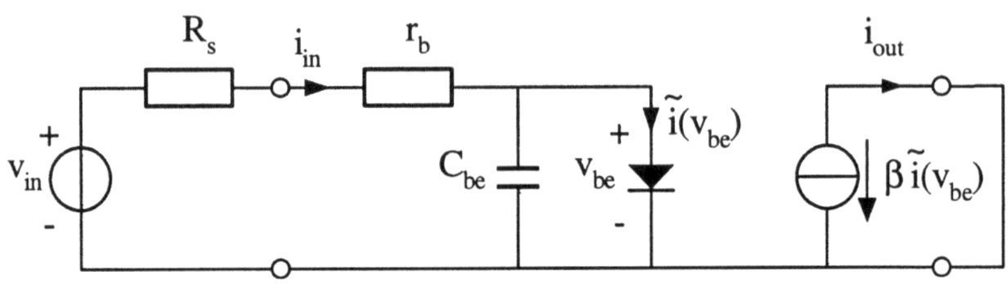


*Figure 5.16.* The nonlinear signal diagram of a CE stage for determining its harmonic distortion.

is loaded by a relatively low impedance, i.e. ideally a short. The driving source is a voltage source with a source resistance $R_s$. By making $R_s = 0\,\Omega$, the CE stage is ideally voltage driven. In the case $R_s = \infty\,\Omega$, the CE stage is ideally current driven. In this way the effect of the relative source resistance can be described. Further, $C_{be}$ is the input capacitance of the CE stage, which is considered here to be constant. The nonlinear behavior of the transistor is modelled by the diode and the output current of the transistor is modelled with the controlled current source.

Deriving the Volterra kernels is considered to be beyond the scope of this book. The interested reader is referred to, for instance, reference [28] and [29].

The expressions obtained for the second and third-order harmonic distortion are given by (in pole-zero format), [30]:

$$HD_2(s) = |K_2| \left| \frac{1 - \frac{2s}{z_1}}{\left(1 - \frac{s}{p_1}\right)\left(1 - \frac{2s}{p_1}\right)} \right| \quad (5.66)$$

$$HD_3(s) = |K_3| \left| \frac{\left(1 - \frac{3s}{z_1}\right)\left(1 - \frac{2s}{z_2}\right)}{\left(1 - \frac{s}{p_1}\right)^2\left(1 - \frac{2s}{p_1}\right)\left(1 - \frac{3s}{p_1}\right)} \right| \quad (5.67)$$

in which

$$K_2 = \frac{\hat{V}_{in}}{4V_T} \frac{1}{(1 + R_s/r_\pi)^2} \quad (5.68)$$

$$K_3 = \frac{\hat{V}_{in}^2}{24V_T^2} \frac{1 - 2R_s/r_\pi}{(1 + R_s/r_\pi)^4} \quad (5.69)$$

$$p_1 = \frac{-1}{(R_s//r_\pi)C_{be}} \quad (5.70)$$

$$z_1 = \frac{-1}{R_s C_{be}} \quad (5.71)$$

$$z_2 = \frac{2R_s/r_\pi - 1}{R_s C_{be}} \quad (5.72)$$

Clearly, the ratio of $R_s$ and $r_\pi$ is important. This is in line with what was found in the instantaneous situation, see table 5.1 and 5.2. The expression for $HD_2(s)$ and $HD_3(s)$ are depicted for two situations: $R_s \ll r_\pi$, see figure 5.17 and $R_s \gg r_\pi$, see figure 5.18. These two situations correspond to voltage and current drive, respectively. Clearly, in the case of voltage drive, the distortion is up to the pole ($p_1$) frequency independent, whereas the distortion in the case of current drive starts very low and increases beyond$z_1$. From the two graphs, specific information can be obtained.

For the voltage driven case, the maximum distortion levels are obtained at the relatively low frequencies already:

$$HD_{2,v} = \frac{1}{4}\frac{\hat{i_c}}{I_c} \quad (5.73)$$

$$HD_{3,v} = \frac{1}{24}\left(\frac{\hat{i_c}}{I_c}\right)^2 \quad (5.74)$$

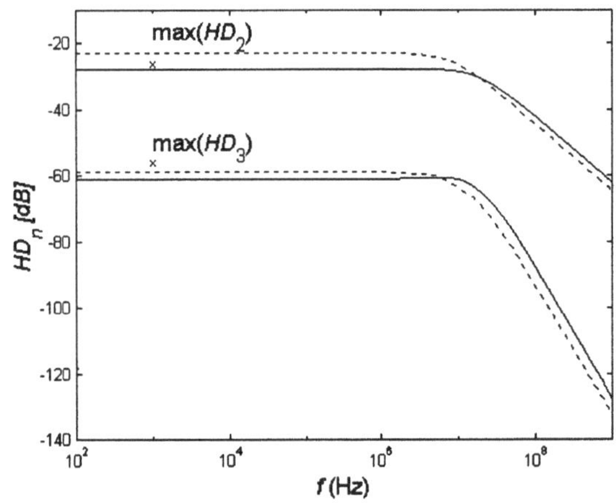


*Figure 5.17.* The second and third-order harmonic distortion of a CE stage which is voltage driven, $R_s \ll r_\pi$. The solid lines are the model predictions and the dotted lines the results obtained from a simulator using a complete model for the transistor.

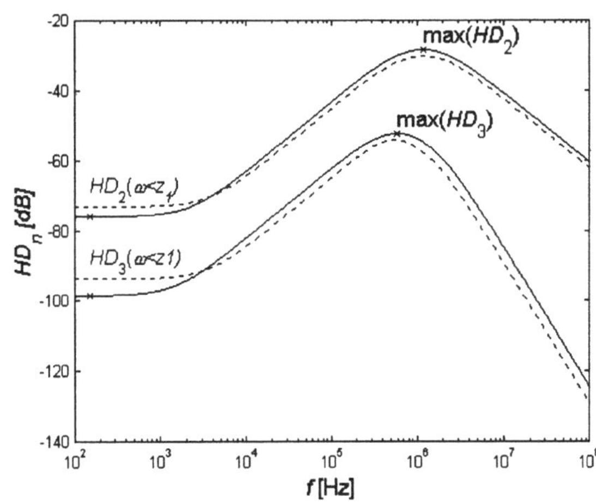


*Figure 5.18.* The second and third-order harmonic distortion of a CE stage which is current driven, $R_s \gg r_\pi$. The solid lines are the model predictions and the dotted lines the results obtained from a simulator using a complete model for the transistor.

in which $\hat{i_c}$ is the maximum of the output current and $I_c$ is the collector bias current. These expressions are in line with equations (5.47)

For the current-driven case, the distortion at relatively low-frequencies is given by ($\omega < z_1$):

$$HD_{2,i}(\omega < z_1) = \frac{1}{4}\frac{\hat{i}_c}{I_c}\left(\frac{r_\pi}{R_s}\right)^2 \tag{5.75}$$

$$HD_{3,i}(\omega < z_1) = \frac{1}{12}\left(\frac{\hat{i}_c}{I_c}\right)^2 \cdot \left(\frac{r_\pi}{R_s}\right)^3 \tag{5.76}$$

Which, indeed, predict zero distortion for$R_s \gg r_\pi$, as was found in the previous sections when considering the instantaneous case. As a worst case number, also the maximum distortion can easily be obtained for the current-driven case. These are given by:

$$\max(HD_{2,i}) \approx \frac{1}{6}\frac{\hat{i}_c}{I_c} \tag{5.77}$$

$$\max(HD_{3,i}) \approx 0.047\left(\frac{\hat{i}_c}{I_c}\right)^2 \tag{5.78}$$

### 5.5.3 Behavior of Differential CE stage

One of the reasons why differential CE stages, or differential pairs, are used, is that because of the balancing the even harmonics disappear and thus lower distortion is obtained. This was also displayed in table 5.1, i.e. $\kappa_2 = 0$. However, as will be seen in this section, as a result of a non-ideal implementation of the tail current source, even-harmonic distortion may still arise. In order to describe both effects, the typical configuration for a differential pair, as depicted in figure 5.19, is used. The relation between the input voltage and the output current of a differential pair is given by:

$$i_{out} = I_{tail}\tanh\left(\frac{v_{in}}{V_T}\right) \tag{5.79}$$

in which $I_{tail}$ is the tail current. When the tail-current source has a non-zero output conductance, $Y(s)$, an additional tail current, $I_t$, is obtained due to a common-mode input voltage, $V_{cm}$, according to:

$$I_t = \frac{Y(s)g_m}{2g_m + Y(s)}V_{cm} \tag{5.80}$$

in which $g_m$ is the transconductance of a single transistor. Including this effect into expression (5.79) yields:

$$i_{out} = I_{tail}\tanh\left(\frac{v_{in}}{V_T}\right) + i_t(v_{cm})\tanh\left(\frac{v_{in}}{V_T}\right) \tag{5.81}$$

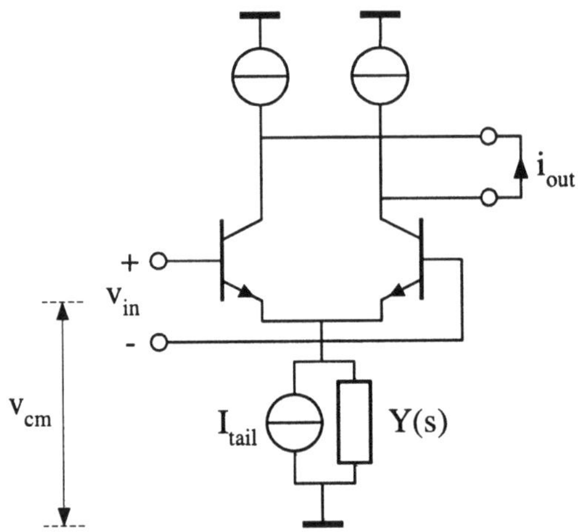


*Figure 5.19.* A typical implementation of a differential pair.

So, when the common-mode signal is zero, no effect is obtained due to the non-ideal output conductance of the tail-current source. However, in many cases the differential pair is used in an asymmetric manner, i.e. one side grounded. In that case, the "common-mode" voltage equals $v_{in}/2$ and thus an effect must be obtained. Having a closer look to expression (5.81) a simplification can be made for determining the resulting distortion. The function $\tanh(v_{in})$ has got only even-order Taylor terms with respect to $v_{in}$. Consequently, the product $v_{in} \tanh v_{in}$ has got only odd Taylor terms with respect to $v_{in}$. Thus, the even-order distortion is due to the (ideal) differential pair, whereas the odd distortion is due to the non-ideal tail-current source.

By using the Volterra series up to the third order, expressions can be found for $HD_2$ and $HD_3$. For the transistors again the nonlinear model as depicted in figure 5.16 is used. Calculations yield [31]:

$$HD_2(s) = |K_2| \left| \frac{1 - \frac{s}{z_1}}{\left(1 - \frac{s}{p_1}\right)\left(1 - \frac{s}{p_2}\right)} \right| \tag{5.82}$$

$$HD_3(s) = |K_3| \left| \frac{1 - \frac{3s}{z_1}}{\left(1 - \frac{s}{p_1}\right)^2 \left(1 - \frac{3s}{p_1}\right)} \right| \tag{5.83}$$

in which

$$K_2 = \frac{\hat{V}_{in}}{2V_T} \frac{1}{(1 + 2g_m r_o)(1 + R_s/r_\pi)} \tag{5.84}$$

$$K_3 = \frac{\hat{V}_{in}^2}{48V_T^2} \frac{1}{(1 + R_s/r_\pi)^3} \tag{5.85}$$

$$p_1 = \frac{-2}{(R_s//r_\pi)C_{be}} \tag{5.86}$$

$$p_2 = -\frac{1 + 2g_m r_o}{r_o C_o} \tag{5.87}$$

$$z_1 = \frac{-2}{R_s C_{be}} \tag{5.88}$$

$$z_2 = \frac{-1}{r_o C_o} \tag{5.89}$$

in which, $r_\pi$, $C_{be}$ and $g_m$ are the small-signal parameters of a single transistor and $r_o$ and $C_o$ model the output impedance of the tail-current source.

Figure 5.20 depicts the second-order harmonic distortion in the case of a voltage drive. The minimum distortion level (at relatively low frequencies), is

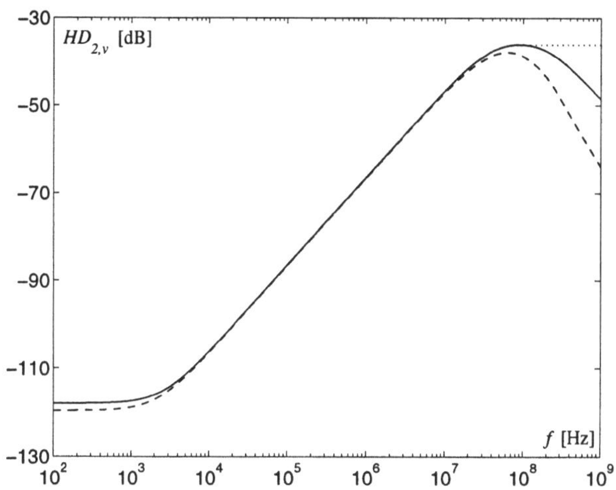


*Figure 5.20.* The second-order harmonic distortion for an asymmetrically voltage-driven differential pair in the case of a non-ideal tail-current source ($r_o =$ 10 MΩ and $C_o = 5pF$). The solid lines are the model predictions and the dotted lines the results obtained from a simulator using a complete model for the transistor.

given by equation (5.84). The maximum distortion level is described by:

$$\max(HD_2)|_{R_s \ll r_\pi} \approx \frac{C_o}{C_o + g_m R_s C_{be}} \frac{\hat{i}_{out}}{I_{tail}} \tag{5.90}$$

Clearly, the lower the output capacitance of the tail-current source, the lower the distortion.

Figure 5.21 shows the third-order harmonic distortion for a voltage and current drive. The same characteristic difference between a voltage drive and cur-

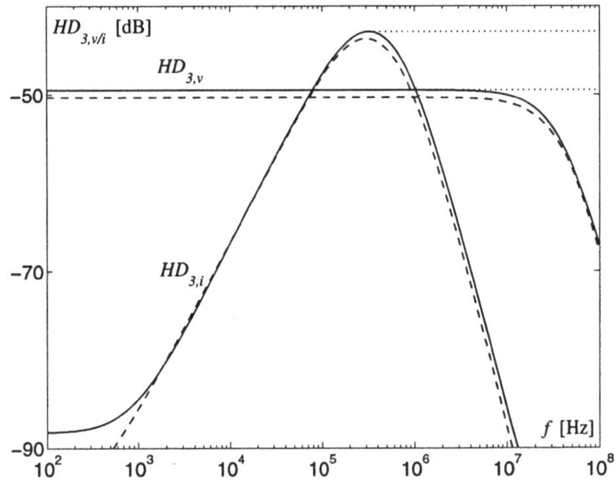


*Figure 5.21.* The third-order harmonic distortion for a voltage-driven and current-driven differential pair, $HD_{3,v}$ and $HD_{3,i}$, respectively. The solid lines are the model predictions and the dotted lines the results obtained from a simulator using a complete model for the transistor.

rent drive is obtained as was found for the CE stage. The third-order distortion for the voltage-driven case at relatively low frequencies is given by:

$$HD_{3,v} = \frac{1}{12} \left( \frac{\hat{i}_{out}}{I_c} \right)^2 \tag{5.91}$$

which corresponds to a double level compared with the single CE stage. The maximum level of distortion in the current-driven case is given by:

$$\max(HD_{3,i}) \approx 0.0558 \left( \frac{\hat{i}_{out}}{I_c} \right)^2 \tag{5.92}$$

## 5.6 Exercises

### Exercise 5.1

Consider the four different output parts of negative-feedback amplifier, as depicted in figure 5.22. The specifications as listed in table 5.3 apply to the four

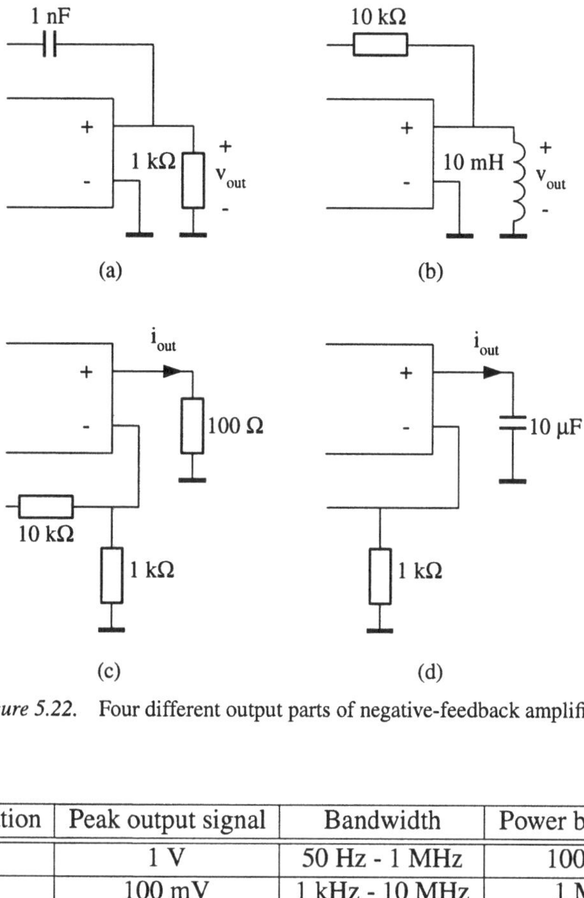


*Figure 5.22.* Four different output parts of negative-feedback amplifiers.

situations.

| Configuration | Peak output signal | Bandwidth | Power bandwidth |
|---|---|---|---|
| (a) | 1 V | 50 Hz - 1 MHz | 100 kHz |
| (b) | 100 mV | 1 kHz - 10 MHz | 1 MHz |
| (c) | 10 mA | 50 Hz - 100 kHz | 100 kHz |
| (d) | 10 $\mu$A | 50 Hz - 20 kHz | 20 kHz |

*Table 5.3.* The specifications for the four outputs depicted in figure 5.22.

1. Design for each of the configurations, depicted in figure 5.22, an output stage.

## Exercise 5.2

Clipping distortion is assumed to be orthogonal to noise and bandwidth performance. This is, however, not always true.

1. In what way can the distortion be made orthogonal to noise and bandwidth?
2. What is the difference/equivalence between clipping distortion and slew rate?

Assume an intermediate stage is required for bandwidth considerations. The biasing conditions for this stages are derived from bandwidth and clipping distortion constraints.

3. In what way should you determine the maximum signal swing for this intermediate stage, by using a simulator?
4. Discuss in this context the effect on the biasing of the intermediate stage when the output stage is a CG or a CD stage.

## Exercise 5.3

Expressions (5.16) and (5.17) describe for a general negative-feedback amplifier the second and third-order nonlinear distortion.

1. What is the effect of an increased loop gain on the nonlinear distortion?

2. When, because of bandwidth consideration, the loop gain needs to be enlarged, what is the effect on the nonlinear distortion?

Apparently, bandwidth and nonlinear distortion are not completely orthogonal.

3. Can the observed relation between bandwidth and nonlinear distortion be a bottleneck in the design methodology?

An other alternative for increasing the bandwidth (LP-product) is to increase the bias current of one of the existing stages.

4. For which stage would you choose for increasing the bias current, taking into account the highest effectiveness in lowering the distortion at the same time?

## Exercise 5.4

Given three amplifiers, for which the nullor implementations are given, as depicted in figure 5.23. Note that, besides CE and CS stages, other stages are

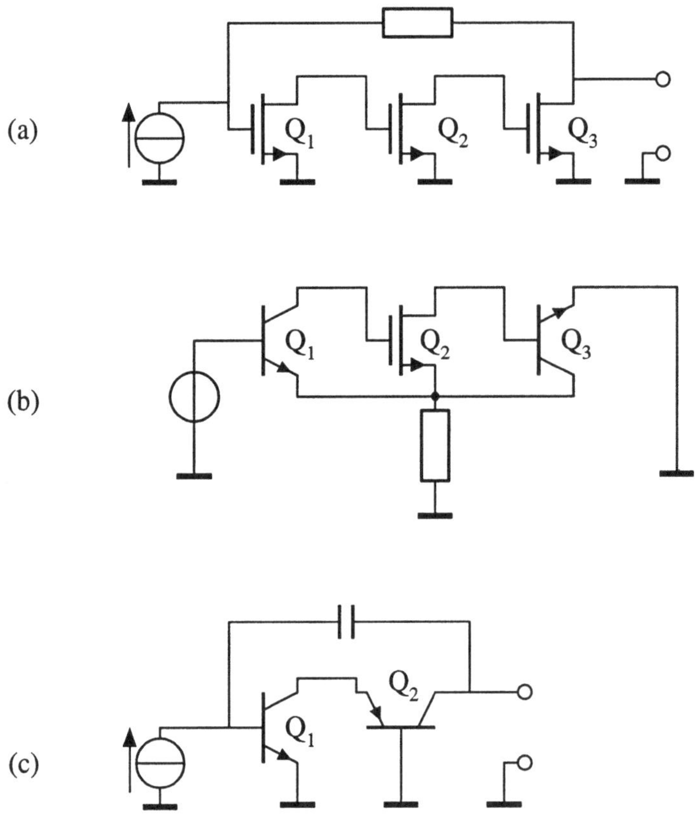


*Figure 5.23.* Three amplifiers for which the nullor implementations are give.

used as well to implement the nullors.

1. Discuss for the three amplifiers, which of the stages is most likely to limit the maximum signal amplitude because of clipping distortion (note that clipping can occur in the voltage and current domain).

2. Indicate for each of the stages of the three amplifiers, for relatively low frequencies, whether they have $\beta$-distortion or $g_m$-distortion.

# 6

# THE LOOP-GAIN-POLES PRODUCT

## 6.1 Introduction

In the previous chapters the input and output stage of the nullor were implemented. These design steps introduced the noise and the distortion to the amplifier behavior, respectively. An example of a (partially) implemented nullor, which can be obtained from the previous design steps, is depicted in figure 6.1. For this nullor, the bandwidth capabilities are still ideal, i.e. an infinite

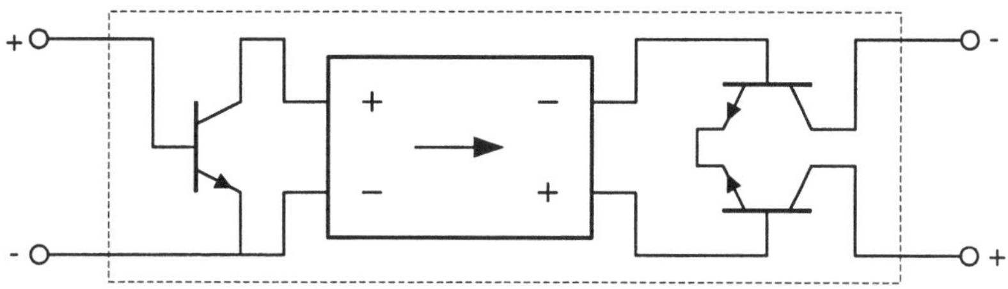

*Figure 6.1.* An example of a (partially) implemented nullor: the input and output stage are implemented.

bandwidth. This is just because of the nullor in between the input and output stage. To design the frequency behavior of the amplifier, an implementation for the remaining nullor has to be made. This implementation has to be guided by demands on the frequency behavior and, of course, the implementation has to be as simple as possible.

The frequency behavior of an amplifier can be split into two separate parts:

- absolute frequency behavior;
- relative frequency behavior.

This distinction is comparable with the distinction that is made when using polar coordinates. The absolute frequency behavior is proportional to the distance between the poles and the origin, i.e. the length of the place vector, and the relative frequency behavior has to do with the relative pole positions, i.e. the angle between the place vector and the negative real axis. The absolute frequency behavior is explicitly determined by the speed capability of the constituent devices. Whereas the relative frequency behavior can be altered easily be adding at appropriate locations some passive components. Therefore, the design of the frequency behavior is split into two steps. First, the absolute frequency behavior has to be derived and made large enough and second, the loop poles have to be moved so that the system poles are at the desired relative positions in the s-plane. The *bandwidth* of a system is closely related to the absolute frequency behavior when the poles *have* the desired relative position. For instance, when the relative frequency behavior is of the Butterworth type, the absolute frequency behavior equals the bandwidth of the transfer. In the remaining discussion, the term bandwidth will be used for the absolute frequency behavior, remembering that for the final transfer the relative positions also have to be realized.

The design of the frequency behavior of the amplifier can take a lot of work. Therefore a simple measure for the bandwidth capability of an amplifier, before starting with the design of the relative frequency behavior, can reduce design time considerably. In this way the chance that one needs to conclude after lengthy calculations that the amplifier cannot reach the desired bandwidth, is reduced significantly.

In this chapter the Loop-gain-Poles product (LP-product) is introduced and it is shown that this LP-product is a very simple measure for the bandwidth capability of an amplifier. When the LP product is, relative to the bandwidth specifications, too low, it is *impossible* to reach the desired bandwidth. Then, the LP product needs to be enlarged. Measures are discussed to increase the LP-product. It is assumed that an all-pole Butterworth characteristic is desired. However, in an analogous way the LP product can be applied to other pole patters as well. Subsequently, in the next chapter, the design of the relative frequency behavior is treated.

## 6.2 The simple transistor model

To obtain a simple criterion for the bandwidth capability of an amplifier, a simple model for the transistor is inevitable. The model used for the bipolar transistor is depicted in figure 6.2. The model comprises the input impedance of the transistor, $r_\pi$ and $C_\pi$, and the voltage-controlled current source, $g_m$. So, the basic components responsible for the gain and the limited speed are taken into account. When this model is used, the DC loop gain and the poles and zeros can be calculated relatively simple. For the FETs, an analogous model is

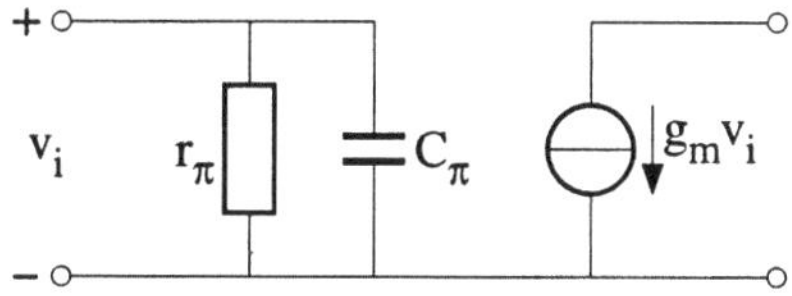


*Figure 6.2.* The simple model for the bipolar transistor, used for the calculation of the LP-product.

used, i.e. $C_{GS}$ and $g_m$. Of course, after the design of the frequency behavior, the validity of this simple model has to be checked. This is discussed in the next chapter.

## 6.3 The LP-product

In chapter 3 the asymptotic-gain model was shown to be the appropriate model for the synthesis of amplifiers. Now, the frequency dependency of the constituent elements is considered. The expression for the asymptotic-gain models then reads:

$$A_t(s) = A_{t\infty} \frac{-L(s)}{1 - L(s)}, \tag{6.1}$$

in which the direct transfer $A_{t0}$ is ignored. This is possible as the term with the direct transfer cannot influence the pole positions; it can only introduce zeros into the system transfer.

Assume a loop gain with $n$ loop poles, $p_{li} = a_i + b_i j$ with $a_i < 0$, as given by:

$$L(s) = \frac{L(0)}{\prod_{i=1}^{n}(1 - s/p_{li})}, \tag{6.2}$$

where $L(0)$ is the DC loop gain. Then the characteristic polynomial, CP, of $A_t(s)$ is given by:

$$CP(s) = s^n + \cdots + [1 - L(0)] \prod_{i=1}^{n} |p_{li}|. \tag{6.3}$$

The zeroth-order term is called the Loop-gain-Poles product, or LP product for short [1]. A more precise name would be the DC-return-difference-poles product, because the term $[1 - L(0)]$ is the return difference as defined in [32]. However, for accurate amplifiers, the magnitude of the loop gain is relatively large and the magnitude of the DC loop gain is approximately equal to the magnitude of the DC return difference. Expression (6.3) is found from the viewpoint of the root-locus method. However, of ultimate interest are the system poles. The characteristic polynomial derived from the $n$ system poles, $p_{s_i}$ =

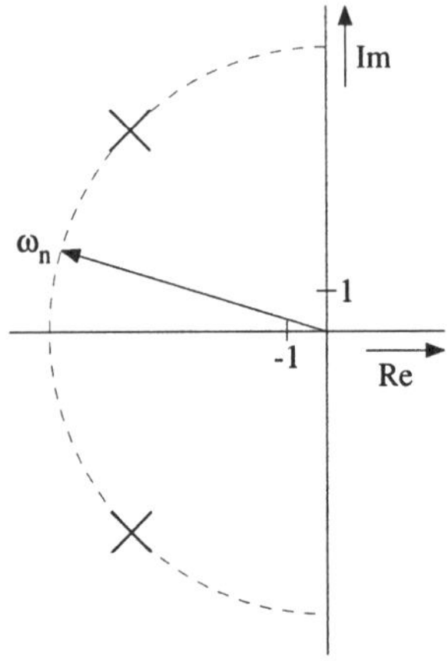


*Figure 6.3.* The characteristic placement of the poles for the case of a Butterworth pattern.

$c_i + d_i j$ with $c_i < 0$, equals:

$$CP(s) = s^n + \cdots + \prod_{i=1}^{n} |p_{si}|. \tag{6.4}$$

Here the zeroth-order term is the product of the moduli of all the system poles, i.e. the product of the lengths of the corresponding place vectors. Thus this term explicitly describes the absolute frequency behavior of the system. Consequently, as equation (6.3) and (6.4) describe both the CP, the zeroth-order term found in equation (6.3) is *also* a measure for the maximum attainable bandwidth of the corresponding system.

In order to get this relation more explicit, a specific relative frequency behavior is considered. For amplifiers, the Butterworth characteristic is a commonly used relative frequency behavior because it results in a maximum-flat-magnitude transfer. For a Butterworth characteristic, the system poles are regularly placed on a half circle in the left half of the s-plane, see figure 6.3. For an nth-order system, the half circle is divided into n equal parts and in the middle of each part a pole is located. For a bandwidth of $\omega_n$, the radius of the circle equals $\omega_n$ and thus the modulus of each pole equals $\omega_n$. Applying this to equation (6.4) yields:

$$CP(s) = s^n + \cdots + \omega_n^n. \tag{6.5}$$

Equation (6.3) and (6.5) both describe the CP, thus the following relation holds:

$$LP_n = [1 - L(0)] \cdot \prod_{i=1}^{n} |p_{li}| = \omega_n^n. \tag{6.6}$$

or

$$\omega_{n_{max}} = \sqrt[n]{LP_n} = \sqrt[n]{\left|[1 - L(0)]\prod_{i=1}^{n} p_i\right|}. \quad (6.7)$$

So, the product of the loop poles and the DC loop gain, is a measure for the maximum attainable bandwidth. The question is, which poles must be used to calculate this LP-product?

Example: What maximum bandwidth can be expected when the loop consists of three poles, $p_{l1} = -1$ kHz, $p_{l2} = -10$ kHz, $p_{l3} = -1$ GHz, and the DC loop gain equals -100?

When the three poles are used to calculate the LP-product, the maximum bandwidth , $B_{s3}$, is found to be:

$$B_{s3} = \sqrt[3]{101 \cdot 1\text{ kHz} \cdot 10\text{ kHz} \cdot 1\text{ GHz}} \approx 1\text{ MHz}. \quad (6.8)$$

With a bit of experience one knows that the pole at -1 GHz is not dominant, i.e. it does not contribute to the bandwidth. The maximum bandwidth, $B_{s2}$, calculated based on $p_{l1}$ and $p_{l2}$ yields:

$$B_{s2} = \sqrt{101 \cdot 1\text{ kHz} \cdot 10\text{ kHz}} \approx 32\text{ kHz}, \quad (6.9)$$

which is about a factor 30 lower than $B_{s3}$.

That the -1 GHz pole does not contribute to the bandwidth is clear; however, what to do when it was at -1 MHz. As was stated, the LP-product only predicts the bandwidth when the poles used can be moved into the required relative positions, for this case the Butterworth positions; these poles then contribute to the bandwidth and are therefore called the *dominant poles*. Thus, only dominant poles should be used to calculate the LP product.

## 6.4 Dominant poles

In principle, one only knows at the end of the design what the maximum attainable bandwidth really is. The LP-product gives a *prediction* of this maximum bandwidth. It is not certain whether or not this bandwidth can be reached. The only thing that is sure is that for the given number of stages the bandwidth cannot be larger than that indicated by the LP-product. An analogous rule can be found for the dominant poles. The following derivation of the dominant poles is not limited to Butterworth behavior, it is generally applicable to other relative frequency behaviors as well. However, the derivation of the dominant poles is limited to loops with only real poles.

To find the dominant poles, the frequency behavior of the system is described again from two points of view. Firstly, the characteristic polynomial is described from the loop point of view, which yields:

$$CP(s) = s^n - s^{n-1}\sum_{i=1}^{n} p_{li} + \cdots, \quad (6.10)$$

with $p_{li} < 0$. Secondly, the characteristic polynomial is described as a function of the system poles, which yields:

$$CP(s) = s^n - s^{n-1}\sum_{i=1}^{n} p_{si} + \cdots, \tag{6.11}$$

with $p_{si} = c_i + d_i j$, $c_i < 0$, the system poles. Now the factor of the (n-1)th-order term is of interest. Comparing the term of equation (6.10) with the corresponding term of equation (6.11) yields:

$$\sum_{i=1}^{n} p_{li} = \sum_{i=1}^{n} p_{si}, \tag{6.12}$$

which states that the sum of the loop poles is equal to the sum of the system poles. From this property a procedure can be derived for the dominant poles, it is as follows.

The LP-product gives a measure of the maximum attainable bandwidth. As the required relative frequency behavior is known, the position of the system poles can be determined and from that their *sum* can be calculated. The *sum* of the loop poles is also given. These sums are generally not equal and frequency compensation has to be used as discussed in the next chapter. All the methods discussed have the property of making the sum of the system poles smaller (i.e. more negative, remembering that the poles are negative). Thus, when the sum of the loop poles is smaller than the sum of the required system poles, frequency compensation will *not succeed*; the loop poles cannot be placed in the desired position; at least one loop pole is too far away from the origin. Such a pole will be called a non-dominant pole. The most negative pole from the loop has to be ignored and the LP-product and the sum of the remaining poles has to be calculated again, et cetera, until the highest number of dominant poles is found. Thus:

When $p_{li}$ are the poles of the loop and $p_{si}$ are the poles of the system, the dominant poles are the largest set of poles for which holds:

$$\sum_{i=1}^{n} p_{li} \geq \sum_{i=1}^{n} p_{si}. \tag{6.13}$$

The sum of the loop poles has to be less negative than the sum of the system poles.

Fulfilling this criterion is necessary but not sufficient. The characteristic polynomial may include more terms which must be given the appropriate values and it must be possible to implement the required frequency compensations in the circuit. In contrast, when the criterion is not fulfilled, it is certain that frequency compensation will not succeed with the given set of poles and loop gain, and the LP-product of the set of dominant poles has to be increased.

Example: For the previous example the LP-product for the third-order system predicted a bandwidth of 1 MHz. For a 1 MHz third-order Butterworth system, the sum of the poles equals -2 MHz ($p_{s1} = -1$ MHz, $p_{s2,3} = -\frac{1}{2}$ MHz $\pm \frac{1}{2}\sqrt{3}j$ MHz). The sum of the loop poles is approximately -1 GHz which is much smaller than -1 MHz and therefore at least $p_{l3}$ is non-dominant. The predicted bandwidth of the second-order system is 32 kHz. The sum of the loop poles equals -11 kHz which is greater than the sum obtained from the system poles, -45 kHz. Thus the system has two dominant poles.

Some remarks have to be made on the dominant pole criterion. The determination of the dominant poles assumes that the poles are real. For complex poles with a high Q, the contribution to the LP product can be very large while the contribution to the sum of the poles is only small (only the real part counts, the imaginary parts cancel). However, when complex poles are present in the loop, the loop comprises a second loop or a resonator (which can be seen as a loop also). The poles of these local loops must be placed at the correct positions, before the overall loop is compensated. Multi-loop techniques are beyond the scope of this book and not discussed here.

Further, the dominant pole criterion is derived with the assumption that only compensation techniques are available which make the sum of the poles smaller. However, techniques exist to increase the sum of poles. These techniques, however, use positive feedback and the risk of instability due to component spread increases. Therefore, these techniques are less favorable.

## 6.5 Increasing the LP-product

For now the focus is on what to do when the LP-product is too low. Because, in that case it is not possible to reach the specified bandwidth. There are two basically different ways for increasing the LP-product:

- Increasing the LP-product without increasing the order;

- Increasing the LP-product by increasing the order.

### 6.5.1 Increasing the LP-product without increasing the order

The LP-product is the product of the DC loop gain and the loop poles. Thus, when the DC loop gain is increased without altering the product of the poles, or vice versa, the LP-product increases.

In figure 6.4 an example of a transimpedance amplifier is given. The nullor implementation is the one from figure 6.1 in which the intermediate nullor is implemented by two wires only. For the DC loop gain and the poles holds,

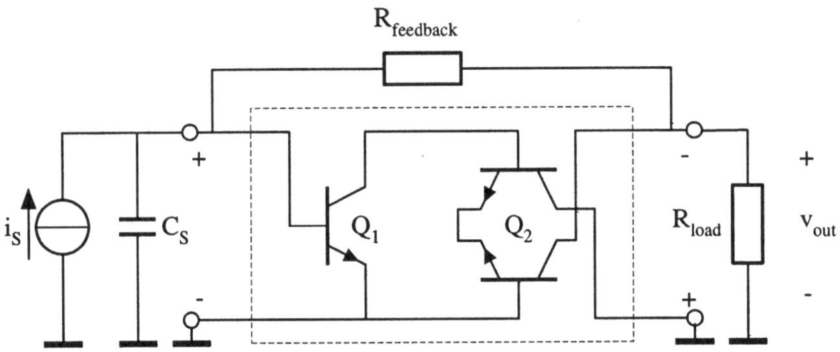


*Figure 6.4.* An example of a transimpedance amplifier.

when $r_{\pi 1} \ll R_{feedback}$:

$$\begin{aligned} L(0) &= -\beta_1\beta_2 \frac{R_{load}}{R_{load} + R_{feedback}}, \\ p_1 &= \frac{-1}{2\pi r_{\pi 1}(C_s + C_{\pi 1})}, \\ p_2 &= \frac{-1}{2\pi r_{\pi 2} C_{\pi 2}}, \end{aligned} \tag{6.14}$$

with the subscripts referring to the corresponding stages. From these expressions the LP-product is found to be:

$$LP_2 = \frac{C_{\pi 1}}{C_{\pi 1} + C_s} \cdot f_{T1} \cdot f_{T2} \cdot \frac{R_{load}}{R_{load} + R_{feedback}}, \tag{6.15}$$

with $f_T = \frac{g_m}{2\pi C_\pi}$. This equation shows some characteristic properties for the LP-product:

- an amplifier stage adds a factor $f_T$ to the LP-product;
- the influence of the source, load and feedback impedance are represented by two factors, a factor concerning the input and a factor concerning the output. Possibly these factors introduce additional poles.

As the source and load impedance are given for a specific design, the LP product can be enlarged, without increasing the order, by increasing the $f_T$ of the used transistors. The $f_T$ of a transistor is determined by:

- the type of transistor;
- its bias current.

At this point of the design the amplifier comprises two stages. Thus the LP product can be enlarged by increasing the $f_T$ of the input and/or the output stage.

#### 6.5.1.1 Increasing the $f_T$ of the input stage

For the input stage the bias current and type of transistor were chosen on noise considerations. For the type of transistor this could have been the transistor with the lowest base resistance, for instance. When it is possible to use a transistor with a higher $f_T$, without degrading the noise behavior such that the specifications are not met anymore, the LP-product can be increased by using this transistor. The increase in LP-product can be less than the increase in $f_T$ because of the first factor of equation (6.15).

In chapter 4 it was found that the noise level as a function of bias current is relatively flat near the noise minimum. Thus increasing the bias current of the input transistor in order to increase its $f_T$ results in a small increase of the noise level only. In contrast, the increase of the LP-product can be considerable. For instance, when the bias point of the transistor is on the left slope of the $f_T - I_C$ curve (see figure 1.11) the $f_T$ increases linearly with increasing bias current.

#### 6.5.1.2 Increasing the $f_T$ of the output transistor

The output transistor was chosen on output capabilities. When the output transistor can be replaced by a transistor with a higher maximum $f_T$ the LP-product can be increased by using this type of transistor. Further, as long as the output transistor is not biased at its maximum $f_T$, the bias current can be increased such that the $f_T$ becomes higher. The increase of the bias current enlarges the output capabilities and thus distortion is lowered. No contradiction between bandwidth and distortion is apparent.

Which from these measures (at the input or output) is the most effective depends on several factors. A transistor with a higher maximum $f_T$ is not always available. For instance, when the circuit needs to be made as an IC, the IC process determines the maximum $f_T$ of the transistors and not much design freedom is left.

Increasing bias currents is often the most effective for the input stage. The bias current of the input stage is often lower than the bias current of the output stage. Consequently, the $f_T$ of the input stage is often lower than the $f_T$ of the output stage and less current is required at the input to increase the $f_T$ by a certain factor.

When with these measures the required LP-product cannot be obtained, the order of the system has to be increased.

### 6.5.2 Increasing the LP-product by increasing the order

In this section the intermediate nullor of figure 6.1 is implemented by a single amplifying stage and consequently, the order is increased by one. Care has to be taken to keep the loop gain negative. When an additional CE stage is used, the two output terminals of the differential pair, $Q_2$, need to be interchanged. Consequently, this differential pair can be replaced by a CE stage. The additional intermediate CE stage introduces an additional pole at:

$$|p| = \frac{1}{2\pi r_\pi C_\pi} = \frac{f_T}{\beta} \tag{6.16}$$

and the loop gain increases by a factor $\beta$. Thus the LP product for the third-order system is ($LP_3$):

$$LP_3 = LP_2 \cdot \frac{f_T}{\beta} \cdot \beta = LP_2 \cdot f_T. \tag{6.17}$$

As long as the $f_T$ of the transistor is larger than the bandwidth given by $LP_2$ product, the bandwidth of the system increases. This option is less favorable with respect to previous ones, because the order increases by one and the frequency compensation of the system becomes more difficult.

#### 6.5.2.1 Biasing the intermediate stage

The bias conditions for the intermediate stage are found from two criteria:

- the $f_T$ of the transistor and
- the output capability of the intermediate stage.

As the transistor is used for increasing the LP-product of the amplifier, the bias current has to be chosen such that the resulting LP-product is large enough. The $f_T$ is, up to its maximum value, more or less proportional to the collector bias current (see chapter 5, figure 1.11) and thus a minimum collector bias current is found.

Further, the intermediate stage has to be able to drive the next stage over the complete frequency range of interest. The input impedance of the next stage can be capacitive for higher frequencies and, consequently, the drive current at high frequencies can be considerably larger than the low-frequency drive current. When the bias current of the intermediate stage is too low, slewing occurs at high frequencies, the *total* bias current meant for the intermediate stage is used to charge the input capacitance of the next stage and the intermediate stage is switched off. As a result, the negative-feedback loop is broken and the complete amplifier behavior (including biasing) may be influenced. When later on the intermediate stage has to be switched on again, the total amplifier has to settle before proper functioning is possible. This settling time may be too large with respect to the frequency of the input signal and the loop remains broken such that no accurate amplification is possible anymore. This switching of stages may cause terrible transients at the output of the amplifier.

The minimal bias current for the intermediate stage is given by the maximum input current of the next stage. This maximum current is easily found from an AC analysis. This analysis linearizes the complete circuit before the transfer is calculated. Consequently, no limitations due to bias sources occurs and the maximum input current of the next stage is found.

## 6.6 The contribution to the LP-product of a (MOS)FET

The previous sections concentrated on the use of bipolar transistors. For a (MOS)FET the situation seems to be different. The contribution to the DC loop gain is approximately infinite because of the absence of a DC input current. Thus the L part of the LP-product is infinite. In contrast, the pole introduced by the (MOS)FET is at the origin. The P part of the LP-product is zero. Thus, the product of L and P is undefined. However, for the maximal attainable bandwidth not the L nor the P separate is what counts, but their product.

In figure 6.5a the small-signal diagram of a (MOS)FET is depicted. In figure 6.5b the corresponding current-gain factor as a function of frequency is depicted. To be able to calculate the contribution to the LP-product, a dummy resistor$R_x$

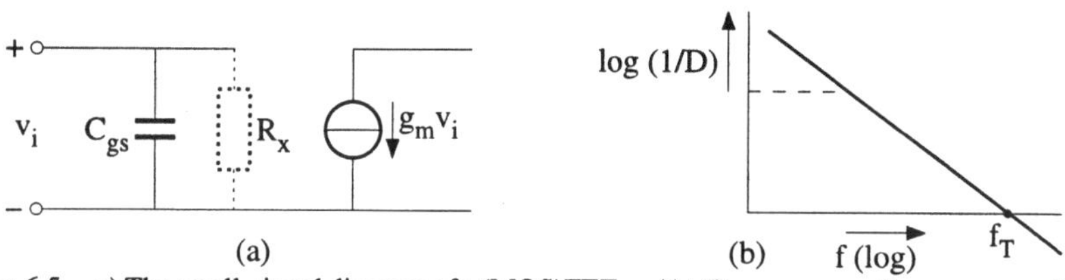


*Figure 6.5.* a) The small-signal diagram of a (MOS)FET and b) The corresponding current-gain factor as a function of frequency.

is used, depicted in fig.6.5a by the dotted-drawn resistor. The low-frequency current-gain factor, including $R_x$, is drawn in fig.6.5b with the dotted line. For the contribution to the L and P now holds, respectively:

$$\begin{aligned} L &= -g_m R_x, \\ P &= \frac{-1}{2\pi R_x C_{gs}}, \end{aligned} \tag{6.18}$$

and thus the contribution to the LP-product equals

$$LP = \frac{g_m}{2\pi C_{gs}} = f_T. \tag{6.19}$$

The contribution to the LP-product is independent of the dummy resistor. Thus, for $R_x$ becoming infinite, the contribution remains equal to the $f_T$ of the transistor.

## 6.7 What to do with a zero in the origin

In some situations it might occur that a loop zero is at the origin. Consequently, the DC loop gain is zero and the calculated LP-product is zero too.

This situation occurs, for example, in a charge amplifier as depicted in figure 6.6. Due to the pure capacitive feedback the DC loop gain is zero. A hint about

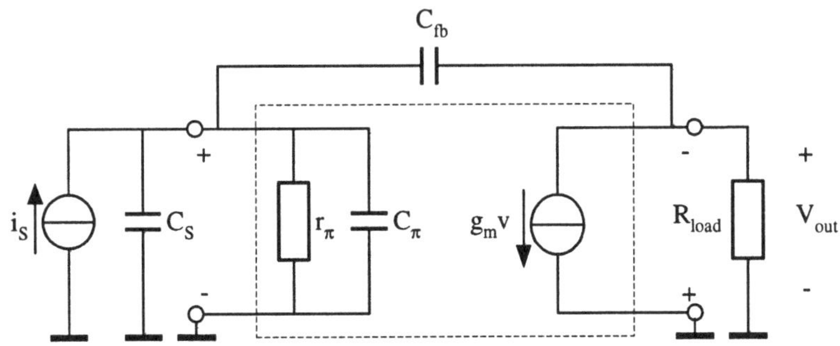


*Figure 6.6.* A charge amplifier with a zero at the origin in the loop.

what to do in such a case, can be found from the root locus. In figure 6.7 a typical root locus of a charge amplifier is depicted, when it is assumed that besides the zero the loop also comprises three poles. In this particular case the zero at

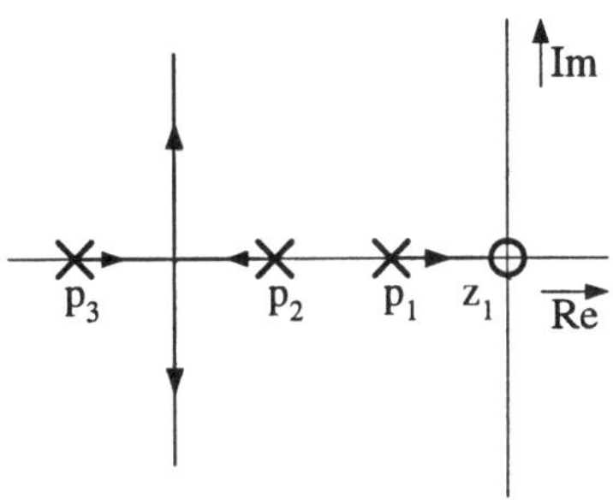


*Figure 6.7.* The root locus of the charge amplifier.

the origin is a phantom zero (to be discussed in the next chapter), representing the desired integrator pole, ($A_{t\infty}$), of the charge amplifier. When the loop is closed, the pole at the lowest frequency moves to the zero at the origin and cancels it, ideally. The other two poles determine the high-frequency behavior of the amplifier. Thus before closing the loop the low-frequency pole and the zero at the origin can already be cancelled and frequency compensation can concentrate on the two high-frequency poles. Then still the question remains, what DC loop gain has to be used. This is found from the loop-gain-versus-frequency plot, figure 6.8. Cancelling the pole and zero means in this figure the extrapolation of the loop gain from higher frequencies to zero (the dotted line). As mostly the pole to be cancelled is at relatively low frequencies, the loop gain of the charge amplifier remains for high frequencies the same. Only for low frequencies, below $p_1$, the loop gain is actually lower.

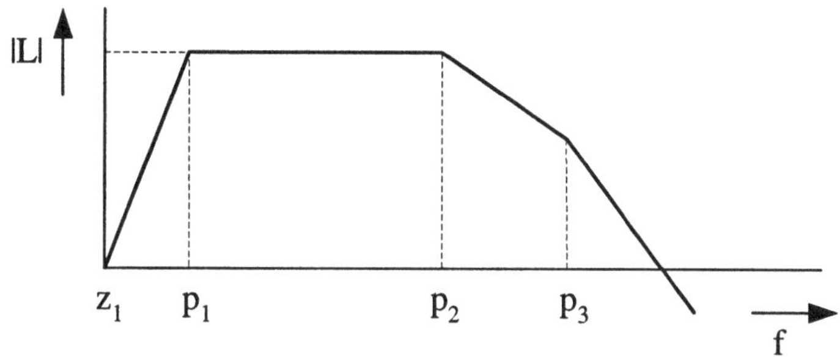


*Figure 6.8.* The loop gain versus the frequency.

Cancelling the pole and zero means that the division in which, in this case, $C_{fb}$ is involved, is made frequency independent. This can be done by removing $r_\pi$ or by placing a resistor in parallel with $C_{fb}$ such that the time constant of the feedback network and the input impedance of the nullor implementation are equal. With the first method a pure capacitive voltage divider is obtained. The pole and zero are cancelled at the origin. Simulation software may have problems with this method as for DC-loop-gain calculations mostly a*resistive* path (loop) needs to be available. With the second method, cancelling the pole and zero at $p_1$, this problem does not occur.

## 6.8 Guaranteeing a negative loop gain

Now the number of stages is determined, the loop gain is not necessarily negative anymore. Some attention was already paid to this in section 6.5.2. For changing a positive loop gain into a negative loop gain, without changing the number of stages, one of the stages could be made differential as discussed in section 2.6.2, or the two output terminals of a differential pair could be interchanged. Another way, which is frequently seen, is replacing one of the CE (CS) stages by a CC (CD) stage. In section 2.6.2, however, it was already made clear that this is not a favorable option.

The CC stage, also called a voltage follower, is a non-inverting stage. Thus when a CE stage is replaced by a CC stage the sign of the loop gain changes indeed. However, the use of a CC stage most probably reduces the performance for one or more of the three design aspects, noise, bandwidth and distortion. When the CC stage is placed at the input, the noise performance is deteriorated as the influence of the voltage noise of the second stage on the noise behavior is increased. Using the CC stage at the output, results in degradation of the distortion behavior. As the voltage gain of the CC stage is just 1, the preceding stage must be able to handle the same output-voltage swing as the output stage itself. Further, due to the local voltage feedback, the input impedance of this stage may become very high, resulting in a larger bandwidth reduction due to parasitic components. All these negative effects are caused by local feedback of the CC stage which makes the parameter A of the chain matrix equal to one. The remedy often used to make this parameter almost zero again, is a CC-CB combination, which is just a differential CE stage. This would just be the long and unstructured way to the solution that should have been chosen in the first place: an asymmetrically driven differential pair. The differential CE stage can be used successfully as a non-inverting stage. The parameters A and D of the chain matrix are equal to that of a single CE stage and the parameters B and C differ only by a factor 2. They can be made equal to the parameters of the CE stage at the expense of some extra power. Consequently, the noise performance, distortion behavior and the bandwidth capabilities remain unaffected.

Summarizing, when a loop gain has to be made negative again, this generally results in the use of the differential CE stage instead of a single CE stage. When the CC stage is the first choice, this stage is nearly always placed at the output because the deterioration of the distortion behavior is less compared to the degradation of the noise performance. Thus, using a CC stage leads to a differential pair at the output. In contrast, when the differential CE stage is used, the designer can still decide whether the differential CE stage is placed at the input or at the output. A differential CE stage at the input is in some cases favorable (temperature behavior of the input offset voltage).

## 6.9 Exercises

### Exercise 6.1

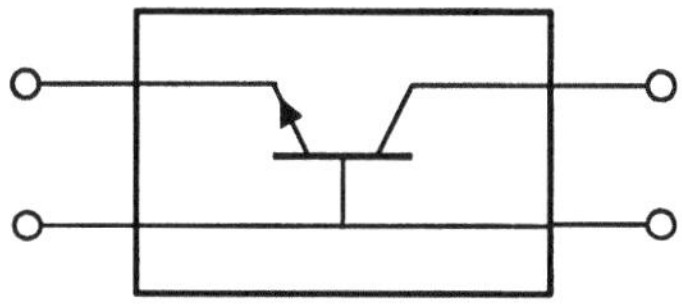

*Figure 6.9.* A CB stage.

1. Calculate the pole of a CB-stage and compare it with the pole of the corresponding CE-stage.
2. Calculate the current-gain factor of a CB-stage and, again, compare the result with the corresponding CE stage.
3. Calculate the contribution to the LP-product of a CB-stage in an amplifier.
4. Is a CB-stage a useful way to increase the LP-product of an amplifier?

## Exercise 6.2

During the design process, i.e. after the noise and distortion minimization, it appears that the LP-product is slightly too low. Increasing the bias current of the input stage can increase it within specs.

1. Why is the input stage a good candidate for increasing the LP-product?

2. Is it allowed to use this option? The input stage was designed to have maximum performance with respect to noise!

3. The addition of an extra amplifying stage is an other option. What can be the considerations to either increase the LP-product via an additional stage or via increasing a bias current?

## Exercise 6.3

Given the $f_T$ of a technology.

1. Up to what frequency can the bandwidth of an amplifier globally be?

2. Is it wise to design an amplifier with a bandwidth equal to that specific frequency?

# 7

# FREQUENCY COMPENSATION

## 7.1 Introduction

In the previous chapter the bandwidth capability was discussed. It was found that the LP-product predicts the absolute bandwidth *capability* of a nullor implementation. Nothing was said about *how* to reach that bandwidth nor about the *possibility* of reaching it. The only thing that can be said is that when the LP-product is too low, it is not possible to reach the specified bandwidth at all.

In this chapter the next design step is discussed: frequency compensation. At this point the LP product is considered to be large enough and now the *placement* of the poles is of concern, i.e. the design of the relative frequency behavior. This is called frequency compensation. The frequency behavior of the transfer function is made such that it meets the specifications. In this chapter it is again assumed that the transfer function has to be of the Butterworth type.

The frequency compensation techniques are not allowed to deteriorate the former design steps, i.e. noise and distortion. Further the reduction of the LP-product due to the frequency compensation must be as small as possible, ideally it should not reduce the LP-product at all.

## 7.2 Model used for the frequency compensation

Frequency compensation tends to be the design step with the largest risk of becoming a tedious way of calculation and trial and error. To prevent this and to make frequency compensation more clear, the model of the transistor is stripped to the relevant part only. Starting with the hybrid-pi model of the bipolar transistor, figure 7.1, some simplifications are made. When the transistor is driven from a relatively high impedance the base resistance, $r_b$, does not influence the transfer and can thus be ignored. When the transistor is loaded

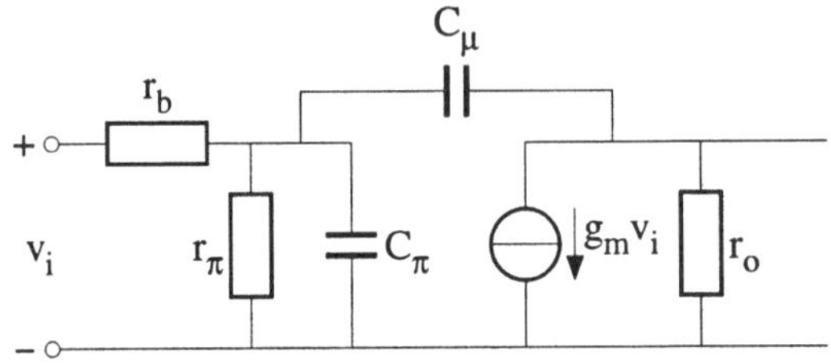


*Figure 7.1.* The hybrid-pi model.

with a relatively low impedance, the output impedance of the transistor,$r_o$, is shorted and can thus be ignored also.

The influence of $C_\mu$ on the input impedance can be described, under certain conditions, by the Miller approximation:

$$C_{input} = C_\mu(1 - A_V). \tag{7.1}$$

in which $A_V$ is the voltage gain from input to output. Due to the assumed low load impedance the voltage gain between the base and collector,$A_V$, is small and $C_\mu$ acts just as it is in parallel with the input of the transistor. Thus when $C_\mu \ll C_\pi$ the influence of $C_\mu$ at the input can be ignored also.

Besides the influence on the input impedance of the transistor,$C_\mu$ introduces a zero in the right half plane at:

$$z_\mu = +\frac{g_m}{C_\mu}. \tag{7.2}$$

At this frequency the transfer from the input to the output of the transistors via $C_\mu$ equals the transfer via the voltage-controlled current source,$g_m$, but with a phase difference of $90^o$. By ignoring $C_\mu$ this zero is ignored too. The simplified model is given in fig.7.2.

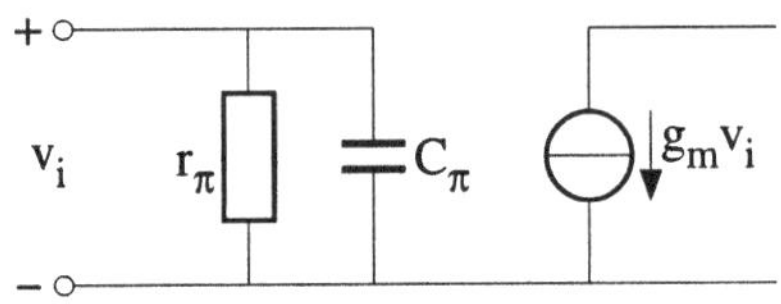


*Figure 7.2.* The simplified hybrid-pi model.

## 7.3 Model validation after frequency compensation

When the frequency compensation using the simplified model of figure 7.2 is finished, the model has to be checked on its validity, for each transistor separately. This can be done by introducing, one by one, the$C_\mu$s, $r_o$s and $r_b$s. When an $r_o$ or a $C_\mu$ introduces an unacceptable influence, the load impedance of the transistor is apparently not low enough. It can be made low by using a current follower cascading the amplifying stage. Remind that a low load impedance was the criterion for ignoring $C_\mu$ and $r_o$. A current follower cascading an amplifying stage is depicted in figure 7.3. With this current follower$r_o$ is short

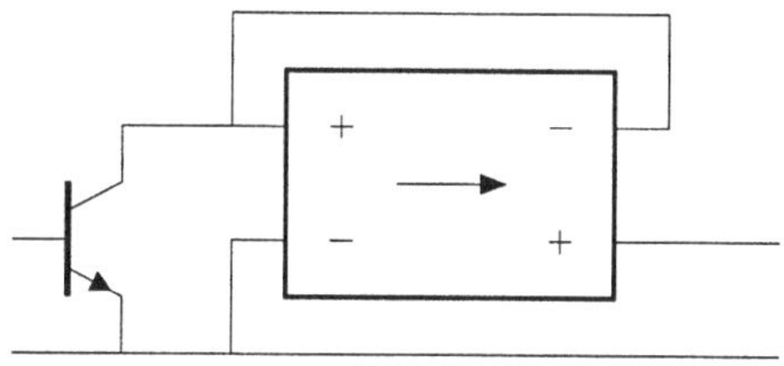


*Figure 7.3.* A transistor cascaded by an ideal current follower.

circuited and $C_\mu$ is in parallel with $C_\pi$. However, the zero due to $C_\mu$ is still at $+g_m/C_\mu$ as the signal current flowing through $C_\mu$ flows through the current follower as well. Thus when an ideal current follower does not reduce the effect of $C_\mu$ significantly, there are two possible causes:

- $C_\mu$ is not negligible with respect to $C_\pi$;
- the zero due to $C_\mu$ is not negligible.

The first problem can be solved by making$C_\mu$ smaller by a higher reverse-bias base-collector voltage ($C_\mu$ is a junction capacitance). When $C\mu$ cannot be reduced enough in this way, $C_\pi$ has to be changed to $C_\pi + C_\mu$ and one iteration for the frequency compensation has to be made. For the second problem there are three possible solution. Firstly, a higher reverse-bias base-collector voltage reduces $C_\mu$ and the zero shifts to a higher frequency. Secondly, a larger bias current can be chosen for the transistor such that$g_m$ increases and the zero also shifts to a higher frequency. In this case the frequency compensation needs some trimming because the pole of the transistor depends on$g_m$ also. Thirdly, just do a frequency compensation in which the right-half-plane zero is taken into account.

After placing all the $r_o$s and $C_\mu$s one by one, the influence of the $r_b$s can be checked. In figure 7.4 the situation is depicted for a transistor that is driven from a current source, i.e. an other transistor, with a parallel impedance of Z(s). This impedance can be due to an output impedance of the preceding stage or due to a frequency-compensation network. The influence of$r_b$ is especially apparent in situations where $Z(s)$ is on the order of $r_\pi$ and $r_\pi$ is relatively small. This can occur in output stages where the bias current can be large. In the case that Z(s) is due to an output impedance, Z(s) can be increased by adding a current follower after the preceding stage. In this way Z(s) is enlarged. Of course, reducing the base resistance by using an other transistor with a lower base resistance, or by taken several transistors in parallel can help also.

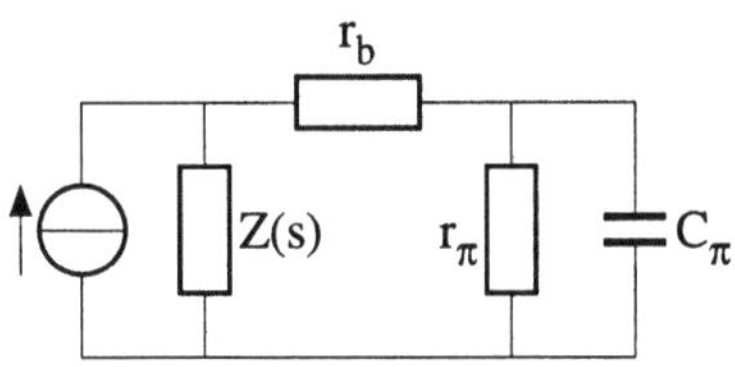


*Figure 7.4.* The influence of a base resistance.

When Z(s) results from a frequency-compensation network, the base resistance could be lowered by taking several transistors in parallel. When this is not possible either the base resistance should be taken into account in the frequency compensation or the frequency-compensation network should be placed somewhere else. The chance on a problem due to the base resistance can be reduced by putting no frequency compensation networks at the input of transistors with a relatively high base resistance and a relatively high bias current.

Sometimes, lowering the bias current is also an option. In that case $r_\pi$ increases and the relative effect of $r_b$ reduces. In the last chapter, in which a design example is discussed, this option is used to reduce the effect of a base resistance. Of course, it must always be checked with respect to the specification whether it is possible or not for the specific transistor to lower its bias current.

When the frequency compensation is made o.k. again with the ideal current followers, implementations for the current followers can be made. The most simple active implementation of the nullor is a single CE (CS) stage. This is depicted in figure 7.5. The single-stage current follower is just the CB stage,

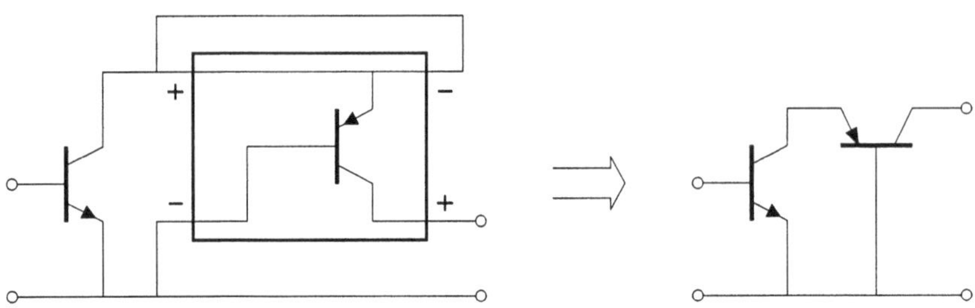


*Figure 7.5.* A single-stage implementation of the current follower.

or in the case of a FET a CG stage. The input impedance of the CB stage equals $1/g_m$. When the bias current of the CB stage is equal to the bias current of the cascoded transistor, the voltage amplification equals $A_V = -1$ and thus the influence of $C_\mu$ on the input is equal to $2C_\mu$. The output impedance of the cascoded stage equals:

$$r_{o_{cascode}} = \beta r_o. \tag{7.3}$$

As a result of the CB stage, the output impedance is increased by a factor $\beta$.

Of course the CB stage adds a pole to the loop. However, ideally this pole is at $f_T$ and is therefor mostly non-dominant.

Analogous to the bipolar transistor the model for the (MOS)FET is as given in figure 7.6. After frequency compensation with this model the model is extended

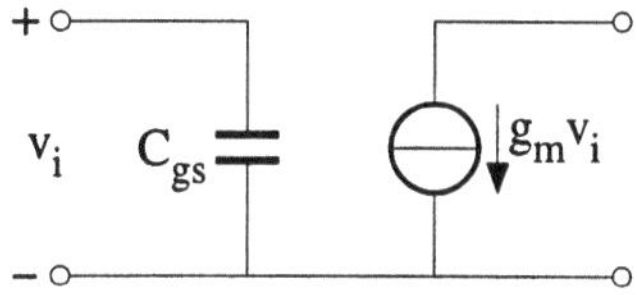


*Figure 7.6.* The model for the (MOS)FET to be used for frequency compensation.

with $C_{dg}$ and $r_d$. The (MOS)FET does not have a gate bulk resistance, the equivalent for the base-resistance. However, in the (MOS)FET technology it is common use to connect the transistor with poly-silicon. This poly-silicon is relatively high ohmic and by that the (MOS)FET can have a series gate resistance of several hundreds of ohms.

So, the start of the frequency compensation is to replace each transistor by its simplified small signal diagram. For the obtained circuit, subsequently, the relative frequency behavior is designed.

## 7.4 Frequency compensation via the root locus

The goal of frequency compensation is to alter the characteristic polynomial of a system such that the poles of the transfer function are placed at the required positions in the s-plane. A general expression for the characteristic polynomial of the system is given by:

$$CP(s) = s^n - \sum_{i=1}^{n} p_i s^{n-1} + \cdots + LP. \tag{7.4}$$

Frequency compensation is equivalent to changing the coefficients of thes terms without changing the LP-product, i.e. the maximum attainable bandwidth is not lowered. The subsequent described method uses a two step relation between the loop poles and the system poles. The first step is the often simple relation between the small-signal diagram and the loop poles. The second step is the straightforward relation between loop poles and the system poles via the root locus method. Thus, in this two-step way, the effect of frequency-compensation elements on the system poles can be rather simple. A*direct* relation between the system poles and the frequency-compensation elements is often not clear enough for design purposes.

In the remaining part of this chapter, frequency compensation is treated for a second-order system. The two poles of the system are moved into Butterworth position. The characteristic polynomial equals:

$$CP(s) = s^2 - s(p_1 + p_2) + \omega_n^2. \tag{7.5}$$

The compensation of a third or even higher-order system is analogous to the compensation of a second-order system. The only difference is the number of compensation networks needed. For a second-order system mostly one compensation network fulfills as for a third-order system mostly at least two compensation networks are required.

It must be noted that third and higher-order systems are inherently unstable for relatively high loop gains, as two or more asymptotes of the root locus are directed into the right half plane. Therefore, these systems are less favorable with respect to second-order systems.

Frequency compensation via the root-locus method makes explicitly use of the asymptotic-gain model. The system poles are found from the loop poles ánd the DC loop gain. Techniques to move the system poles to their specified position concentrate on influencing the starting point of the root locus, i.e. moving the loop poles, and influencing the shape of the root locus, i.e. adding a zero to the loop. These two methods are depicted in figure 7.7. Figure 7.7a

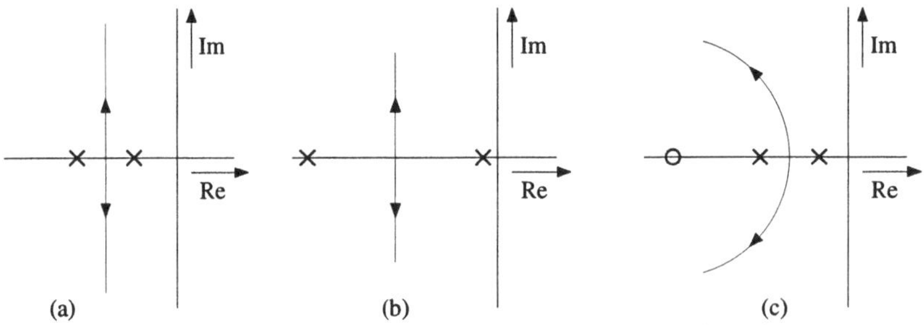


*Figure 7.7.* a) The original root locus, b) changing the starting points of the root locus by moving the loop poles and c) altering the shape of the root locus by adding a zero.

depicts the root locus for the non-compensated system. The system poles do have a relatively high Q, the sum of the poles is too high (remind the poles are negative), and thus frequency compensation has to be done.

Figure 7.7b depicts the frequency compensation by moving the loop poles. Several techniques are available for altering their position without degrading the LP-product significantly. These are:

- pole-splitting;
- pole-zero cancellation;
- resistive broad-banding.

The same result can be obtained by adding a zero to the loop. This is depicted in figure 7.7c. However, the zero must be of a special type. As the system transfer was assumed to be all pole, the zero may not be visible in the system transfer. Zeros that are visible in the loop and *not* in the system transfer and thus can be used for this type of frequency compensation are the:

- phantom zeros.

As phantom zeros show to be the most elegant way of frequency compensation this method is discussed first.

### 7.4.1 Phantom zeros

The characteristic property of a phantom zero is that it is visible in the loop and not in the system transfer. To obtain this, a zero has to be realized in the feedback, $\beta$, of the loop gain $L$. In that case a phantom zero is obtained. Assume a zero at $n_1$ in the feedback factor $\beta$:

$$\beta(s) = \beta(0)(1 - s/n_1), \tag{7.6}$$

then the total transfer is given by (the direct transfer is ignored):

$$A_t(s) = A_{t\infty}(s)\frac{-L(s)(1 - s/n_1)}{1 - L(s)(1 - s/n_1)}. \tag{7.7}$$

in which the zero is made explicit. For the ideal transfer holds:

$$A_{t\infty}(s) = A_{t0}(s) - \frac{\nu(s)\zeta(s)}{\beta(s)} \approx \frac{1}{\beta(s)} = \frac{1}{\beta(0)(1 - s/n_1)}. \tag{7.8}$$

When it is assumed that the direct transfer, $A_{t0}$, is negligible and $\nu$ and $\zeta$ can be approximated by 1, the total transfer is found to be:

$$A_t(s) = \frac{1}{\beta(0)} \cdot \frac{-L(s)}{1 - L(s)(1 - s/n_1)}. \tag{7.9}$$

As can be seen, the zero in $\beta$ is only visible in the loop, i.e. the denominator of the transfer, and not in the total transfer. The phantom zero has only an indirect effect on the total transfer via an additional pole in the system transfer. However, mostly this pole is negligible.

Thus, a zero is a *phantom* zero when it is realized in the feedback network. Three principle places exist where a phantom zero can be realized in the feedback network:

- in its ideal transfer;
- at its input, which is at the output of the amplifier;

- at its output, which is at the input of the amplifier.

With a phantom zero, a reduction in the loop gain, present in $\beta$, is being cancelled beyond the frequency corresponding to the zero. When this reduction, $\delta$, is only small the phantom zero is not very effective. In the pole-zero plot, this effectiveness is represented by an additional pole. This pole and the reduction $\delta$ have a tight relation:

$$p = n \cdot \delta. \tag{7.10}$$

Assume that at the output of $\beta$, by means of a compensation network, a reduction of a factor $\delta$ is cancelled beyond 1 MHz. The corresponding phantom zero is, of course, at -1 MHz. Then the additional pole is a factor $\delta$ apart from the phantom zero. For a relatively small reduction factor $\delta$, the pole is relatively close to the phantom zero and the zero is not very effective; it is almost cancelled by the pole. Thus, the effectiveness of the phantom zero is determined by how close the pole is to the phantom zero and thus how large the reduction was that is cancelled. An example is given in figure 7.8. Originally, the current from the

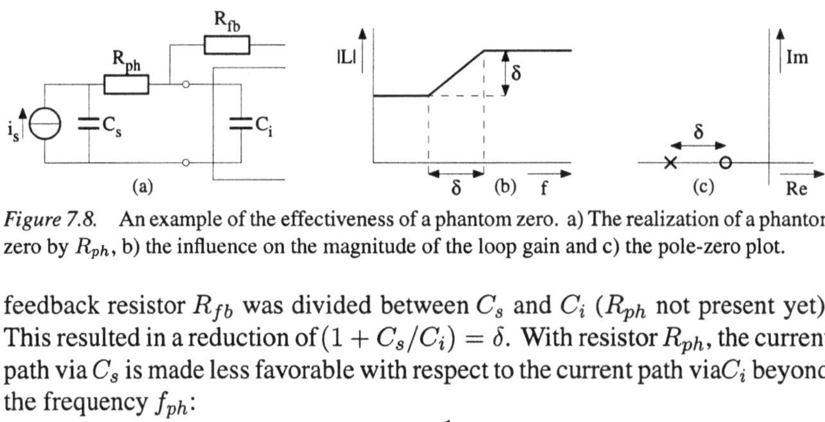


*Figure 7.8.* An example of the effectiveness of a phantom zero. a) The realization of a phantom zero by $R_{ph}$, b) the influence on the magnitude of the loop gain and c) the pole-zero plot.

feedback resistor $R_{fb}$ was divided between $C_s$ and $C_i$ ($R_{ph}$ not present yet). This resulted in a reduction of $(1 + C_s/C_i) = \delta$. With resistor $R_{ph}$, the current path via $C_s$ is made less favorable with respect to the current path via $C_i$ beyond the frequency $f_{ph}$:

$$f_{ph} = \frac{1}{2\pi R_{ph} C_s}. \tag{7.11}$$

At the frequency $f_p$:

$$f_p = \frac{1}{2\pi R_{ph} \frac{C_s C_i}{C_s + C_i}} \tag{7.12}$$

a pole is found. This pole is a factor $\delta$ apart from the phantom zero.

Phantom zeros are mostly placed near the band edge. Consequently, the influence of the components used to realize the phantom zero, is only apparent at relatively high frequencies. Thus the influence of a phantom zero on the noise and distortion performance of the amplifier is only apparent near the band edge and of course, beyond that frequency. That makes this compensation technique favorable. On top of that, it will be seen that most of the other compensation techniques reduces the loop gain in parts of the frequency band. This is, of course, a negative side effect. In contrast, the phantom zero even*adds* some loop gain to the loop. This is, as the zero is placed often near the band edge, also around the band edge.

#### 7.4.1.1 The influence of a phantom zero on the root locus

In figure 7.7c the influence of a phantom zero on the root locus was qualitatively illustrated. But were has the phantom zero to be placed exactly in order to obtain correct frequency compensation? To find this, the characteristic polynomial is examined. When the loop comprises two poles,$p_1$ and $p_2$, and one phantom zero, $n_1$, the characteristic polynomial is given by:

$$CP(s) = s^2 - s(p_1 + p_2 + \frac{LP}{n_1}) + LP. \tag{7.13}$$

The first-order term increases by $LP/n_1$. Via the required sum as given by the system poles, $n_1$ can now easily be calculated. The sum of the system poles$p_a$ and $p_b$ needs to be equal to:

$$p_a + p_b = -\sqrt{2}\omega_2 = -\sqrt{2LP} \tag{7.14}$$

for a Butterworth transfer. This has to be equal to the first order term of equation (7.13) and thus:

$$n_1 = \frac{-LP}{\sqrt{2LP} + (p_1 + p_2)} \tag{7.15}$$

It must be noted that$p_1$ and$p_2$ are negative. For higher-order systems equivalent expressions can be found.

#### 7.4.1.2 Phantom zeros at the input of the amplifier

The phantom zero can be realized at the input of the amplifier. Which component has to be used depends on the type of source impedance, i.e. resistive, capacitive or inductive, and the type of feedback, i.e. parallel or series. In figure 7.9 a general input is depicted with parallel feedback. The type of component, $Z_{ph}$, to realize a phantom zero is given in table 7.1 as a function of the type of source impedance. The number of zeros obtained is given also. The effectiveness of the phantom zero can be checked by making$Z_s$ infinite (ideal current source). The larger the increase in loop gain the more effective the phantom zero is. In the case of a capacitive source, the reduction in the loop gain can be removed by a series resistor or inductor. For a resistor a single phantom zero is found. In contrast, with an inductor two (complex) phantom zeros are found.

In figure 7.10 the situation is depicted for a series feedback at the input. The effectiveness of this compensation can be checked by making the source impedance zero (ideal voltage source). The corresponding compensation com-

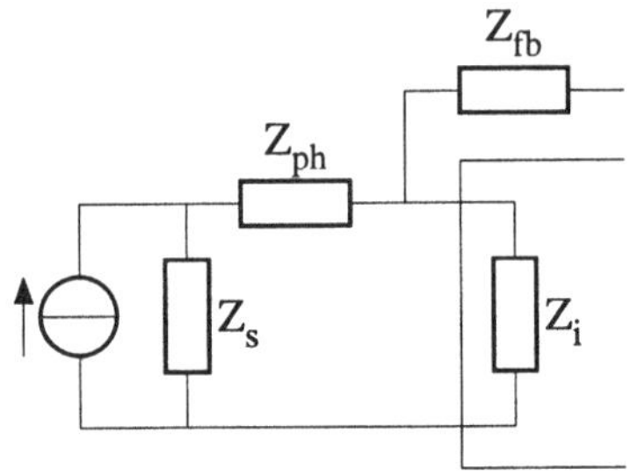


*Figure 7.9.* A general input with parallel feedback.

| type $Z_s$ | type $Z_{ph}$ | order |
|---|---|---|
| C | R and/or L | 1, 2 |
| R | L | 1 |
| L | - | - |

*Table 7.1.* Type of compensation component required to realize a phantom zero at the input of the amplifier with parallel feedback.

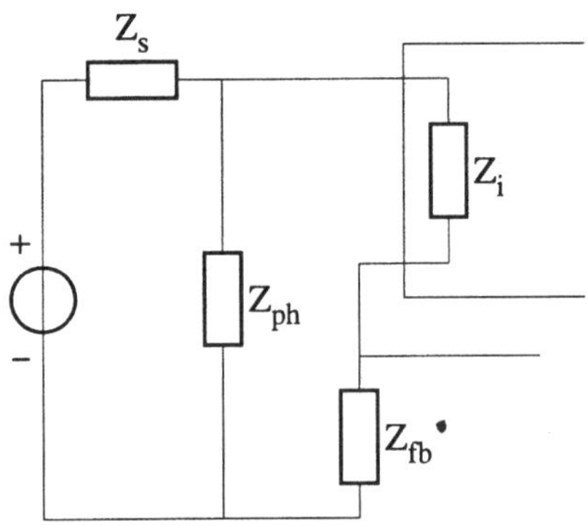


*Figure 7.10.* Realization of phantom zeros with series feedback at the input.

ponent and the number of realized phantom zeros are given in table 7.2.

| type $Z_s$ | type $Z_{ph}$ | order |
|---|---|---|
| C | - | - |
| R | C | 1 |
| L | R, C | 1, 2 |

*Table 7.2.* The type of component required to realize a phantom zero at the input with series feedback.

When realizing phantom zeros at the input, care has to be taken with respect to noise. The compensation components introduce an increase of the noise by their own noise and/or by transforming noise from other components.

#### 7.4.1.3 Phantom zeros at the output of the amplifier

For the realization of phantom zeros at the output, analogous tables can be derived with respect to those for the input. In figure 7.11 the output part of an amplifier is depicted with a parallel feedback network. The corresponding type for the compensation component and the number of realized phantom zeros are given in table 7.3. The effectiveness of the phantom zero is checked by making $Z_L$ infinite (ideal voltage load).

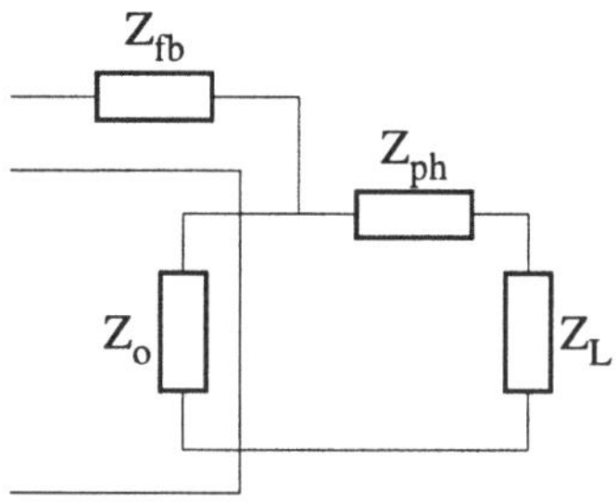


*Figure 7.11.* A general output with parallel feedback.

| type $Z_l$ | type $Z_{ph}$ | order |
|---|---|---|
| C | R and/or L | 1, 2 |
| R | L | 1 |
| L | - | - |

*Table 7.3.* The type of component required to realize a phantom zero at the output with parallel feedback.

The output with series feedback is depicted in figure 7.12. The corresponding type of component and the number of zeros obtained are given in table 7.4. Now the effectiveness is found by making the load impedance zero (ideal current load). As the load is only slightly coupled to the loop, mostly the phantom zero is not effective.

Phantom zeros at the output degrade the distortion performance due to the higher load. However, this can be negligible. For instance, the output depicted in figure 7.11 has to supply more current to realize the original output voltage, consequently, the distortion increases. However, due to the phantom zero the loop gain increases also and the increase of distortion is counteracted.

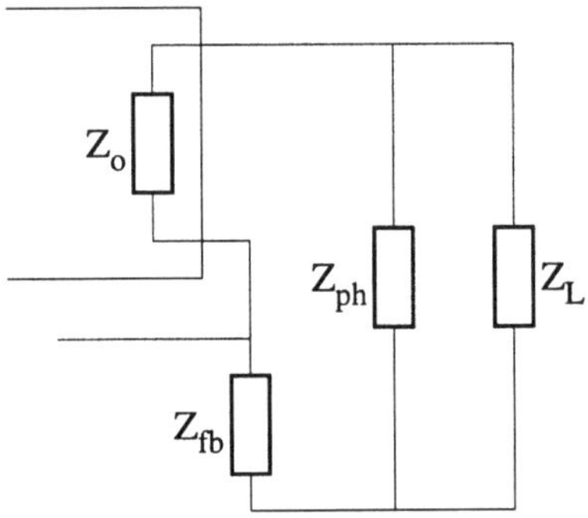


*Figure 7.12.* Realization of phantom zeros with a series feedback at the output.

| type $Z_l$ | type $Z_{ph}$ | order |
|---|---|---|
| C | - | - |
| R | C | 1 |
| L | R, C | 1, 2 |

*Table 7.4.* The type of component required to realize a phantom zero at the output with series feedback.

#### 7.4.1.4 Phantom zeros in the feedback network

To be able to realize a phantom zero in the feedback network, the feedback network has to introduce a reasonable reduction of the loop gain. Otherwise the effectiveness of the phantom zero is low. In contrast to the phantom zeros at the input and output, the phantom zeros in the feedback network can have an influence on both, distortion and noise.

In figure 7.13 the V-I and I-V feedback networks are depicted. For the V-I

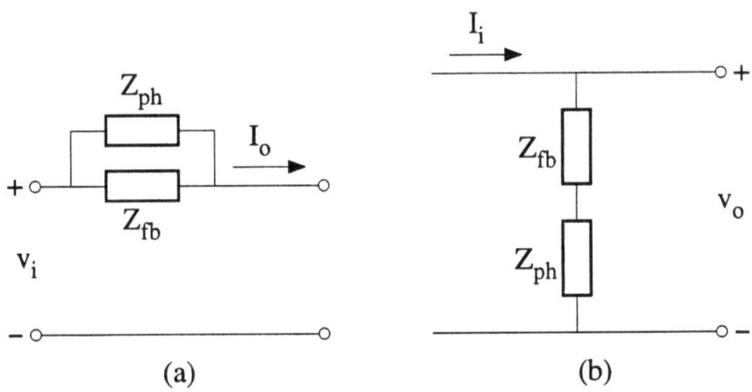


*Figure 7.13.* a) The V-I and b) the I-V feedback network

feedback network the impedance has to be made lower to obtain an increase of loop gain. Thus the effectiveness is checked by making the feedback impedance zero. The different types of compensation components are given in table 7.5. The effectiveness of the of I-V feedback network is found when the feedback

| type $Z_{fb}$ | type $Z_{ph}$ | order |
|---|---|---|
| C | - | - |
| R | C | 1 |
| L | R, C | 1, 2 |

*Table 7.5.* The type of component required to realize a phantom zero in a V-I feedback network, including the number of zeros obtained.

impedance is made infinite. Table 7.6 gives the corresponding type of components to be used.

| type $Z_{fb}$ | type $Z_{ph}$ | order |
|---|---|---|
| C | R, L | 1, 2 |
| R | L | 1 |
| L | - | - |

*Table 7.6.* The type of component required to realize a phantom zero in a I-V feedback network. The number of zeros which are obtained is given also.

For the voltage-to-voltage and current-to-current feedback network the situation is a bit different as they consists of two impedances. In figure 7.14 the V-V feedback network is depicted. To obtain an increase in the loop gain the

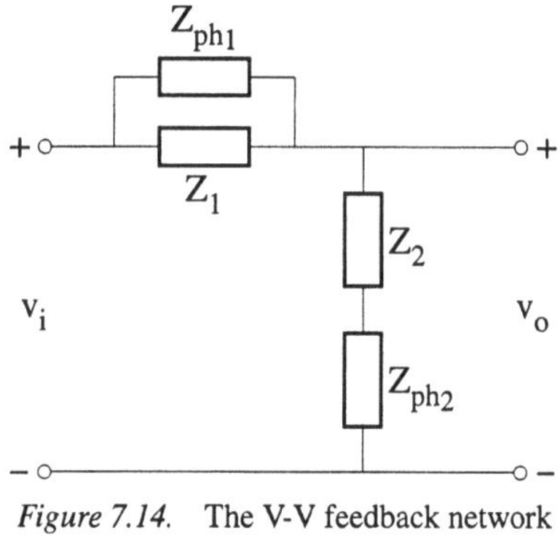


*Figure 7.14.* The V-V feedback network

impedance of $Z_1$ has to be made lower and/or the impedance of $Z_2$ has to be made higher. Effectiveness is found by making $Z_1$ zero and/or $Z_2$ infinite. The corresponding component types are given in table 7.7. For the I-I network the

| type $Z_1$ | type $Z_{ph1}$ | type $Z_2$ | type $Z_{ph2}$ |
|---|---|---|---|
| C | - | C | R, L |
| R | C | R | L |
| L | R, C | L | - |

*Table 7.7.* The type of component required to realize a phantom zero in a V-V feedback network.

phantom zeros can be realized in an analogous way. The feedback network is given in figure 7.15 and the compensation components in table 7.8.

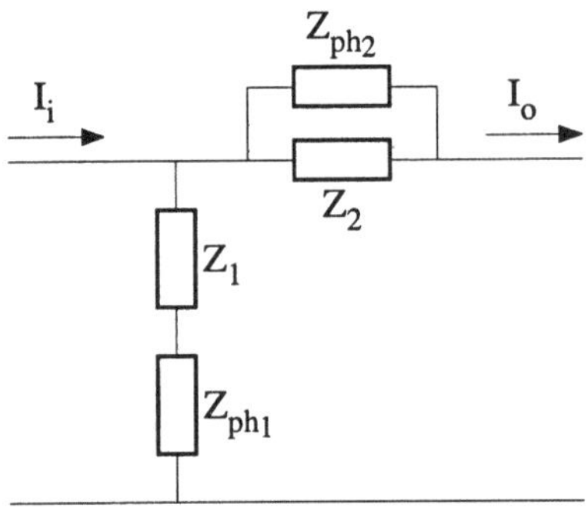


*Figure 7.15.* The I-I feedback network.

| type $Z_1$ | type $Z_{ph1}$ | type $Z_2$ | type $Z_{ph2}$ |
|---|---|---|---|
| C | R, L | C | - |
| R | L | R | C |
| L | - | L | R, C |

*Table 7.8.* The type of component needed to realize a phantom zero in a I-I feedback network.

From the foregoing treatment of the phantom zero, it can be concluded that the phantom zero is always realized *outside* the nullor implementation. As the feedback network is realized with accurate components, frequency compensation using the phantom zero is often more accurate to design than the methods to be discussed now, which are realized *inside* the nullor implementation.

### 7.4.2 Pole-splitting

Pole-splitting is the technique that introduces an interaction between two poles such that they split. The principle is depicted in figure 7.16. The poles are split apart while their product remains constant such that the LP-product is not changed, ideally. Due to this splitting the *sum* of the poles decreases (poles

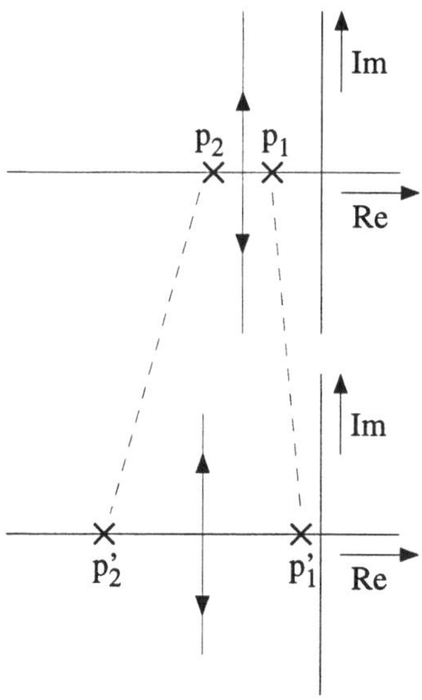


*Figure 7.16.* The principle of pole-splitting.

are negative). Assume a part of a loop is as given in figure 7.17. The pole at the

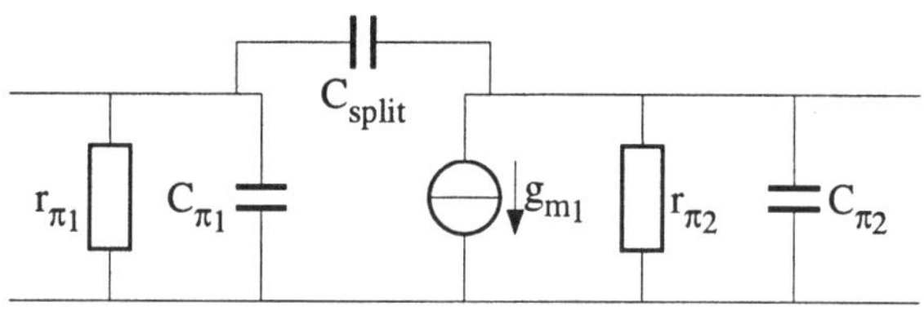


*Figure 7.17.* An example of pole-splitting.

input and output of the transistor are given by, respectively ($C_{split}$ is ignored for the moment):

$$p_{in} = \frac{-1}{2\pi r_{\pi 1} C_{\pi 1}},$$
$$p_{out} = \frac{-1}{2\pi r_{\pi 2} C_{\pi 2}}. \qquad (7.16)$$

Further, assume that for obtaining a Butterworth characteristic the sum of these poles has to be decreased by ΔP. With capacitor $C_{split}$ a local loop around the transistor is created and $p_{in}$ and $p_{out}$ interact via this loop. For exact compensation the poles can be calculated from the schematic, however by means of inspection a good estimation can be made already.

It can be said that the pole at the input shifts to a lower frequency by a certain factor, due to the Miller effect. The parallel feedback at the output, introduced by $C_{split}$, causes the output impedance of the transistor to lower and pole $p_{out}$ shifts to a higher frequency. Pole $p_{out}$ must be at a higher frequency than $p_{in}$ otherwise the Miller approximation does not hold. As the LP-product remains almost constant, $p_{out}$ shifts the same factor upwards (higher frequency) as $p_{in}$ shifts downwards (lower frequency). The upward shift of $p_{out}$ is the dominant cause of the decrease of the sum of the poles. From $\Delta$P can be found what factor $p_{out}$ has to shift. Pole $p_{in}$ is lowered the same factor as $p_{out}$ and the required split capacitor for realizing this factor, K, can be found by the Miller approximation. For $p_{in}$ holds:

$$p_{in} = -\frac{1}{2\pi r_{\pi 1} C_{\pi 1}[1 + \frac{C_{split}}{C_{\pi 1}}(1 + g_{m1} r_{\pi 2})]} = -\frac{1}{2\pi r_{\pi 1} C_{\pi 1} \cdot K}. \quad (7.17)$$

From this equation $C_{split}$ can easily be found.

When exact calculations are made it appears that the LP-product lowers due to the splitting according to:

$$\frac{LP_{before}}{LP_{after}} = 1 + C_{split}\left(\frac{1}{C_{\pi 1}} + \frac{1}{C_{\pi 2}}\right). \quad (7.18)$$

The more the poles are split, the more the LP is lowered. From equation (7.17) can be seen that $C_{split}$ is multiplied by the voltage gain of the stage, $g_{m1} r_{\pi 2}$. For a higher voltage gain the split capacitor can become lower and consequently, less LP-product is lost. Thus effective places for pole splitting are between nodes with a high voltage gain. Of course, the choice is limited to the places where interaction can be realized between the right dominant poles.

Besides the effect pole splitting has on the loop poles, it also introduces a zero, compare with a zero due to $C_{\mu}$. Because of $C_{split}$, the stage is no longer unilateral, which can be a severe problem for the stability. Luckily, some simple measures exist to counteract this effect and even to change it into a positive effect. The effect of the zero could be cancelled by making the capacitor unilateral, i.e. using a buffer in series with the capacitor such that a current can only flow from one side of the capacitor to the other side.

A more simple method is using a resistor, $R_{split}$, in series with the capacitor, see figure 7.18. Now the zero due to $C_{split}$ is found at:

$$n_l = \frac{+1}{2\pi C_{split}(\frac{1}{g_{m1}} - R_{split})}. \quad (7.19)$$

Clearly, three situations can occur, depending on the value of $R_{split}$ relative to $g_m$:

- $R_{split} < \frac{1}{g_{m1}}$ : zero in the right-half plane (RHP);
- $R_{split} = \frac{1}{g_{m1}}$ : zero at infinity;

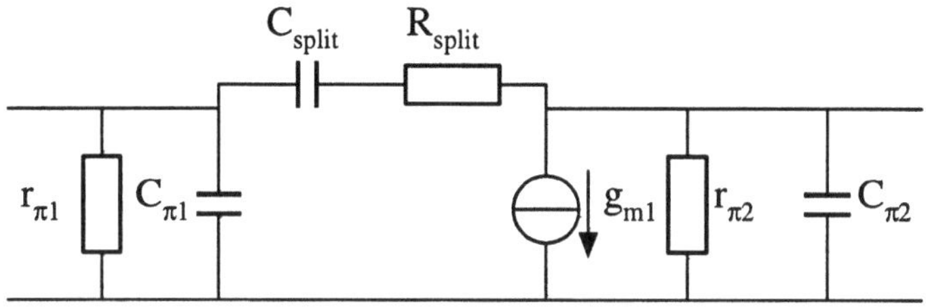


*Figure 7.18.* The compensation of the right-half plane zero by a resistor.

- $R_{split} > \frac{1}{g_{m1}}$ : zero in left-half plane (LHP).

In some situations this third case can be used advantageously. Then, a combination of poles-splitting and a LHP zero is obtained which can be a very powerful technique for frequency compensation. Again, like when discussing the phantom zero technique, this LHP zero has got a certain effectiveness. This means that besides this LHP zero also an additional pole is obtained. When calculating the characteristic polynomial, the additional pole is found at:

$$p_{l3} = \frac{-1}{2\pi R_{split} C_{split}} \cdot \left[1 + C_{split}\left(\frac{1}{C_{\pi 1}} + \frac{1}{C_{\pi 2}}\right)\right], \qquad (7.20)$$

with the last factor equal to the factor of equation (7.18). Thus, for $R_{split} > 1/g_{m1}$ and $C_{split}$ being relatively large, this pole can be relatively far away from the LHP zero. The ratio of the pole and zero is equal to the factor that was lost in the LP-product due to this pole-splitting!

In contrast, for a relatively small split capacitor, the pole is found at:

$$p_{l3} = \frac{-1}{2\pi R_{split} C_{split}} \qquad (7.21)$$

This means, that with an exact compensation of the RHP zero, $g_{m1} = R_{split}$, the zero in the RHP vanishes, but an additional pole is found at the same position in the LHP. The additional loop gain due to the zero is removed but the additional phase shift remains.

Figure 7.19 depicts a situations for which $R_{split} > 1/g_{m1}$ applies and $C_{split}$ is relatively small. The zero is indeed found in the LHP, closer to the origin than the third pole. However, as the pole is relatively close to the zero, it is probably not very effective.

As presented in the beginning of this chapter, after the frequency compensation with the simple models, the next step is to introduce one by one the $C_\mu$s, $r_o$s and $r_b$s. When pole-splitting is used, a special situation might occur. The pole-split capacitor is in many cases in parallel with $C_\mu$ or $C_{dg}$. When $C_\mu$ ($C_{dg}$) is larger than $C_{split}$ it seems that a problem arises. However, with a current

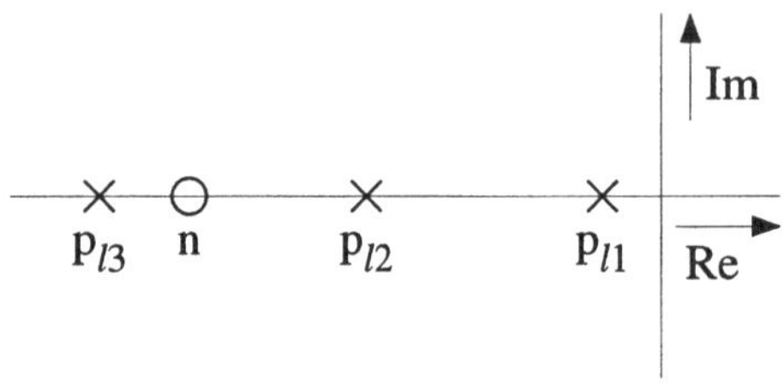


*Figure 7.19.* The pole-zero plot of pole splitting with $R_{split} > \frac{1}{g_{m1}}$.

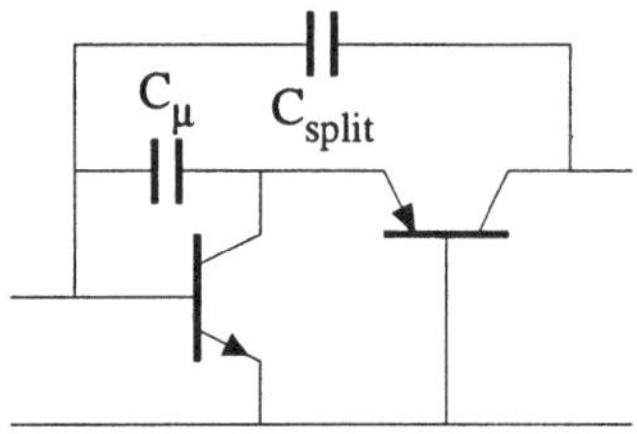


*Figure 7.20.* Pole-splitting which is insensitive for $C_\mu$.

follower the stage may be made unilateral again, and the split capacitor can be placed over the total stage, see figure 7.20.

Generally, introducing $r_o$ to the transistor does not influence the pole-splitting much. The base resistance however, can introduce an additional pole and some extra reduction of the LP-product.

### 7.4.3 Pole-zero cancellation

With pole-zero cancellation two poles can be split, just like pole-splitting does. But with pole-zero cancellation the stages involved remain unilateral. The principle is depicted in figure 7.21. With a compensation network a pole ($p_1$) is shifted to a lower frequency. When at higher frequencies the influence of this network is removed again, a zero ($n$) is obtained. With this zero an other pole ($p_2$) in the loop can be cancelled. Of course, an additional high-frequency pole ($p_3$) is obtained at a frequency determined by the effectiveness of the zero (cf. the phantom zero). This effectiveness is equal to the factor the low-frequency pole is shifted downwards, assuming that the LP-product is not changed. In figure 7.22 a straightforward method for implementing a pole-zero cancellation is depicted. The stages can be part of an arbitrary loop. Without the pole-zero cancellation network, $R_{pz}$ and $C_{pz}$, the poles of the circuit are

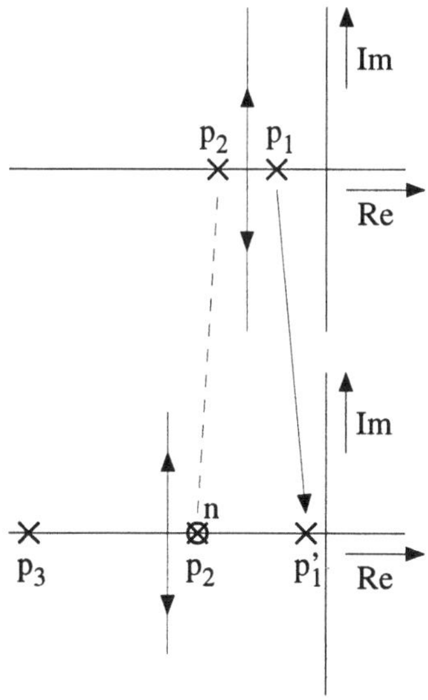


*Figure 7.21.* The principle of pole-zero cancellation.

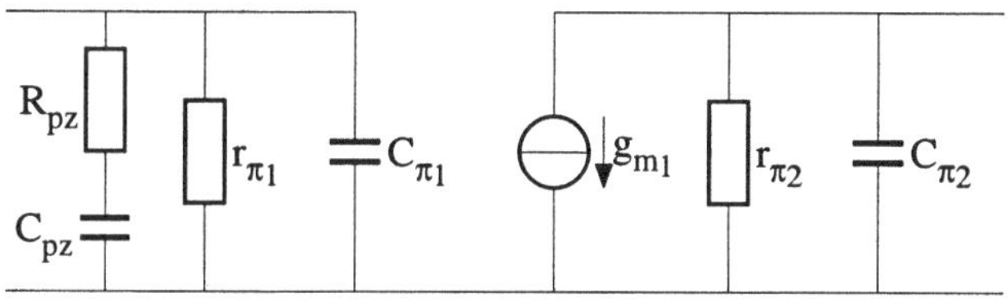


*Figure 7.22.* A straight forward implementation of pole-zero cancellation.

given by:

$$p_1 = \frac{-1}{2\pi r_{\pi 1} C_{\pi 1}},$$
$$p_2 = \frac{-1}{2\pi r_{\pi 2} C_{\pi 2}}. \tag{7.22}$$

Next, the pole-zero cancellation network is introduced. For relatively low frequencies the capacitor of the network is dominant and causes $p_1$ to shift downwards to:

$$p'_1 = \frac{-1}{2\pi r_{\pi 1}(C_{\pi 1} + C_{pz})}. \tag{7.23}$$

Beyond a certain frequency the influence of $C_{pz}$ is cancelled by $R_{pz}$ resulting in a zero. This zero is at

$$n = \frac{-1}{2\pi R_{pz} C_{pz}}. \tag{7.24}$$

When this zero is located at the same frequency as pole $p_2$, the zero cancels the pole. For the higher frequencies the resistor is the dominant part of the compensation network, resulting in a pole at

$$p_3 = \frac{-1}{2\pi R_{pz} C_{\pi 1}}. \tag{7.25}$$

The resulting pole-zero pattern is depicted in figure 7.21. The product of the two new poles equals:

$$p_1' p_3 = \frac{-1}{2\pi r_{\pi 1}(C_{\pi 1} + C_{pz})} \cdot \frac{-1}{2\pi R_{pz} C_{\pi 1}} = \frac{1}{2\pi r_{\pi 1} C_{\pi 1} \cdot 2\pi r_{\pi 2} C_{\pi 2}}. \tag{7.26}$$

As the DC loop gain is not changed, the LP-product remains exactly the same. However, when the base resistances are introduced a reduction of loop gain equal to a factor:

$$\frac{LP_{before}}{LP_{after}} = 1 + \frac{r_b}{R_{pz}} \tag{7.27}$$

is found. This is caused by the remaining high frequency current division between the compensation network and the input impedance of the transistor. This reduction can be counteracted by adding an inductor in series with the compensation network such that for relatively high frequencies the compensation network is disconnected from the circuit by that inductor.

The values for the compensation components can be found as follows. $C_{pz}$ is determined by $C_{\pi 1}$ and how much the poles have to split. How much has to be split is easily found when it is assumed that the difference of $p_2$ and $p_3$ is the dominant part for the decrease of the sum of poles. Finally, $R_{pz}$ is given from the fact that the zero has to cancel $p_2$.

Compared to pole-splitting, pole-zero cancellation is easier to design, less reduction of LP-product is obtained and the stages remain unilateral. However, this way of implementing a pole-zero cancellation has two disadvantages. Firstly, when with pole-zero cancellation the poles are split a factor X, the capacitor $C_{pz}$ needs to be X-1 times as large as $C_\pi$. In contrast, for pole-splitting the required capacitor is a factor equal to the voltage gain between the nodes where the splitting capacitor has to be placed smaller than $C_{pz}$. This can be very advantageous in the case of IC design. The second disadvantage is made clear with the help of figure 7.23. The thick and the thin line depict the loop gain versus the frequency of the original circuit and the compensated circuit, respectively. The figure holds for both methods, pole-zero cancellation and pole-splitting. The dashed part in the figure is the change in loop gain. In the case of a pole-zero cancellation implemented as in figure 7.22, this part of the loop gain is not used, it is totally wasted. In contrast, pole-splitting uses this part in a local loop for linearization purposes. Consequently, the effect of a

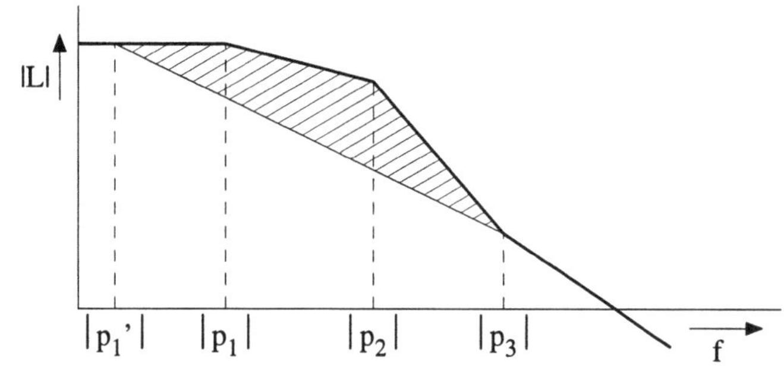


*Figure 7.23.* The influence of splitting poles on the loop gain.

lower overall loop gain on the distortion performance is reduced. However, sometimes pole-zero cancellation can also be used by introducing a local loop, which is depicted in figure 7.24. The influence on the loop gain is depicted

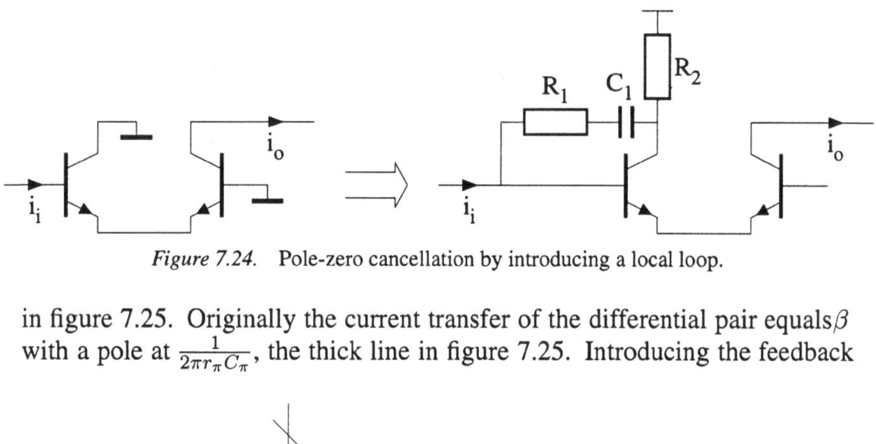


*Figure 7.24.* Pole-zero cancellation by introducing a local loop.

in figure 7.25. Originally the current transfer of the differential pair equals $\beta$ with a pole at $\frac{1}{2\pi r_\pi C_\pi}$, the thick line in figure 7.25. Introducing the feedback

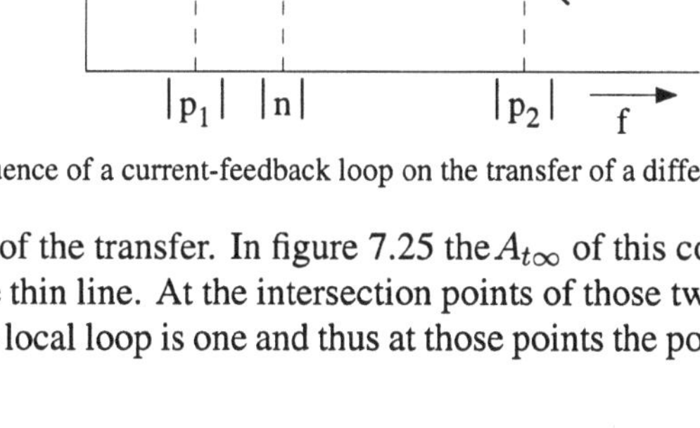


*Figure 7.25.* The influence of a current-feedback loop on the transfer of a differential pair.

results in a change of the transfer. In figure 7.25 the $A_{t\infty}$ of this configuration is depicted with the thin line. At the intersection points of those two functions, the loop gain of the local loop is one and thus at those points the poles of the I-I transfer are found. Below $p_1$ and beyond $p_2$ the transfer is given by the thick line as the loop gain is smaller than one. Between those two poles, the feedback network sets the transfer and a zero is found at

$$n = \frac{-1}{2\pi C_1(R_1 + R_2)}. \tag{7.28}$$

With this zero, a pole of an other stage can be cancelled and a pole-zero cancellation is realized.

The influence of the base resistance, for this type of pole-zero cancellation is significantly reduced with respect to the pole-zero cancellation without a local loop. For the high frequency behavior of the local loop the total resistance of the pole-zero cancellation, $R_1$+$R_2$, is in parallel with the two $r_b$s. As the impedance level of the feedback network is free to choose, higher values can be chosen. The decrease in the LP product is given by:

$$\frac{LP_{before}}{LP_{after}} \approx 1 + \frac{2r_b}{R_1 + R_2}. \tag{7.29}$$

### 7.4.4 Resistive broad-banding

In contrast to the previous two methods, resistive broad-banding acts only on one pole. To keep the LP-product constant, the DC loop gain has to change also. The principle is given in figure 7.26. With a compensation network an

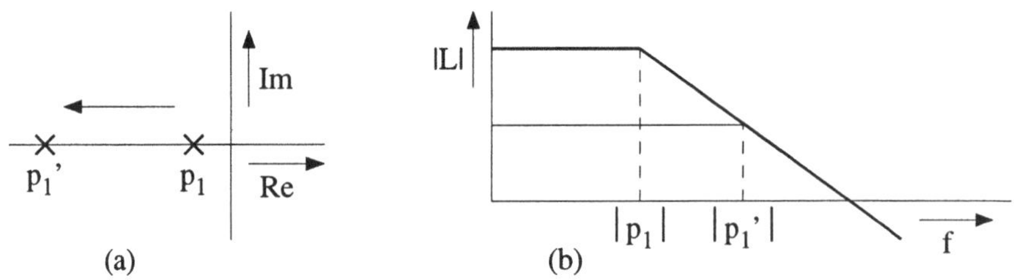


*Figure 7.26.* The principle of resistive broad-banding. The influence on a) the pole-zero plot and b) the loop gain.

upward shift of a single pole is realized. The factor the DC loop gain changes is equal to the factor the pole shifts. Consequently, the LP-product remains the same.

For this type of compensation technique also two different ways of implementation are possible, as it is for pole-zero cancellation. In figure 7.27 an implementation is given which just wastes the excess overall loop gain. The original pole is shifted a factor $1 + r_\pi/R_{br}$ upwards and the DC gain is reduced by the same factor. As was expected, the LP-product remains the same. However, as it is for pole-zero cancellation, the LP-product changes when the base

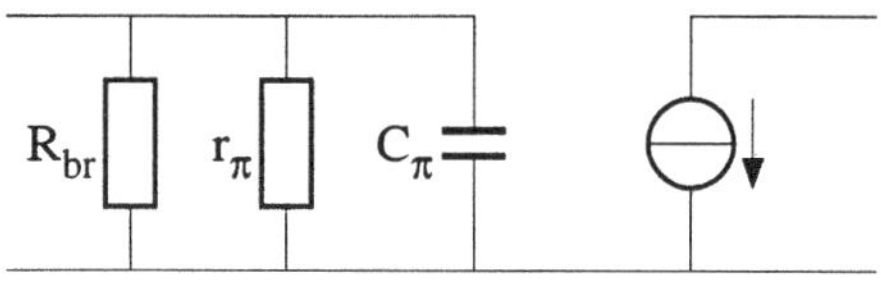


*Figure 7.27.* An implementation of resistive broad-banding.

resistances are introduced. The reduction equals a factor:

$$\frac{LP_{before}}{LP_{after}} = 1 + \frac{r_b}{R_{br}} \tag{7.30}$$

and again this reduction can be removed with an inductor in series with$R_{br}$.

Resistive broad-banding that uses the fraction of overall loop gain, by which the overall loop gain is reduced, for linearization purposes, is depicted in figure 7.28. The differential stage is locally fed back by a current-feedback network.

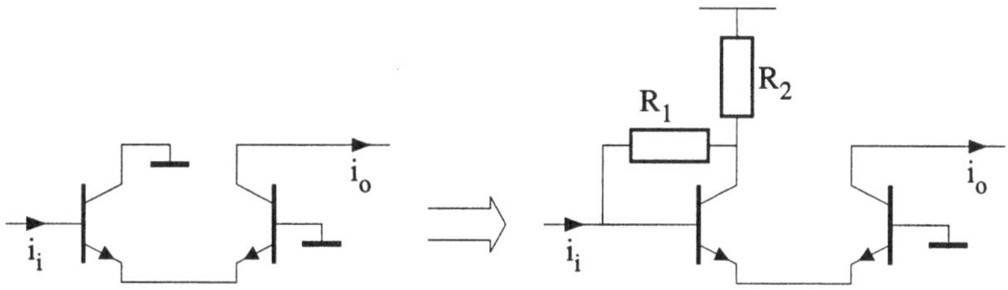


*Figure 7.28.* Resistive broad-banding implemented with a local loop.

In figure 7.29 the influence on the gain of the stage is depicted. In the figure

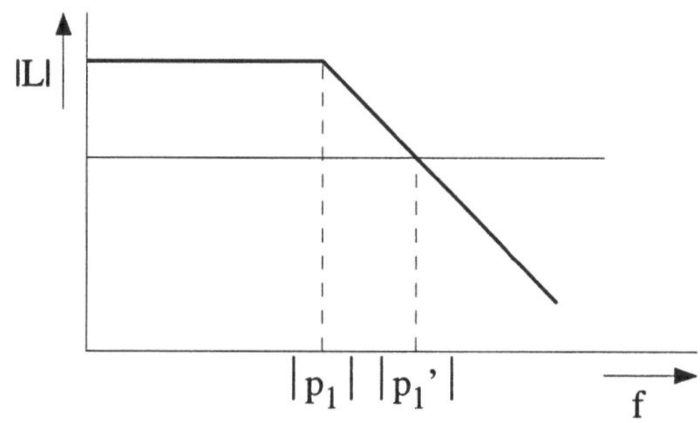


*Figure 7.29.* The influence of the local feedback on the gain of the stage.

the thick line is the original transfer and the thin line is the ideal transfer,$A_{t\infty}$. Again the intersection point gives the new pole position.

The reduction of the influence of the base resistance is analogous to that for pole-zero cancellation and equals approximately:

$$\frac{LP_{before}}{LP_{after}} \approx 1 + \frac{2r_b}{R_1 + R_2} \tag{7.31}$$

### 7.4.5 Changing the contribution to the LP-product

A frequency compensation method that does not fit in the list of previous discussed methods, is frequency compensation by changing the contribution of a stage to the LP-product. Basically, this method changes the LP-product. However, when the LP-product is high enough, it can be a very convenient method.

The contribution of a stage to the LP-product can be changed by choosing an other value for its bias current. This influences the $f_T$ of the stage. As the contribution of a stage to the LP product is governed by its $f_T$, less or no additional frequency compensations are required. Of course, the influence of the changed bias, on noise, bandwidth and distortion, has to be checked.

## 7.5 Conclusions

A method for performing a frequency compensation was discussed. The method is based on the asymptotic-gain model and it aims on influencing the behavior of the total system via the root-locus method. Four different types of compensation techniques, which do not change the LP-product significantly, were discussed. These are in order of preference:

- phantom zero;
- pole-splitting;
- pole-zero cancellation;
- resistive broad-banding.

Depending on the situation this order of preference may be different.

Further, one type of frequency compensation technique that alters the LP-product was discussed:

- changing the contribution of a stage to the LP-product.

Using this technique can result in a reduction of the number of frequency compensations required.

## 7.6 Exercises

### Exercise 7.1

Given the current-to-voltage amplifier as depicted below:

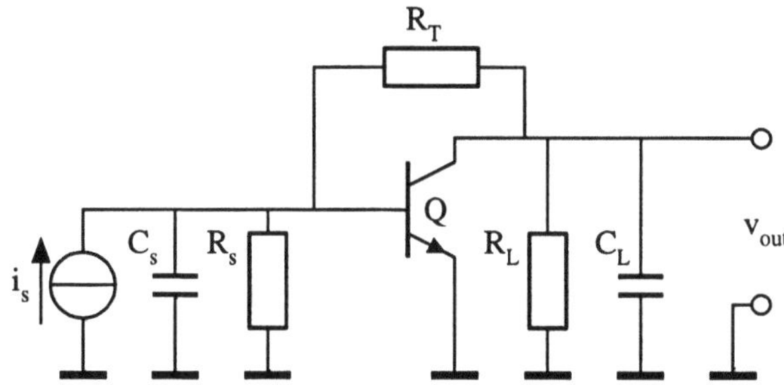


*Figure 7.30.* A current-to-voltage amplifier.

For the transistor, Q, holds:

- $g_m = 40$ mA/V
- $C_\pi = 4$ pF
- $C_\mu = 0.5$ pF
- $r_o = 50$ k$\Omega$
- $\beta = 100$

For the other components hold:

- $C_s = 100$ pF
- $R_s = 10$ k$\Omega$
- $R_T = 10$ k$\Omega$
- $R_L = 10$ k$\Omega$
- $C_L = 10$ pF

1. Determine the DC-loop gain.
2. Determine the poles and the root locus.
3. At which locations can a frequency-compensation network be introduced?
4. Perform the frequency compensation such that the poles of the closed-loop amplifier are in Butterworth position.

## Exercise 7.2

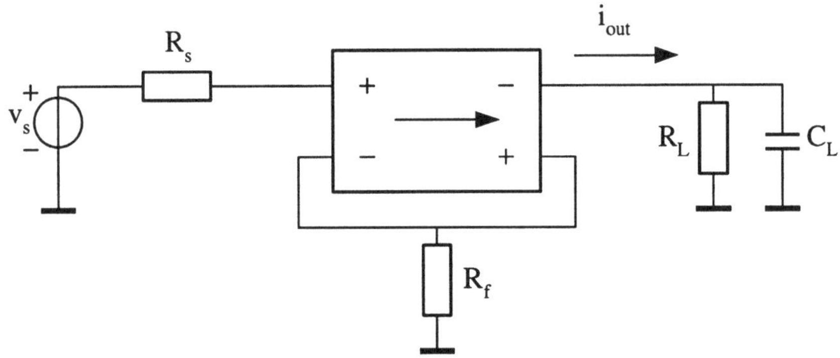


*Figure 7.31.* A transconductance amplifier.

Given the transconductance amplifier as depicted above. The nullor is implemented with bipolar transistors. The load is modelled as a resistor in parallel with a capacitor.

1. What is the effect of the pole due to $R_L$ and $C_L$ on the LP-product?

2. Determine the LP-product of the amplifier when the active part comprises two active stages (transistors). Use symbolic notation like $g_m$, $r_\pi$ etc. for the transistors.

3. Indicate whether it is possible or not to realize at the input, output and feedback network a phantom zero. If possible, what component is required and how must it be connected to realize the phantom zero?

4. Suppose a student has measured with an oscilloscope the transfer function of the amplifier. It appears that the measurement result differs from his calculations and simulations: an additional pole is found. What is causing this extra pole and how can it be prevented?

## Exercise 7.3

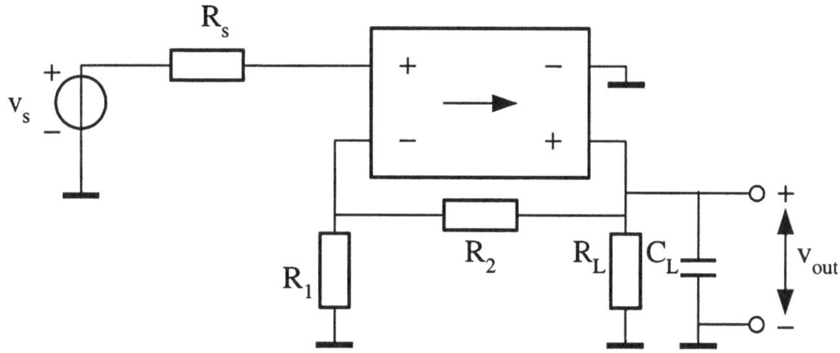


*Figure 7.32.* A voltage amplifier.

The above depicted voltage amplifier has a voltage gain of 100. The active part is implemented with bipolar transistors. Its gain is high enough such that up to a certain frequency the active part can be approximated as a nullor. Three amplifying stages where used.

1. The amplifier is optimized for a low noise level. What does this mean for the value of $R_1$?
2. Determine the expression for the LP-product.
3. For what values of $R_s$ is the effect of $R_s$ on the maximum attainable frequency considerable? What is the consequence for the value of$R_1$?
4. At which location can a phantom zero be realized by means of a capacitor?
5. When the gain of the amplifier is reduced from 100 to only 2, what is the consequence for the maximum attainable bandwidth?
6. Do the values of$R_s$ ,$R_L$ ,$C_L$ or the gain (2 or 100) influence the effectiveness of the phantom zero chosen in question 4?
7. For most operational amplifiers (OpAmps) the output stage is a CC stage instead of a CE stage.
   (a) What is the contribution of a CC-stage to the LP-product?
   (b) The use of a CC-stage at the output may cause problems for the preceding stage. Evaluate what these problems can be by considering the three quality aspects: noise; distortion and bandwidth.
   (c) What can, despite these problems, be the reason for using a CC-stage at the output of amplifier?

## Exercise 7.4

Given four different amplifier output parts, as depicted in figure 7.33.

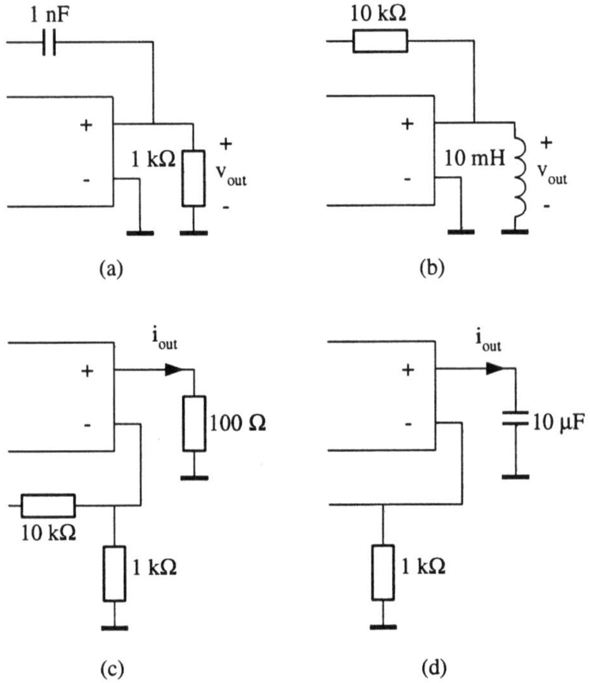


*Figure 7.33.* Four different output parts of negative-feedback amplifiers.

1. For each of the amplifiers as depicted in figure 7.33, indicate where a phantom zero can be made. What type of component is required? How much zeros are obtained?

2. Discuss the effectiveness for each of the phantom zeros from the previous question. Give the ratio of the phantom zeros and its accompanying pole.

## Exercise 7.5

Given the small-signal diagram of a current amplifier (figure 7.34). The

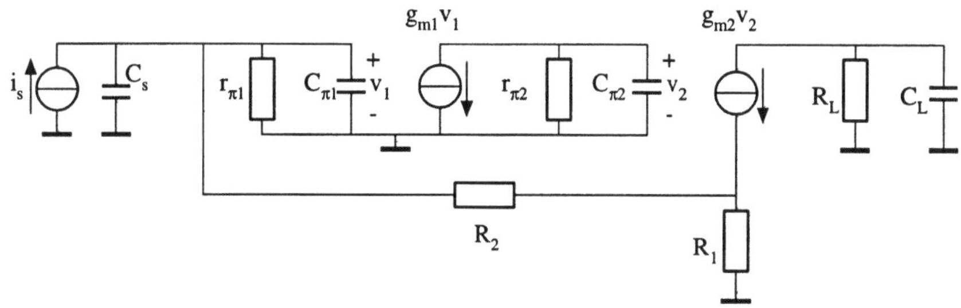


*Figure 7.34.* The small-signal diagram of a current amplifier.

active part is implemented by two amplifying stages. The values for the various (small-signal) components are listed in table 7.9.

| Component | Value | Component | Value |
|---|---|---|---|
| $r_{\pi 1}$ | 20 kΩ | $r_{\pi 2}$ | 1 kΩ |
| $c_{\pi 1}$ | 20 pF | $c_{\pi 2}$ | 100 pF |
| $g_{m1}$ | 10 mA/V | $g_{m2}$ | 100 mA/V |
| $R_1$ | 1 kΩ | $R_2$ | 100 kΩ |
| $R_L$ | 10 kΩ | $C_L$ | 10 nF |
| $C_s$ | 1 nF | | |

*Table 7.9.* The values for the (small-signal) components of figure 7.34.

1. Calculate the loop gain as a function of the complex frequency,$s$.
   (a) What is the DC loop gain L(0)?
   (b) What are the loop poles and loop zeros?
2. Determine the LP-product. Which of the poles found in the previous question are dominant?
3. What is the maximum attainable bandwidth in the case of a Butterworth transfer?
4. What phantom zero is required for the frequency compensation?
5. Where can you implement the required phantom zero? Which option do you choose and why?

## Exercise 7.6

Given the small-signal diagram of a transimpedance amplifier (figure 7.35). The active part is implemented by one amplifying stage. The corresponding

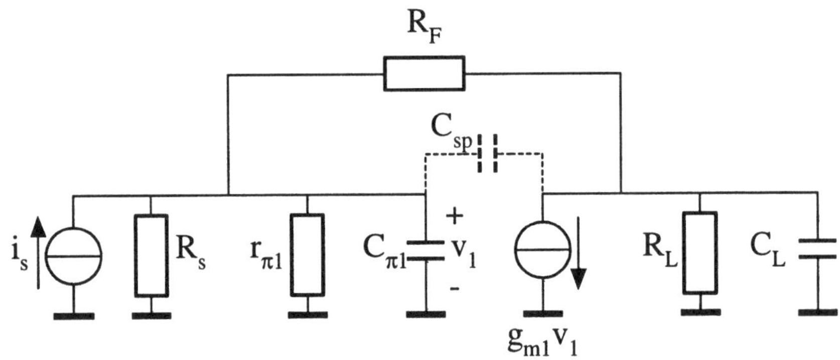


*Figure 7.35.* The small-signal diagram of a transimpedance amplifier.

element values are listed in table 7.10.

| Element | Value |
|---|---|
| $R_s$ | 10 kΩ |
| $r_{\pi 1}$ | 10 kΩ |
| $C_{\pi 1}$ | 100 pF |
| $g_m$ | 10 mA/V |
| $R_L$ | 10 kΩ |
| $C_L$ | 5 pF |
| $R_F$ | 1 kΩ |

*Table 7.10.* The element values corresponding to the circuit as depicted in figure 7.35

1. Calculate the loop gain as a function of the complex frequency, $s$ (ignore $C_{sp}$ for the moment).

    (a) What is the DC loop gain L(0)?

    (b) What are the loop poles and loop zeros?

2. Determine the LP-product. Which of the poles found in the previous question are dominant?

3. What is the maximum attainable bandwidth in the case of a Butterworth transfer?

The frequency compensation of this amplifier is performed by a pole-split capacitor, $C_{sp}$ (indicated in the figure by the dotted-drawn capacitor).

6. What factor do the two poles need to split for a Butterworth closed-loop transfer?
7. How much does the total input capacitance need to increase to reach this?
8. Calculate the voltage gain for the amplifying stage.
9. Determine the value of the split capacitor.

Alternatively, the frequency compensation of the amplifier is performed by using a phantom zero in the feedback network.

10. What component is required for a phantom zero in the feedback network?
11. What is the value for the component?
12. When you compare the two performed frequency compensations, what do you conclude?

## Exercise 7.7

Consider a negative-feedback amplifier for which the frequency compensation, using the simple small-signal models, is performed. The system poles are located at

$$p_{s1,2} = -1\text{ MHz} \pm j\text{ MHz} \tag{7.32}$$

Subsequently, all the parasitic elements, like $C_{\mu}$ and $r_o$, are added to the amplifier circuit. It appears that only $C_{\mu 1}$ deteriorates the frequency performance. In order to repair the frequency behavior, an ideal current follower is placed in cascade with the corresponding amplifying stage. As a result, the poles are at their original location again. As a final step, the ideal current follower is implemented by means of a CB stage. The obtained poles and zeros are given in table 7.11.

| Situation | Poles $p_{s1,2}$ |
|---|---|
| No parasitics | $-1$ MHz $\pm\, j$ MHz |
| All parasitics but $C_{\mu 1}$ | $-1$ MHz $\pm\, j$ MHz |
| All parasitics inclusive $C_{\mu 1}$ | $-100$ kHz $\pm\, 1.4j$ MHz |
| All parasitics plus ideal current follower | $-1$ MHz $\pm\, j$ MHz |
| All parasitics plus CB stage | $-500$ kHz $\pm\, 1.2j$ MHz |

*Table 7.11.* The poles of the frequency-compensated amplifier in different situations.

1. What goes wrong and how can it be improved?

## Exercise 7.8

Sometimes it is required that frequency compensation can be controlled via a current or a voltage. This could be required in the case of a completely integrated circuit. Components like trimmers and potentiometers cannot be integrated.

1. Give some options by which a frequency compensation ca be controlled electronically, i.e. via a voltage or a current.

# 8

# BIASING

## 8.1 Introduction

When we arrive at the biasing step, we are at the end of the "small-signal design". A circuit has been designed using small-signal models that has optimal noise behavior, maximum bandwidth, with the dominant poles and zero's at the desired position and with an acceptable amount of distortion. So, in principle the circuit design for meeting the specifications is finished. This also means that, at this point in the design procedure, all the bias quantities are known. To obtain the desired small-signal parameters, operating points have been chosen for every transistor. For example, the small-signal quantity $g_m$, the transconductance of a bipolar transistor, follows from the collector bias current $I_c$ as:

$$g_m(I_c) = \frac{I_c}{\frac{kT}{q}} \tag{8.1}$$

So, to obtain a certain value for $g_m$, a certain value of $I_c$ is required. At this point in the design procedure, the biasing quantities are nothing more than parameters that influence the values of the small-signal parameters in the used small-signal models, they are not implemented as real bias sources yet.

The problem that has to be solved now is that, using transistors, practical two-ports have to be designed that show a behavior at their input and output ports that sufficiently matches to the behavior of the initial small-signal model.

The biasing procedure has to be such that it does not influence the performance, since this was optimized during the small-signal design. In this chapter a method that meets this requirement will be described.

First, section 8.2, introduces some general bias rules. Which is followed in section 8.3 by a discussion of the differences between the small-signal design and a practical transistor circuit. In this section, also a method to bias a single transistor stage in such a way that the desired small-signal behavior is seen at the input and output port, and, on top of that, that the bias is not visible at those ports. Then every "small-signal two-port" can be replaced by a new one that contains a "real" biased transistor, but has the same small-signal parameters.

Section 8.4 treats how to handle linear devices in the presented bias procedure. Subsequently, section 8.5 - 8.10 generalizes the bias methodology for a single transistor to circuits with more transistors and linear devices. During the first bias steps, for every transistor four bias sources are added. This yields a correctly biased circuit, but probably there are to many separate bias sources in the circuit. Section 8.9 will show methods to reduce the number of bias sources while keeping, of course, the overall biasing of the circuit unaltered. It will be shown that by using the same shifting rules for voltage and current sources as used for noise in section 4.2.2 and section 4.2.3, the number of bias sources can be reduced. Voltage sources are shifted as much as possible into branches that have one end connected to ground. These voltage sources become the supply voltage sources. So the supply voltage *follows* from the small-signal design. Sometimes it is possible to optimize the supply voltage needs by changing device types from N to P or vice versa. This will be discussed in section 8.9.3.

Also a number of bias control loops is necessary in every biased circuit. Some of them are trivial loops, some of them are very complex. Section 8.5 will deal with this.

In section 8.11 the method will be discussed by which the ideal bias sources that have been used so far are replaced by practical transistor circuits. The practical implementation of bias sources is discussed in chapter 9.

Differential amplifiers will be treated separately in section 8.12, since some extra possibilities exist to bias these circuits. Especially there are more ways in which the bias loops can be implemented.

## 8.2 General biasing rules and floating nodes

In a correctly biased circuit, all nodes should have a well defined DC-voltage relation to each other. So, if nodes exist to which only elements with a current-source character are connected, something should be done about this. Figure 8.1 shows an example of a floating node. Neither the transistor nor the following stage define the DC bias voltage on the node. A control loop is necessary that measures the DC-voltage on the floating node and controls the DC component of at least one of the currents that flows into or out of the node, in order to fix the DC voltage on that node. The result of this is that both the voltage offset and the current offset at that node are nullified. Voltage offsets are easy to measure and it is not difficult to implement a loop that controls them. Current offsets pose a larger problem because there are no true current sensors available, except for exotic elements like Hall-sensors, but these are not practical solutions in standard circuit design. The only practical way to measure a current offset is

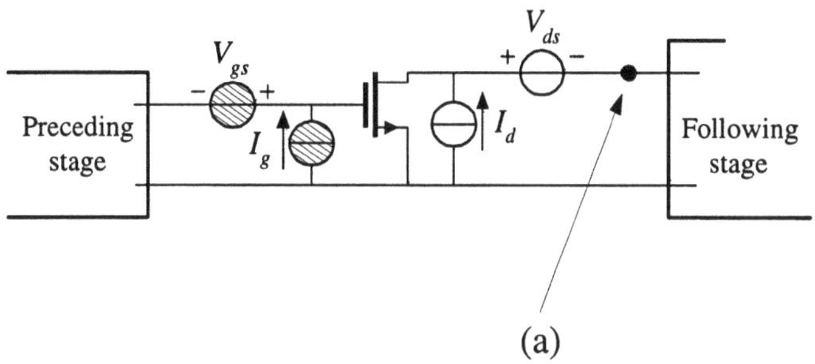


*Figure 8.1.* Example of a floating node (a).

to convert it into a voltage offset via an impedance (like in a multi-meter) . Fortunately, it is rarely necessary to specially introduce impedances in a circuit for this purpose because in most cases a circuit will have one or more floating nodes at which the voltage already is a good measure for current offsets in the circuit. Sometimes floating nodes are even created for this purpose. The proper ways to use these nodes to reduce the current offsets will be discussed later in section 8.6.1.

## 8.3 The ideal transactor and the real transistor

At the end of the small-signal design, usually a circuit results as depicted in figure 8.2. It consists of transistor symbols that actually represent a small-signal model and not a "real-life" transistor and for the rest a feedback network, in this case $R_f$, a source $I_{source}$ with impedance $R_{source}$, a load $R_{load}$ and possibly some frequency compensation components, like $C_{comp}$. To be more clear in

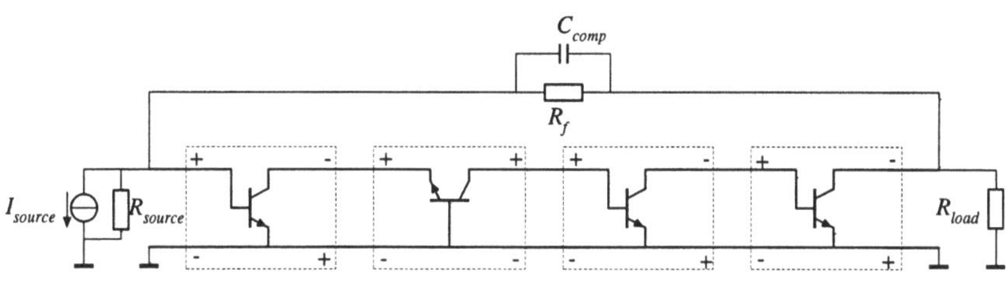


*Figure 8.2.* A finished small-signal design.

what the circuit diagram *really* shows, in figure 8.3 a better, but a little more uncommon, circuit diagram is depicted.

The first thing to do now is to find a two-port containing a real transistor that can replace the small-signal two-port containing the ideal controlled source without changing the small-signal behavior. So, the first step is to start the

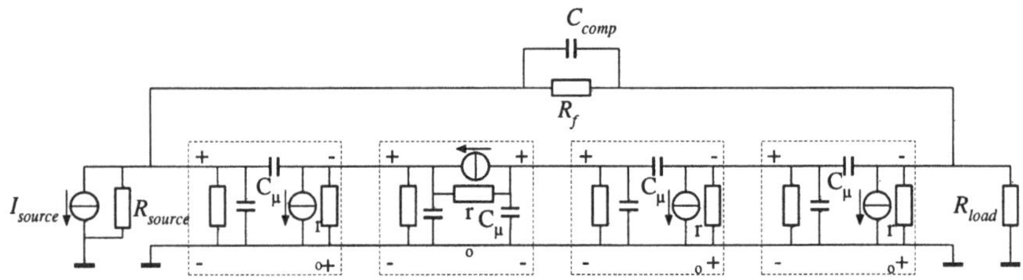


*Figure 8.3.* A finished small-signal design with the proper small signal models shown.

biasing procedure locally, within the bounds of one two-port. This will be discussed in the next section.

### 8.3.1 Biasing one transistor

The essential difference between the small-signal model and the real-life transistor is that the small-signal model contains a linear controlled*source* and the transistor does not. The circuit in the small-signal model can generate power, a separate transistor cannot. So, a transistor can not be the only element in the two-port that has to replace the small-signal two-port. There should be at least one source in there to deliver the power. This can be a DC source and the transistor can convert this DC power into signal power at other frequencies. Only non-linear elements can convert power from one frequency to another. Obviously, transistors are non-linear elements, so in combination with a DC source they can generate power at any desired frequency.[1]

The combination of DC sources and a non-linear element being the transistor can behave like the small-signal model in a certain signal range. The size of this range is determined during the distortion analysis in the small-signal design phase. In that phase, the magnitude of some of the bias quantities has been chosen. So, linearization and defining small-signal models are distortion problems. A result from the distortion design phase is the selection of the part of the device characteristic that is used in the information transfer. The signal is translated to that part of the device characteristic and the result is translated back again as shown in figure 5.6 on page 156. A prescription for the bias quantities follows from that. But even in non-linear circuits, translation of signals may be required, resulting in the need for biasing.[2]

[1]Sometimes this matter is used to argue that amplifiers are "by definition" non-linear circuits. However, one should be careful with this statement because it is not a fundamental property of amplifiers. For example, the circuit depicted in figure 8.3 is truly linear. It is only the present practical unavailability of the controlled sources which makes that the non-linear alternative is chosen.

[2]Actually, this translation of the information carrying signal is the*essence* of biasing: shifting the operating point of the transistor to the origin. So linearization and biasing (translation) should not be confused.

Apart from delivering the required power, the second task of the DC sources used in the transistor-source combination is the*translation.* In figure 8.4 it can

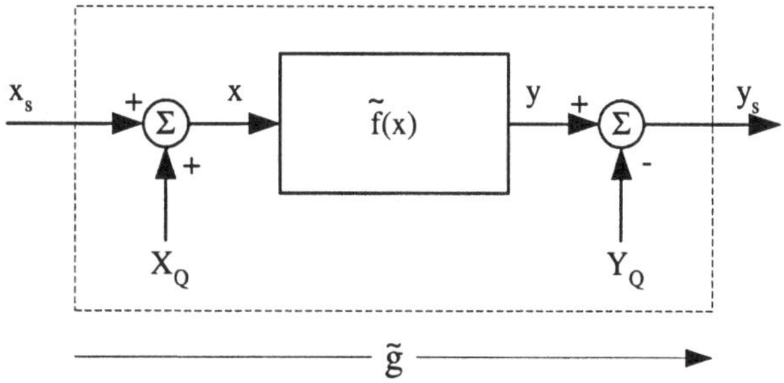


*Figure 8.4.* The principle of translation with bias signals.

be seen how the operating point of a non-linear transfer function $\tilde{f}(x)$ can be translated to the origin by adding and subtracting the proper bias signals. For the new function $\tilde{g}(x)$ holds:

$$\tilde{g}(0) = 0 \tag{8.2}$$

When this strategy is applied to a transistor, both the voltages and the currents need to be translated at the input and at the output. So, two DC-voltage sources and two DC-current sources are inserted. Figure 8.5 shows the results for a bipolar transistor and a FET. It can be seen that the transistors are surrounded

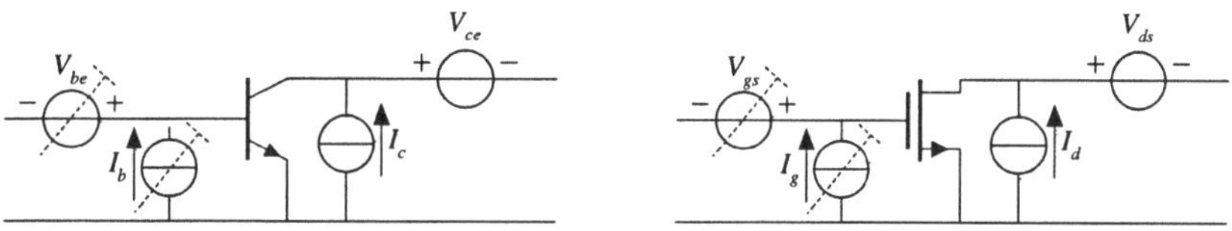


*Figure 8.5.* Transistor-DC-source combinations as a practical substitute for small-signal models.

by two independent DC sources and two controlled DC sources. At the output an independent voltage source and an independent current source are found, because they usually define the operating point the most accurate. Most small-signal parameters have a more direct relation to the drain/collector current than to the currents or voltages at the input, see for example equation (8.1). Via the large-signal two-port relations, that are for example described by a chain matrix, the input voltage and current are completely defined when the bias sources at the output have been given their value as independent sources. Therefore, the two sources at the input need to be controlled in such a way that their values are matching the large-signal transistor equations. This means that their value is controlled in such a way that at the input and output of the two-port there is no voltage or current offset.

As a result, the first step in replacing the small-signal two-port by a more practical circuit results in a two-port comprising:

- one transistor;
- two DC-voltage sources;
- two DC-current sources;
- two control loops.

Note that even for a FET the current source at the input is present. Though its nominal value may be zero, it cannot be deleted just like that since it is a controlled source and deleting the source would also delete the control. This controlled current source is needed to generate the temporary current that has to flow into the gate to get it at the right potential, for example, during startup of the circuit. When the control is not explicitly considered, the practical circuit may find a "less attractive" way to control the current and voltage at the gate, perhaps resulting in a slow startup behavior or even latch-up.

## 8.3.2 Examples of control loops for a single transistor

The control loops are needed to adjust the two bias sources at the input such that the offsets at the input and output of the two-port are equal to zero. This means that all offsets have to be measured and then reduced to zero by the loops. There is one current and voltage offset, so one voltage and one current sensor are needed. For now the loops containing these sensors will be inserted in the circuits in a symbolic way. The actual implementation of the loops is done in a later phase. For now, in the examples below that show how the loops could work, it will be assumed that both voltage and current sensors are available.

Depending on the environment in which the two-port is placed, there are different ways in which the control loops are implemented. In the cascade topology that is used to implement the nullor circuit, there are three different environments for a stage:

- an intermediate stage;
- the output of the amplifier;
- the input of the amplifier.

### 8.3.2.1 The control loops for an intermediate stage

An intermediate stage is usually preceded by either a CE/CS stage or a CB/CG stage. In both cases, the preceding stage has a "current source behavior", with a nominal bias current equal to zero because it can be assumed that this stage is biased properly too. This means that a simple model for the preceding stage can be an open circuit.

The following stage usually is either a CE/CS stage or a CB/CG stage. When this following stage is properly biased, it actively keeps its input voltage offset to zero via its own bias loops. Therefore it has a "voltage source behavior" at its input with a nominal value of zero. This means that a simple model for the following stage can be a short circuit.

Figure 8.6 shows the intermediate stage in this environment. It can be seen

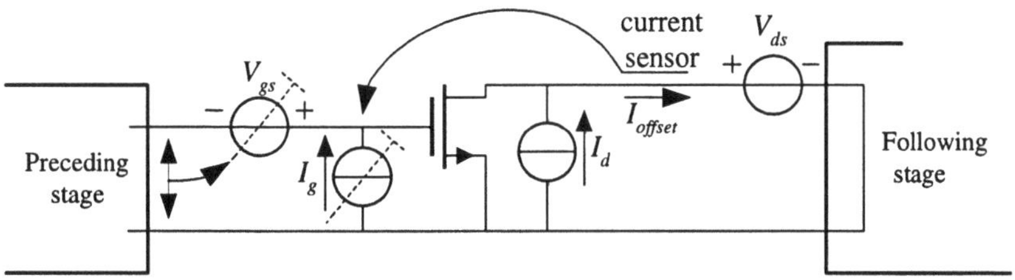


*Figure 8.6.* An intermediate stage placed in its environment.

that the drain-source voltage is correctly set via $V_{ds}$ and the short circuit formed by the following stage.

The drain current is made equal to the bias current $I_d$ via the control loop that measures the difference between the drain current and this bias current via the current sensor. The measurement result is used to control $I_g$. With $I_g$ it is possible to set the gate voltage at the right value. It is not possible to do so via $V_{gs}$ since the preceding stage is an open circuit. The voltage at the input is therefore not defined by the preceding stage, but has to be set by the intermediate stage itself. So, $V_{gs}$ has to be controlled such that the offset voltage at the input equals zero.

To summarize:

- the drain current is set via $I_g$;
- $I_g$ is controlled via a current sensor that measures the current offset at the output of the stage;
- the input offset is controlled via $V_{gs}$;
- $V_{gs}$ is controlled via a voltage sensor that measures the voltage offset at the input of the stage.

For a bipolar transistor the same biasing scheme holds. The question on how to implement the current sensor or any other part of the loops is not important yet, it will be dealt with later. For now the only thing that matters is to know which quantities are measured and how they are used to control the bias sources.

#### 8.3.2.2 The control loops for an input stage

For the input stage two different situations can occur.

- The signal source may have a voltage-source character, which is usually the case for the voltage and the transadmittance amplifier. For the biasing this can be modelled as a short circuit at the input.
- The signal source may have a current-source character, which is usually the case for the current and the transimpedance amplifier. This can be modelled as an open circuit at the input.

(See also figure 2.13 at page 45 and observe the similarity.) When the input has the current-mode character, this is comparable to the case of the intermediate stage and is dealt with in the same way.
The exception that can occur for the first stage is the possible voltage-source character of the signal source. At the output always a similar situation can be expected as for the intermediate stage.

Figure 8.7 shows the situation with the voltage signal source, the short-circuit at the input of the amplifier. It can be seen that the drain voltage is correctly set

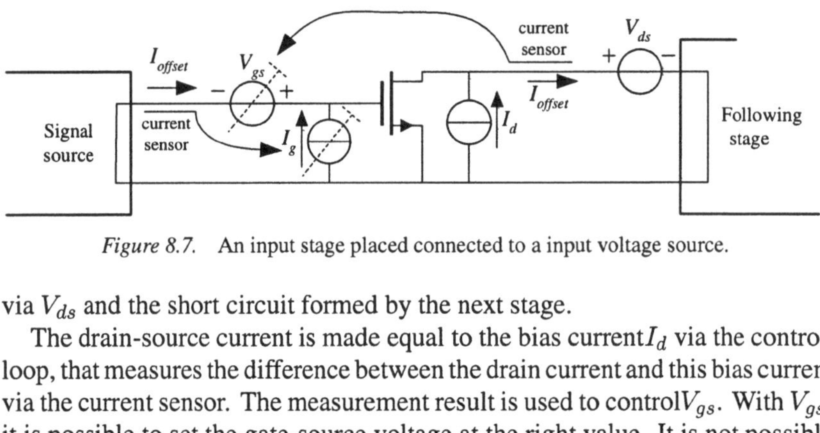


*Figure 8.7.* An input stage placed connected to a input voltage source.

via $V_{ds}$ and the short circuit formed by the next stage.

The drain-source current is made equal to the bias current$I_d$ via the control loop, that measures the difference between the drain current and this bias current via the current sensor. The measurement result is used to control$V_{gs}$. With $V_{gs}$ it is possible to set the gate-source voltage at the right value. It is not possible to do so via $I_g$ since the signal source is a short circuit. Controlled source$I_g$ is used to reduce the offset current at the input to zero.

To summarize:

- the drain current is set via $V_{gs}$;
- $V_{gs}$ is controlled via a current sensor that measures the current offset at the output of the stage;
- the input offset is controlled via $I_g$;

- $I_g$ is controlled via a current sensor that measures the current offset at the input of the stage.

Again, for a bipolar transistor a similar biasing scheme holds.
(Sometimes, having a drain or collector bias current controlled via the voltage source is not desired. In that case a coupling capacitor may be introduced, see section 8.4.2.1.)

#### 8.3.2.3 The control loops for an output stage

For the output stage also two different situations are possible.

- The loading circuit may have a voltage input, resulting in an open-circuit character, which is usually the case for the voltage and the transimpedance amplifier.
- The loading circuit may have a current input, which is the case for the current and the transconductance amplifier. This results in a short-circuit character.

The second (short-circuit) situation is the same as the situation for the intermediate stage. The difference is in the first situation with the open-circuit character of the load.

Figure 8.8 shows the output stage in this situation. It can be seen that the

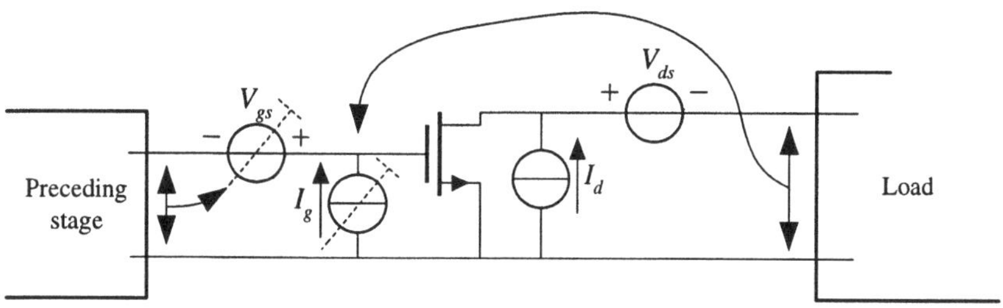


*Figure 8.8.* An output stage with a load with an open-circuit character.

drain-source voltage is *not* correctly set via $V_{ds}$ since the load is an open circuit and it does not define the voltage across it. The node to which $V_{ds}$ and the load are connected and also the drain node are floating nodes. A floating node is a node of which the DC voltage with respect to the other nodes in the circuit is undefined, which is of course intolerable and must be corrected. The two nodes in figure 8.8 form a *floating node-cluster.* Now the DC-voltage relation between this cluster and the other nodes of the circuit is undefined. This must be corrected. The "standard" bias loop to make the drain current correct, using a current sensor would indeed make the drain current correct, but the node-cluster remains floating. Another bias loop is needed to control the DC-bias voltage on this floating node-cluster. Since the two nodes are connected via a voltage

source, a loop equating the DC-bias voltage at the output to zero also makes the voltage at the other node correct.
Fortunately, there is no need to implement a separate loop to do this. The offset voltage at the output can be measured and then used to control the drain current of the FET via the gate current. When the offset voltage is equated to zero, this also implies that the drain current is equal to the bias current $I_d$. So, the bias loop that nulls the DC-bias offset voltage at the output also nulls the offset current, making the drain current correct. (The floating node is used as an "indirect" current sensor.) As with the intermediate stage the drain current is controlled by controlling $I_g$.

To summarize:

- the drain current is set via $I_g$;
- $I_g$ is controlled via a voltage sensor that measures the voltage offset that results from the current offset at the output of the stage;
- the input current offset is controlled via $I_g$;
- $V_{gs}$ is controlled via a voltage sensor that measures the voltage offset at the input of the stage.

Again, for a bipolar transistor the same biasing scheme holds.

### 8.3.3 The influence of the bias loops on the signal behavior

When the bias loops are introduced as described above, the transistor is properly biased. However, it does not show the proper signal behavior. Introducing the bias loops just like that, changes the circuit topology because they also influence the signal behavior. Of course, the loops should only function for the bias signals and not for the information carrying signals. So, each bias loop has to contain a loop filter that separates the bias signals from the information carrying signals. The loop must be made ineffective for the information carrying signals. So, bias filters form an essential part of the biasing loops. Still we will wait to the very end of the biasing procedure before actually inserting a filter into the bias loop. This is because many changes will be made to the biasing topology e.g. by shifting bias sources or even by skipping bias loops. At all times proper biasing can be checked without installing bias filters via a simulator, but the signal transfer will be gone until the proper filters are installed. This should not pose a problem (perhaps except making the designer nervous) since the small-signal design was correct already and the biasing procedure will (should) not change this. It is most efficient to wait with inserting the filters to the last. There are many ways to implement these filters, that this topic will be dealt with in a separate section (section 8.10.5).

### 8.3.4 Biasing of differential stages

When differential stages are used, like the stage depicted in figure 8.9, each transistor is surrounded by four bias sources just like it is done for any other transistor. The differential nature of the stage only has influence on the number of ways the control loops for the controlled sources can be implemented. This is something that is dealt with on the circuit level and not on the level of the separate stages.

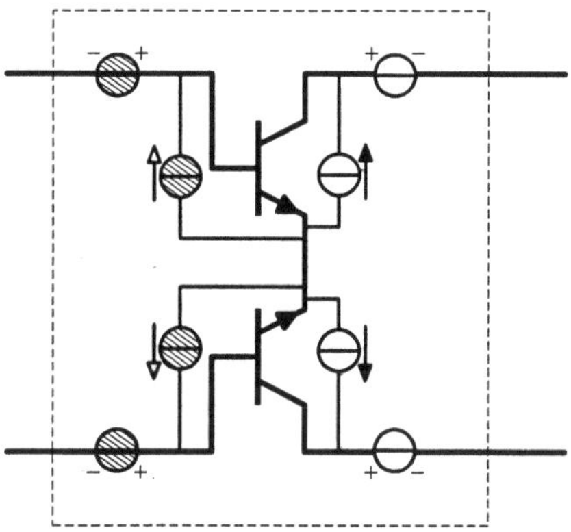

*Figure 8.9.* A biased balanced stage after the first biasing step: placement of the sources. The shaded sources need control.

When the biasing is done as shown in figure 8.9, not making use of the fact that the circuit is a differential circuit at all, this results in a perfectly biased circuit. Using the standard "common-mode/differential-mode biasing strategies" does not result in any extra performance for the circuit. This strategy does not even make the circuit a "DC amplifier" although many designers think it does. Still, there is much to say about the way to make special use of the differential character for biasing purposes. Therefore a separate section (8.12 on page 281) is dedicated to these circuits.

## 8.4 Linear components

Linear components like resistors, inductors and capacitors already have a transfer function that is centered around the origin. No translations via DC sources are needed, so no sources are added around them. *Linear components are not biased.*

When the proper bias sources are added around all nonlinear components in a circuit and all bias loops function properly, it is guaranteed that there are no offset voltages across the linear components and no offset currents through them. This is very good because it is never sure if a linear (passive) component is able to handle the offsets. For example for a $1\Omega$ resistor, a small offset current is not expected to cause much trouble, whereas the same offset current could cause serious trouble when the resistor would have a value of $1\text{G}\Omega$. Note that all linear components in the circuit obtained their value during the small-signal design, and there their value can not ne changed for biasing purposes.

Apparently the linear components play no role in the biasing of a circuit. Therefore, during the first biasing step, they can be replaced by a simpler model: either a short circuit or an open circuit. This results in a simpler circuit which makes the design of the bias circuit easier.

- **Inductors** are be replaced by short circuits.
- **Capacitors** are replaced by open circuits.
- **Resistors** are also replaced by open circuits. The reason for this will become clear in the next section. Replacing the resistors by open circuits makes detecting floating nodes much easier. A floating node is a node of which the DC-voltage relation to the other nodes in the circuit is undefined. Floating nodes (for biasing) cannot be tolerated in a circuit. A resistor may solve this problem when it connects the floating node to a node with a well defined bias voltage, but it also may not. The $1\Omega$ resistor is likely to do the job, the $1\text{G}\Omega$ most likely not. The safe way is to first detect the floating node and then to evaluate if resistors that are already present can solve the problem. Also the time-constant involved should be evaluated. A large resistor can take a long time to charge the floating node to the desired voltage, making the amplifier a "slow starter".

Working this way always yields a properly biased circuit. However, sometimes a linear component can play a role for biasing. A closer inspection for each component is necessary to evaluate this.

### 8.4.1 Resistors

Resistors relate the voltage across them to the current through them via the well known Ohm's law. So, an offset current leads to an offset voltage across it and vice-versa. When an offset current is allowed through the resistor the resulting offset voltage must be evaluated to see if it is acceptable. The same holds when an offset voltage is allowed across a resistor for the resulting offset current.

Usually, bias sources related to the nonlinear (active) components appear around the resistor as shown in figure 8.10a. The resistor operates without offsets. It may happen that by *pure coincidence* holds:

$$R \approx \frac{V_{bias}}{I_{bias}} \tag{8.3}$$

In that case a designer may decide to remove the two sources and let the resistor be involved in the biasing as shown in figure 8.10b. This of course relates$I_{bias}$ with $V_{bias}$ via $R_2$, whereas they may have been unrelated before.

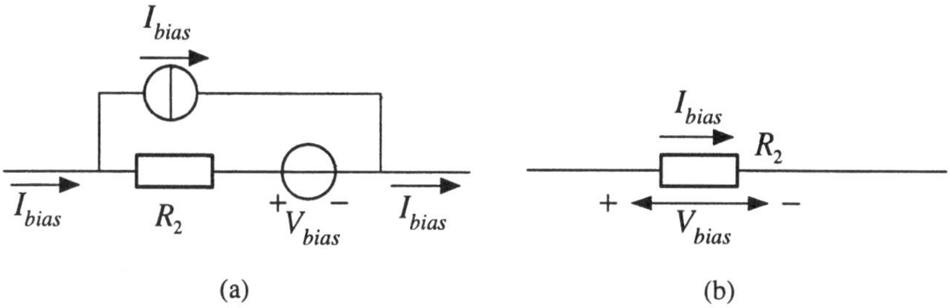


*Figure 8.10.* An unbiased resistor (a) and a resistor supporting a bias current and voltage (b).

Bias quantities are related to each other via the device equations and via the control loops. When the resistor is getting involved in the biasing it must be part of (or just be) the implementation of one of those bias loops.

In practice a resistor used in this way is very often not seen as a bias loop implementation, but it should. This is because keeping track of the number of bias loops gives the designer important information on the biasing behavior of a circuit. When not all loops are implemented, there is an extra degree of freedom for the circuit that is dealt with by the circuit via "parasitic" and unpredictable loops, perhaps resulting in latch-up situations. When there are too many loops there may be no "normal" bias solution but the practical circuit will find one anyway via the non-linearity of the active components and again it may end up in an unexpected operating point.

Situations like shown in figure 8.10a and figure 8.10b usually occur in the feedback network, that reduces the signal and contains no sources that generate power. There dissipative combinations of a voltage and a current source can be found of which the quotient leads to a positive resistor value.

In the active part signal reduction is avoided. Therefore it should not be possible to find a suitable dissipative combination of a voltage and a current source[3]. For example, in the situation of figure 8.11, the signs of the bias sources around $R_{load}$ are such that $R_{load}$ should have a *negative* value to generate the same biasing conditions for the transistor as the two sources. This resistor would need to generate and not to dissipate power.

When is is possible to find a suitable dissipating combination the designer must have had special small-signal considerations (e.g. for frequency compensation) to put it there.

[3]If you can find one, you most probably made a design error. There are not so many good reasons. The least favorable frequency compensation method, resistive broad-banding may be a "good" reason.

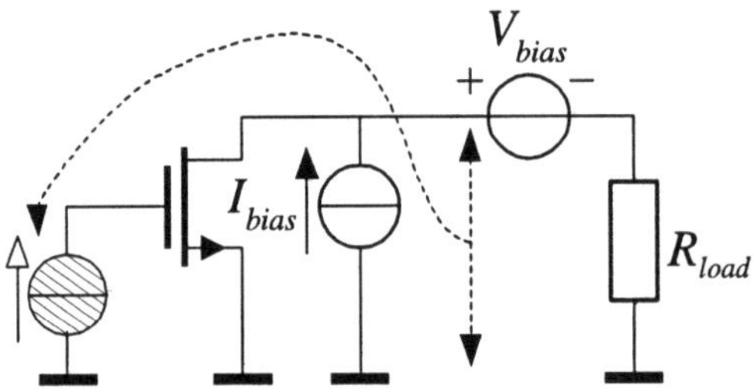


*Figure 8.11.* A biased stage.

When a resistor creates biasing problems, adding a coupling capacitor in series solves this problem. This can also be done when absolutely no DC offsets are tolerated for the resistor. Some resistors show increased $\frac{1}{f}$ noise when biased and potentiometers may create excess noise when varied.

## 8.4.2 Capacitors

Capacitors can be in a circuit for two reasons:

- because of their influence on the signal behavior of the circuit. Among them are for example the capacitors used in frequency compensation.
- as biasing components used for:
  - decoupling, to reduce the impedance at a node for signals in the information band.
  - coupling, to connect two nodes for signals in the information band frequencies while keeping them separated for biasing.

Capacitors of the first class, the "signal capacitors" can only cause problems when they are sensitive to the polarity of the offset voltage across them (like electrolytic capacitors). It would be possible to describe all kinds of very sophisticated measures to reduce the offset to zero, but the most practical solution is to just have the capacitor connected with the right polarity.

Of the capacitors used for biasing purposes, the application of the decoupling capacitor is rather trivial. The most interesting one is the coupling capacitor. There are two different tasks they can perform for biasing. They will be extensively discussed in this chapter. The next section will already give a preliminary overview.

### 8.4.2.1 Coupling capacitors

There are two different situations in which coupling capacitors have a useful application. They can be used to:

- create a voltage or current-controlled DC voltage source, without interfering with the small-signal behavior. The most likely place is at the input of the amplifier.

- create a floating node for DC, without braking the connection for the signal. The most likely place is at the output of the amplifier.

**Coupling capacitors at the input**

In figure 8.5 on page 245, it can be seen that for correct biasing of a transistor one controlled voltage source is necessary. Depending on the environment this source is current controlled via a measurement of the offset current at the output (figure 8.7) or voltage controlled via an offset voltage measurement at the input (figures 8.6 and 8.8).

In section 8.7.1, the implementation of these controlled voltage sources will be discussed in detail. For now, the role a coupling capacitor can play in this matter will be discussed. Sometimes the short-circuit behavior as shown in figure 8.7 at the input of a stage is undesirable. Reasons for this can for example be not wanting to use the exponential relation between voltage and current of a bipolar transistor for bias control or the difficulty to implement a floating controlled voltage source. A coupling capacitor changes the situation to the open-circuit variant as described for the intermediate stage (figure 8.6), in which there is a voltage-controlled voltage source ($V_{be}$ or $V_{gs}$) that is controlled via a loop that measures the bias offset-voltage. The voltage across the coupling capacitor that replaces this controlled voltage source adapts such that it perfectly nullifies the input offset voltage. The (temporary) current needed to adapt the capacitor voltage is supplied by the bias current source$I_g$ or $I_b$. So, for DC the capacitor plays the role of *controlled* source that removes the input offset. For the information carrying signal it is a short-circuit, so even the bias-loop filter is implemented in this way.

Note that placing a coupling capacitor at the input of an amplifier when the preceding circuit has a current output, results in an undefined bias voltage at the output of that preceding circuit. Measures have to be taken to define the DC-bias voltage at that node.

**Coupling capacitors at the output**

There are two reasons for placing a coupling capacitor between the output stage and the load:

- because it must be made absolutely sure that no DC flows through the load.

- to create a floating node to be able to use the voltage at that node as a measure for the offset current flowing into or out of that node. In other words, to create a practical implementation of a current sensor.

Placing a coupling capacitor between the output of a stage and the load (or the next stage), leads to the behavior and the strategies described for the output stage that is connected to a load circuit that has open-circuit character for biasing (figure 8.8). The bias loop nullifies both the offset current and offset voltage at the output of the stage. The capacitor voltage makes sure that the load is without DC bias.

### 8.4.3 Inductors

In most cases, inductors can be considered short circuits. However, sometimes the behavior of an inductor changes when a DC current is allowed to flow through them e.g. due to saturation of the core material. Then the safest way may be to consider them open circuits and guarantee this with a coupling capacitor. Of course this decision can be taken very quickly by the designer.

Because of their frequency dependent behavior, also inductors can be used as coupling and decoupling devices for biasing. They can be used to couple two nodes for biasing while keeping them separate for the signal. Like for example a capacitor can be used to reduce the impedance of a supply voltage source, an inductor can be used to increase the impedance of a current source. In standard amplifier design, inductors are not very frequently used for this purpose yet, but taking in mind the way capacitors are used it will not be difficult to imagine the opportunities for biasing that are available when the proper inductors are available.

## 8.5 Biasing step 1: Identification and first implementation of the bias loops

The first step is to place the four bias sources around each transistor. Two of them need control. So, starting with the circuit depicted in figure 8.2 on page 243, after placement of all sources, the circuit in figure 8.12 results. The

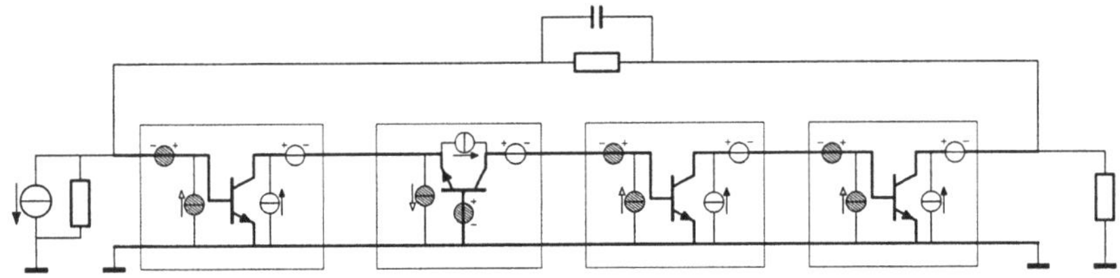

*Figure 8.12.* The circuit of figure 8.2 after the first biasing step: placement of the sources. The shaded sources need control.

shaded sources need control. Following the theory of section 8.3.1, for each controlled source the origin of the controlling signal can be indicated. The control loops are only symbolically indicated in figure 8.13. The resistors are

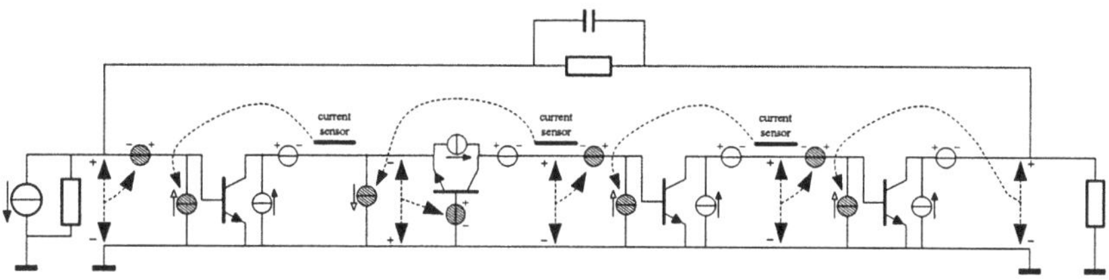


*Figure 8.13.* The circuit of figure 8.12 including the (symbolic) control loops.

considered to be open circuits. As a result, the output node of the last stage is a floating node as described on page 242. So, to both measure the error in the collector bias current and define the node voltage, a voltage sensor is used instead of a current sensor. The use of a current sensor does make the collector bias current correct, but the DC-bias voltage on the collector node remains undefined.

In a simulator like PSPICE, it is very easy to implement every controlled source described in figure 8.13 with the four controlled sources that are available in the simulator. Doing so, without implementing the bias-loop filters, produces simulation result that can be used to verify the proper biasing of the circuit. Of course, lacking the bias-loop filters, the small-signal behavior of the circuit will be gone, but that is of no concern at this design stage. It is just to check if the bias sources have been given their correct value and if no control loops have been forgotten. Also from the operating point information the small-signal parameters can be read that the simulator calculates under the actual bias conditions. It can be verified if they indeed match to the small-signal parameters used sofar.

## 8.6 Biasing step 2: The bias-current loops

Before starting with the implementation of the bias loops, designing loop filters and going into the trouble of creating floating nodes to be able to measure offset currents, first the number of explicit bias loops should be made as small as possible. Preferably, only those loops should be left over that can be controlled via offset voltages at existing floating nodes. Special current sensors are to be avoided. The other loops are skipped with their related controlled source being fixed at the nominal value.

### 8.6.1 Measuring DC-bias offset currents

Measuring bias-current offsets poses a large problem because there are no true current sensorcurrent sensors available, except for exotic elements like Hall-sensors, but these are not practical solutions in standard circuit design. The only practical way to measure a current offset is to convert it into a voltage offset via an impedance (like in a multi-meter). Strict requirements are imposed on these impedances, since they may not interfere with the small-signal behavior of the circuit. In section 2.6.1 on page 40 it has already been made clear that passive devices can only reduce a signal (especially resistors that dissipate signal), so their presence in a nullor circuit would only degrade its performance. Unfortunately it is precisely in this active circuit where the offset currents need to be measured.

There are several ways to deal with a bias offset current:

- The first choice would be *not* to control the offset current but to accept it. Not all offset currents lead to a disfunction of the circuit. They may just lead to a small acceptable variation in the small-signal parameters. In this chapter a method will be discussed to reduce the number of current control loops and to evaluate the resulting current-offset errors.

- The second choice is to make use of the DC voltage variations on a floating node already present in the circuit to obtain an (indirect) indication of the offset current that needs to be controlled to zero. The strategy is to reduce the number of current control loops in such a way that only loops remain that can make use of an existing floating node to measure the offset.

- The last resort is to *create* a floating node by inserting for example a coupling capacitor. Then proceed as above.

- The *dangerous* method is to skip the explicit control of current sources injecting into a floating node and accept the biasing error in the node*voltage* that results. If a parasitic resistor like the output impedance of a transistor is connected to the node the sum of currents at the floating node can be equated to zero via this impedance. It works as a "poor-mans" voltage controlled current source. The risk is the uncertainty in the value of the impedance and its probably high value. Small offsets lead to large voltage swings.[4]
  The only "reliable" resistors may be found at the input, the output and the feedback network. Sometimes such a resistor can indeed be used to deal with the offset current and control the voltage. For example, skipping the bias-current control for the last transistor stage in figure 8.13 may result in an acceptably small current through the load resistor and a small offset voltage across it.

*Note that if a control is skipped, the result inevitably is incorrect biasing of at least one of the transistors. The designer has to decide if this is acceptable.*

When there is no floating node at which offset currents can be measured, the most convenient approach is to create one. However, there are cases in which

[4]Usually this biasing method is not an intentional design decision, but just a forgotten loop. An unreliable and EMC sensitive circuit usually results.

this is not desired or not possible. An example of this is the push-pull class-AB output stage. Controlling the bias currents in stages like this is a notorious problem, that is not solved yet via a practical easy-to-use current sensor. Very often, via a translinear loop (current-mirror like approach), a copy of the bias current is made that is converted to a voltage via a resistor. This resistor is not in the signal path and therefore acceptable.

### 8.6.2 Changing the topology of the current control loops

In section 8.6.1 it was already mentioned that in the original bias scheme it is not possible to skip bias-current loops just like that. To reduce the number of control loops, loops must be found that, when skipped, leave an acceptable error. A method to *create* such loops will be discussed now.

The topology of the control loops is changed from one containing just current loops around a single stage to a topology that contains one more global loop around more than one stage. Starting from the circuit of figure 8.13 on page 257, the circuit of figure 8.14 can be found by exchanging the sensing locations controlled current sources of transistor $Q_3$ and $Q_4$. So, now the base current of $Q_3$ is controlled by measuring the bias offset voltage at the floating node at the output of the fourth stage. This is loop (a). The base current of$Q_4$ is based on the offset current at the output of the third stage. This is loop (b). Close inspection shows that the collector bias currents are still completely correct in both transistors. In this bias topology it is possible to skip loop (b). Bias loop

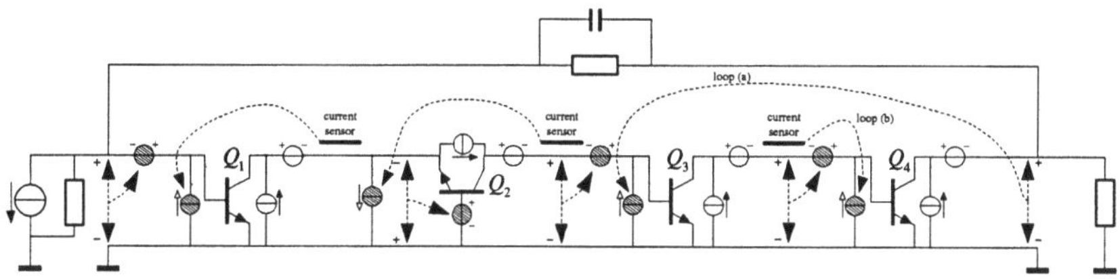


*Figure 8.14.* The circuit of figure 8.13 after exchanging the control locations of two loops.

(b) is a local loop, there is no transistor in the loop. When the controlled source is set at its nominal value and the control is skipped, this results in an error in the collector current of transistor $Q_3$. The error is just the difference between the nominal value of the current source $I_{b4}$ and the actual base current of $Q_4$. This difference can be very small. When this error is acceptable, the loop can indeed be skipped and the circuit of figure 8.15 is found. The formerly controlled source $I_{b4}$ is now set at a fixed value. This process of creating a local current control loop and then skip it can be repeated several times until the number of current controlling loops is at a minimum. Very often the minimum can be just one overall loop over the complete amplifier, figure 8.16 shows this. Of course,

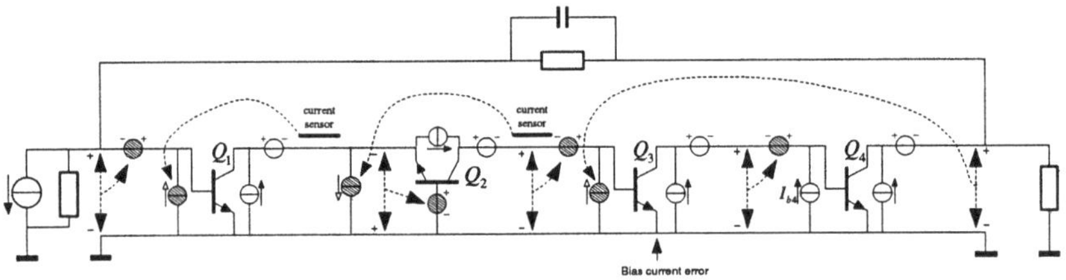


*Figure 8.15.* The circuit of figure 8.14 after skipping loop (b). The formerly controlled source $I_{b4}$ is now set at a fixed value. The collector current of $Q_3$ now has a small error.

then all transistors have a small error in the collector bias current except for the transistor of which the bias offset-current is still measured. This transistor still has a perfectly correct bias current. In this case it is transistor $Q_4$.

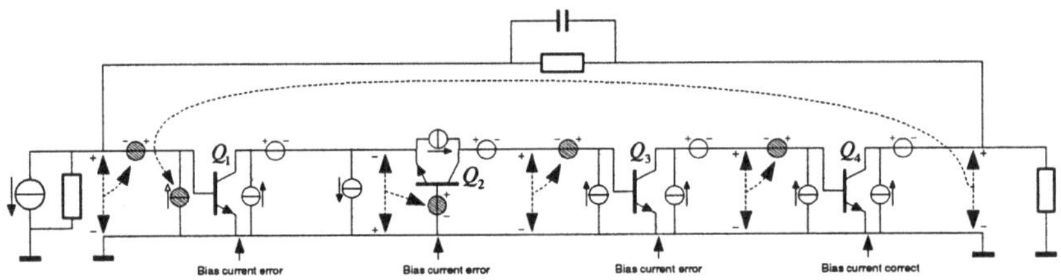


*Figure 8.16.* The circuit of figure 8.14 after skipping three out of four bias-current loops. The formerly controlled sources are now set at their nominal value. Only the collector bias current of $Q_4$ is without error.

This enlargement of the control span of a loop only works when there is a complete DC path. A coupling capacitor breaks this DC path. In figure 8.17 it can be seen how a coupling capacitor $C_{couple}$ breaks the DC-path between $Q_2$ and $Q_3$. The node at the collector of $Q_2$ has become a floating node now. This node is used as measurement node for a bias control loop as shown in the figure. Now $Q_2$ and $Q_4$ are correctly biased and $Q_1$ and $Q_3$ have a small error

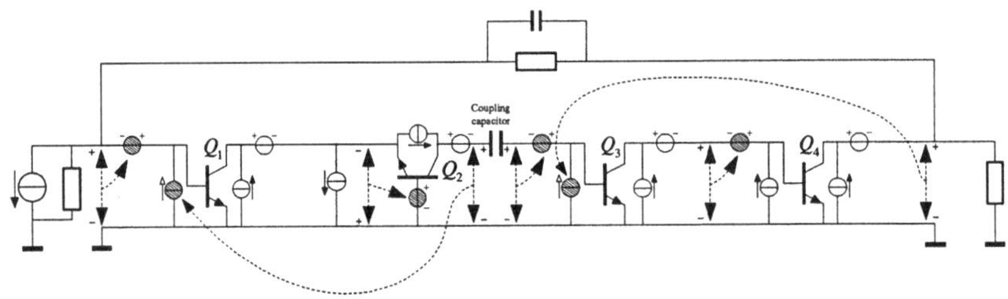


*Figure 8.17.* The effect of a coupling capacitor.

in the collector bias current.

There are a number of reasons why a designer may decide to break a DC path. They will not be discussed here because they normally are very design specific but they are usually related to unacceptably large biasing error when some of the controlled sources are fixed to their nominal value. This may, for example, happen when the bias currents in the transistors differ several orders of magnitude.

The sum of the number of errors in the collector current and the number of current control loops is constant and equal to the number of transistors. So, after implementation of a limited number of current control loops, the designer knows how many collector currents are completely correct and how many have an error. It is easy to find where the errors are and to check if they are indeed acceptable.

### 8.6.3 Floating nodes

In the circuit in the previous section all offset-current measurements were done via floating nodes. No special current sensors were needed. When floating nodes are available, usually it is possible to rearrange the current controls in such a way that no more loops are needed than there are floating nodes. Then the implementation of the current control loop(s) becomes rather simple. So, preferably the situation is the following:

- when there are $N_Q$ transistors
- and $N_{float}$ floating nodes
- it is easy to implement $N_{float}$ current control loops
- and it should be acceptable to have $N_Q - N_{float}$ transistors with a small error in the collector bias current.

Apparently, a lot depends on the number of floating nodes. So, the first thing to do, is to find the floating nodes. Their presence will guide the bias implementation process e.g. to find out what is the best way to change the bias-current loop topology, or what is the best place to create a floating node if there are not enough floating nodes.

There are a number of ways in which two nodes can have a well-defined DC-voltage relation to each other:

- via a voltage source (the ground node and all supply nodes are related in this way);
- via the base-emitter connection of a bipolar transistor or via the gate-source connection of a FET;
- via a two-terminal non-linear element like a diode;

- via an inductor when this is allowed to support the DC-offset current;
- via a resistor of which the value is low enough and that is allowed to support the DC-offset current.

Of every node, the bias voltage must be well defined with respect to all other nodes in the circuit via one of the connections described above. A floating node is a node that does not have this well defined voltage relation.

Figure 8.18 shows the way transistors define a voltage relation between two nodes. Since the collector/drain current of every transistor is well-defined, via

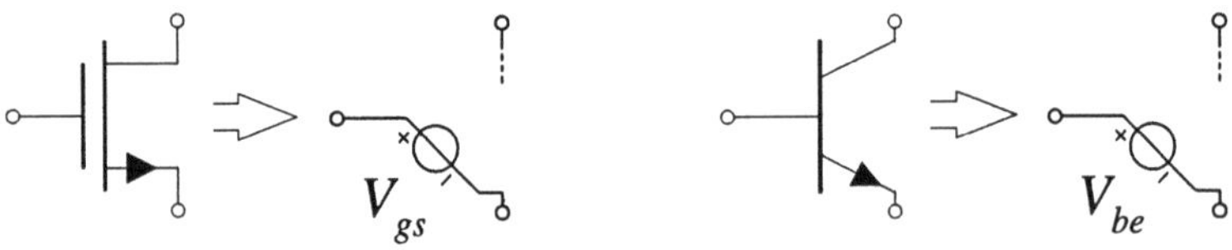


*Figure 8.18.* The way a transistor defines the voltage relation between two nodes. Note that the drain/collector node is floating.

the device input-output relations the base-emitter or gate-source voltage is well-defined too. So, it is not because the base-emitter junction of a transistor is a forward biased diode that the base-emitter voltage is well defined. Therefore, for any device, even for MOSFETs and vacuum-tubes this voltage relation is well-defined.

Resistors are a special case. Of course, they do define a DC voltage between their nodes, but it may be that the relation is not the required one for correct biasing of the particular circuit. For example a10GΩ resistor does not establish much of a relation, but very small resistors may even be considered a short circuit, since the voltage across them, generated by the biasing currents may be negligible. Therefore, to have a quick and easy way to find floating nodes resistors are considered open circuits. Very often, a floating node remains usable as current sensor even when it is completely floating anymore because some resistors are connected to it. This depends on the values of these resistors. For resistors with such a low value that offset-current measurements become impossible, putting a coupling capacitor in series solves this problem.

#### 8.6.3.1 Floating node *clusters*

Inspecting figure 8.13, it can be seen that there are two interconnected floating nodes. They are the collector of the last transistor and the output node, which are interconnected via a voltage source. They form a floating node cluster. Both DC node voltages are good indirect measures for the correctness of the DC bias currents. Then the question is which node is preferably used to control the bias current of that stage. The safest and most correct way is to work with the floating nodes that are at the inputs and the outputs of the biased devices.

They are the nodes that are correctly biased at DC zero volts. It is, for example, the floating node that is used in figure 8.13. Then, when the control connection is shifted to another node in the cluster, it is sure that the proper voltage offsets are taken into account. This is shown in figure 8.19.

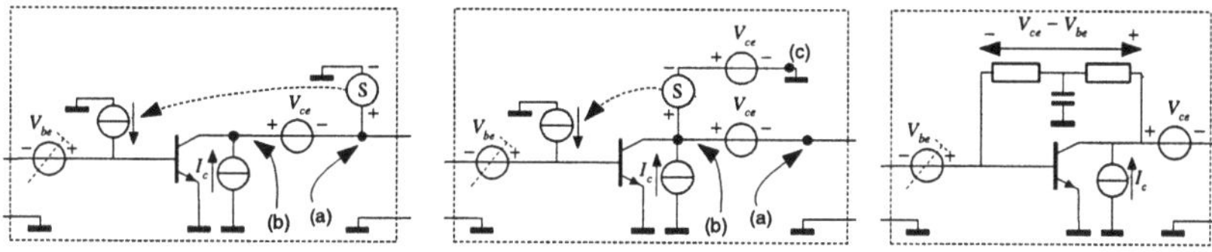


*Figure 8.19.* Controlling the bias current of a transistor stage using a floating node.

The circuit contains a cluster of floating nodes (a) and (b). Node (a) is the appropriate node to start with, since this is a floating node at the output of the biased device. This is done in the left-hand circuit. The DC bias voltage at node (a) is equated to the voltage at the other output node, so the output DC offset becomes zero and the transistor bias currents becomes correct. The connection of the comparator can become connected to internal floating node (b), by shifting the voltage source $V_{ce}$ through the comparator[5]. This results in the center circuit. Actively keeping the DC bias voltage at node (b) at$V_{ce}$ also keeps the DC bias voltage at node (a) equal to zero.

The voltage-to-current converter used for biasing can be implemented with for example a differential pair and some fixed current sources. This yields very accurate biasing.

## 8.7 Biasing step 3: The bias-voltage loops

### 8.7.1 Practical controlled DC-bias voltage sources

Active devices that come close to controlled voltage sources are not readily available. This implies that there are only indirect ways to implement them. In practice, there are four approaches towards the implementation of the controlled voltage sources:

- via a controlled current source and an impedance that is already in the circuit for the small-signal behavior. This is shown in figure 8.20(b). Of course, the DC current should cause no problems for the signal behavior of the resistor, like flicker noise.

- via a controlled current source and an extra impedance specially inserted in the circuit to create the voltage across. This is shown in figure 8.20(c).

[5]This is done according to the voltage shift method also used for noise sources, see section 4.2.2 on page 93.

The value of this impedance at signal frequencies should be low enough to keep it from interfering with the signal behavior. This can be done with a decoupling capacitor.

- by inserting a coupling capacitor.
- by skipping the control completely and accepting the voltage offset.

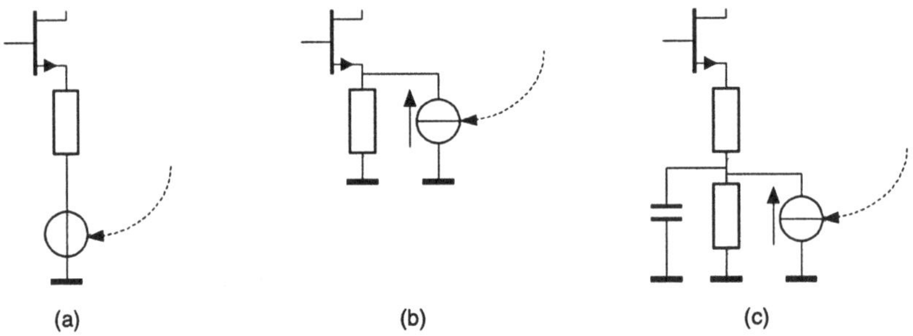


*Figure 8.20.* Implementation of a controlled voltage source. (a) the ideal source, (b) using an existing (signal)resistor, (c) using an additionally inserted impedance.

## 8.7.2 Capacitors as controlled voltage sources

The voltage control loops can be local and "not local". In a local loop the offset measurement is done between a node of the controlled source itself and another node. In this respect a resistor was already introduced as a voltage controlled current source. In a similar way, a capacitor can be seen as a current controlled voltage source and when a finite DC impedance exists between the measurement nodes, even as a voltage controlled voltage source. Figure 8.21 shows a detail of the circuit of figure 8.13. At the input there is a bias-voltage source that is controlled via a local loop. The controlled voltage source has to

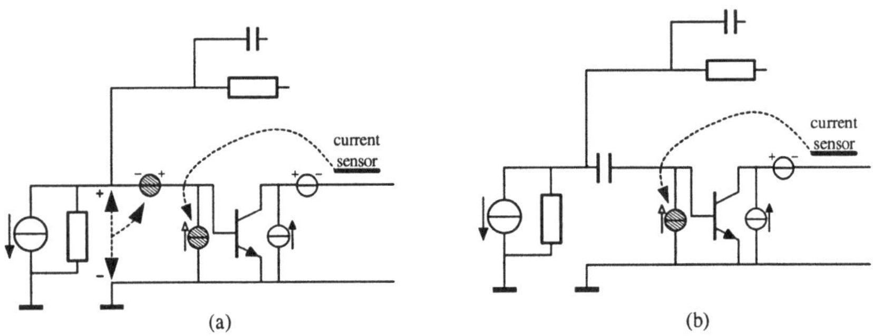


*Figure 8.21.* A capacitor used as a current controlled voltage source (b) to implement the voltage controlled voltage source in (a).

nullify the DC bias voltage at the input of the transistor stage. The direct method is to measure the offset voltage and to control the source with that information. In this case, the source impedance is resistive. A non-zero offset voltage leads to an offset current through the controlled source. When the controlled source is replaced by a capacitor and the correct voltage is not across its terminals, also an offset current flows. This charges the capacitor exactly to the required value that is needed to nullify the input offset. This sounds (is) trivial, but it is discussed here in this detail to clearly show that a coupling capacitor can be used to implement a voltage or current controlled current source when the control loop is *local.*

Since the coupling capacitor is a short-circuit for the signal—so it does not interfere with the signal behavior—even the bias-loop filter action is included. Even when a more elaborate circuit would be used to implement the controlled voltage source, across its terminal is would show the behavior of a capacitor. Of course, such a circuit can be a good option when too extreme values are needed for direct use of a coupling capacitor.

### 8.7.3 Skipping the control of a bias voltage source

For all controlled voltage sources in figure 8.16, except for the one at the input, skipping the control and just setting them at their nominal values results in an error in the collector bias voltage of one of the transistors close to the source. Usually, the collector bias voltage is not very critical and the error can be accepted without problems. Figure 8.22 shows a typical example of this case. The controlled source is in series with a collector. The accuracy of

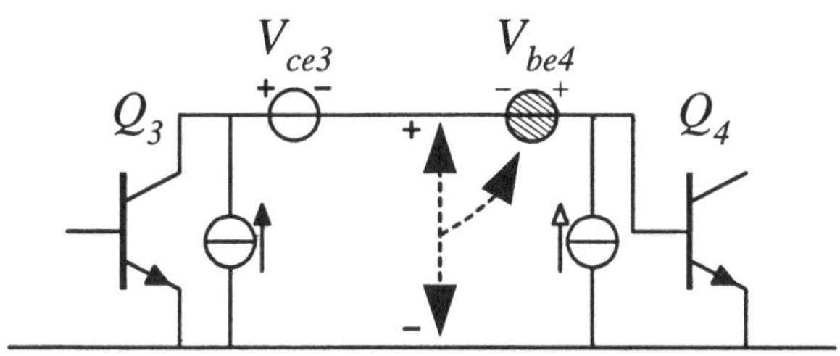


*Figure 8.22.* A typical situation in which the control of a source may be skipped. The collector voltage of $Q_3$ get a small voltage error.

the parameters that depend on the value of the collector-to-emitter (or actually collector-to-base) voltage is not much affected when the bias source$V_{be}$ is set at its nominal value instead of being controlled. So, for every controlled voltage source, the sensitivity of the circuit for variations of the source must be checked.

If the sensitivity is low enough, the control can be skipped. Typically this is when controlled sources determine a drain or a collector voltage.

Figure 8.23, shows the circuit of figure 8.13 on page 257, in which the control of three out of four voltage control loops is skipped. It can be seen that:

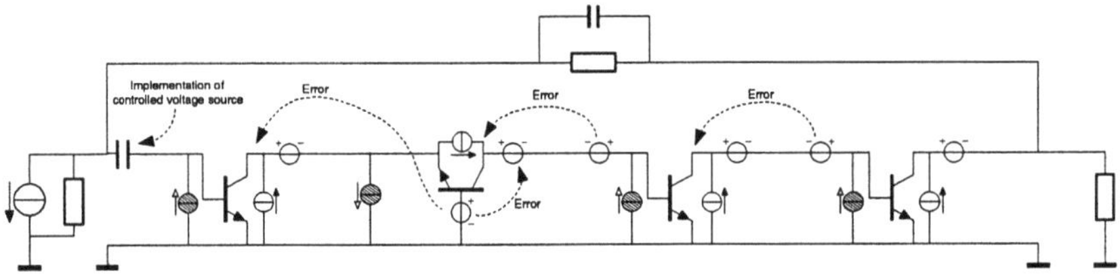


*Figure 8.23.* Location of the voltage errors when three controls are skipped.

- the controls of the voltage sources belonging to $Q_3$ and $Q_4$ are skipped
- variations of the controlled voltage source belonging to $Q_2$ affect the collector voltages of $Q_1$ and $Q_2$: the control can be skipped
- the controlled bias voltage source belonging to $Q_1$ is connected to a node to which the signal source is connected, not just collectors or drains. Therefore the control of this source *cannot be skipped* just like that.

Skipping the control of the controlled voltage source at the input in figure 8.23 may lead to an offset current through the signal source and an offset voltage across it. The current control loop keeps the currents in the input transistor correct, so the biasing of the transistor remains correct. It is up to the designer to decide whether a DC-offset voltage is allowed across the source or not, and if the offset current, that is generated due to this voltage offset, can be supplied by the controlled current source that is connected to the same node. In other cases, a controlled source must be implemented.

### 8.7.4 The voltage offset control at the input

When control is needed, depending on the source properties, two different situations can occur:

- the impedance at the source node is high for DC. So, offset currents would not become un-tolerably high. The source just does not allow DC offsets. The simplest model for the impedance is an open circuit. This is the situation shown in figure 8.6 on page 247
- the impedance at the source node is low for DC. The simplest model for the impedance is a short circuit. This is the situation shown in figure 8.7 on page 248

#### 8.7.4.1 Source with open-circuit model

When the simplest model for the source impedance is an open circuit, one way to deal with this is to insert a coupling capacitor. This breaks the DC path from the source to the amplifier input. Then the voltage across the capacitor is the exact value that is necessary to have the offset at the input nullified. It can be seen as one of the most simple ways to implement the control that nullifies the input offset voltage. this was described in section 8.4.2.1 on page 255.

The coupling capacitor introduces an extra finite impedance in series with the input. This implies that it will play a role in noise behavior of the circuit. *The value of the capacitor is therefore not only determined via the frequency behavior of the circuit, but also via the noise behavior!* Frequently the value that follows from the noise analysis is higher than the value that follows from bandwidth considerations (see exercise 3 on page 144).

#### 8.7.4.2 Source with short-circuit model

When the simplest model for the source is a short circuit, it is the controlled voltage source that sets the collector or drain current of the first stage. The controlled current source at the input reduces the offset current at the input. This is the situation in the circuit shown in figure 8.24.

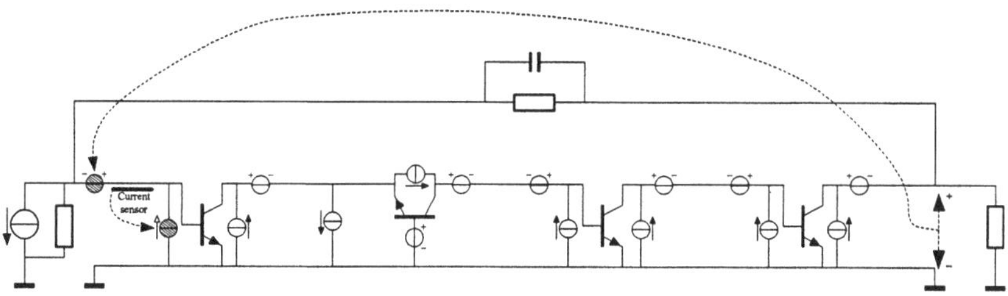


*Figure 8.24.* The bias loop for the controlled voltage source that sets the collector current of the first transistor.

If the source accepts small offset currents, the control for the controlled current source can be skipped.

Note that for this way of biasing, the load has to be present in order to have a correctly biased circuit. When removed, it should be replaced by a short circuit. If this is not guaranteed and the amplifier needs to remain biased correctly even when the source is detached, the bias topology has to be changed.

The alternative is to create the "open-circuit situation". This can be done by inserting a coupling capacitor. Then the bias topology of figure 8.25 is found. This topology keeps the circuit correctly biased even when the source is detached. This alternative is also a good choice when it is not possible to implement a good quality controlled voltage source or when a designer does not want to make use of the exponential relation between base-emitter voltage and collector current for biasing. Figure 8.25 shows the result. In this circuit all necessary bias sources have been placed and all controls are established. The

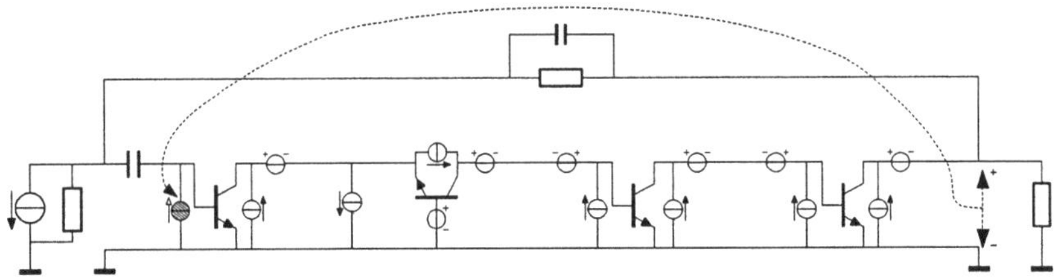

*Figure 8.25.* The circuit of figure 8.24 with an alternative biasing scheme, making use of a coupling capacitor.

number of loops has been reduced by skipping some loops and accepting the small offsets that result. After this reduction two controls remain. One loop is explicitly present, the other is established by placing a coupling capacitor.

## 8.8 Summary after biasing steps 2 and 3

After biasing step 2, the circuit shown in figure 8.16 on page 260 resulted. Most bias currents are correct enough and only one current control loop is implemented. The small errors in the bias currents of $Q_1$, $Q_2$ and $Q_3$ that result from skipping three bias currents loops are accepted. In biasing step 3, the controls for the voltage sources were evaluated. This resulted in the circuit shown in figure 8.23 on page 266. These voltage loops nullify a "true" voltage offset, they do not control the voltage at floating nodes. This is done by the current control loops that use the voltage on these floating nodes to indirectly sense current offsets in the circuit.

In figure 8.26, the combined result of both steps is depicted. For this amplifier

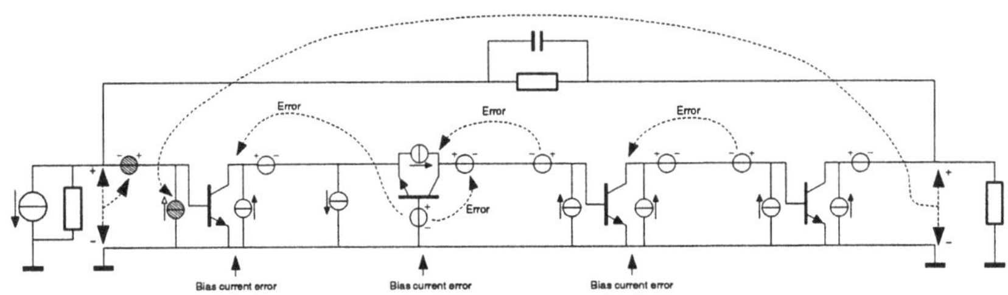


*Figure 8.26.* The circuit of figure 8.16 with the minimal amount of control loops.

two control loops remain and six bias errors are accepted. This is in accordance with the number of transistors. Four transistors require eight loops for perfect biasing.

The order in which the manipulation of the loops is done is not important. It is possible to do biasing step 3 (the voltage loops) before biasing step 2 (the current loops). However, sometimes a floating node is created in biasing step 2 and this may lead to different choices made in step 3. This is why there is a slight preference to manipulate the current loops first.

## 8.9 Biasing step 4: Reduction of the number of bias sources

Having found all the bias sources (current and voltage) and controls, one can now implement each source and end up with a correctly biased circuit with lots of bias sources. But as may be clear, this is not the most efficient way of biasing; there are too many sources.

Fortunately, it is easy to reduce the number of sources. For this some of the same transformations can be used that are also used to transform noise sources. They are the Voltage-Source Shift discussed in section 4.2.2 at page 93 and the Current-Source Shift discussed in section 4.2.3 at page 94.

### 8.9.1 Using the Voltage-Source Shift

Voltage sources are shifted through the circuit preferably into branches that are grounded. The sum of the sources in these branches form the supply voltages. Sources that cannot be shifted into a grounded branch become level-shifts. When two or more sources combine into a level-shift, they may cancel each other. So careful shifting of the sources may reduce the number of level-shifts. Also it is possible to place the level-shifts at the places where they cause the least inconvenience. When the type of one of the devices is changed, for example, an N-type FET is replaced by a complementary (in order not to change the small-signal behavior) P-type FET, the related bias sources change sign. This is a useful method to have floating voltage sources cancel each other, thus removing the need for a level-shift.

After completion of the Voltage-Source Shift, the voltage supply sources are found. The convenience of this method is that:

- the minimal supply voltage needed to achieve the required circuit performance is found.
- when a lower supply voltage is required than found here, it is possible to select which of the original bias sources can be reduced to achieve this and what the consequences on the performance are. So the designer can decide to give in on either noise, distortion or bandwidth since it is clear to what performance aspect each of the bias sources is related.
- it offers the possibility for an optimal selection of the type of each device (P or N) to find the minimally required supply voltage while maintaining performance.

Figure 8.27 shows the result of the shifting operation. To keep the figure simple, a source $V_{cc4}$ is introduced with:

$$V_{cc4} = V_{ce1} + V_{ce2} - V_{be3} + V_{ce3} - V_{be4} + V_{ce4} \tag{8.4}$$

A floating source can be seen in series with the output with a value $V_{cc4}$; it

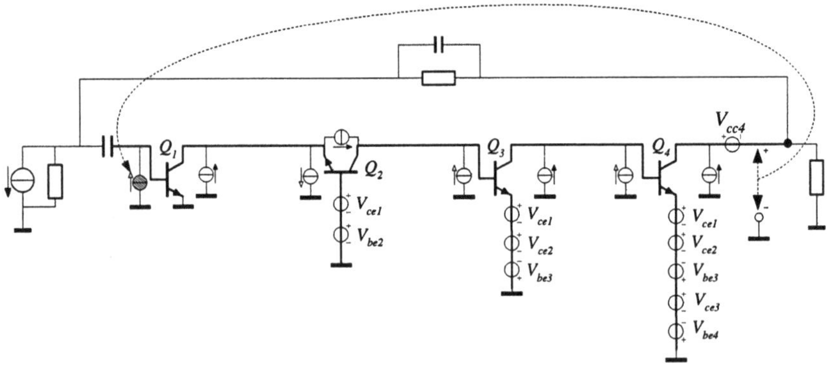


*Figure 8.27.* The result of the Voltage Source Shift.

is a level shift. In this case the simplest way to get rid of this floating source is to shift it through the voltage sensor. This results in a grounded source in series with the voltage sensor and a floating source in series with the output. The latter source can be replaced by a coupling capacitor. The node at the right side of this capacitor is a floating node, but no bias currents are injected into that node, so no control for the voltage is added. The definition of the voltage of the node is left to the resistors connected to it[6]. It is easy to calculate the settling time of the node voltage.

For this particular circuit the end result is shown in figure 8.28. For the supply voltages, apart from equation (8.4), the following equations hold:

$$V_{cc1} = V_{ce1} + V_{ce2} - V_{be3} + V_{ce3} - V_{be4} \tag{8.5}$$

$$V_{cc2} = V_{ce1} + V_{ce2} - V_{be3} \tag{8.6}$$

$$V_{cc3} = V_{ce1} + V_{be2} \tag{8.7}$$

## 8.9.2 Using the Current-Source Shift

To reduce the number of bias current sources and to make it more easy to implement them, first the Current Shift Transform is used to replace all floating

[6]The resistor as a "poor-mans" implementation of a voltage controlled current source: there *is* control at the node, albeit not a very explicit one.

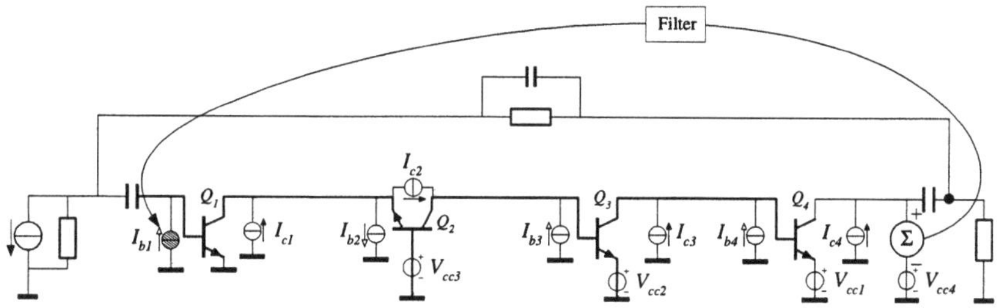


*Figure 8.28.* The end-result of the Voltage Source Shift and implementation of the level shift via a capacitor.

current sources by two sources that have one terminal connected ground. Figure 8.29 shows this transform. Two extra conditions, compared to the transform when used for noise, are introduced:

- the third node to which the new sources are connected is ground (figure 8.29a);
- a controlled source is introduced that symbolizes the fact that the two new current sources should be exactly equal. When they are not exactly equal, an offset current would flow, that is compensated by the controlled source (figure 8.29b).

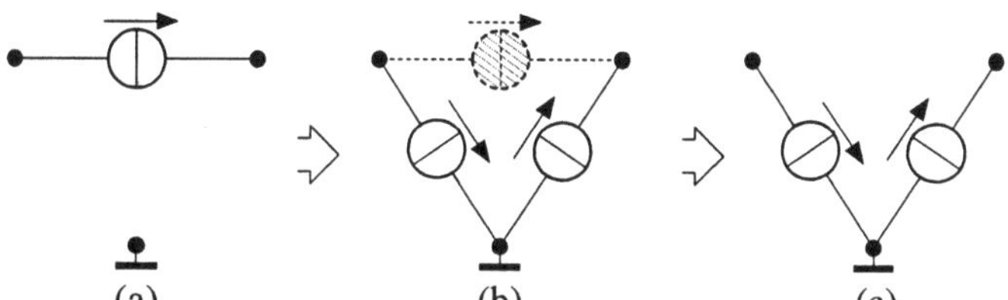


*Figure 8.29.* The Current Source Shift for bias sources.

The situation in figure 8.29b is rather "symbolic". In practice the matching error between the sources is small and the resulting offset current is easily dealt with in the circuit via the other controlling mechanisms and bias sources. However, it is good to be able to trace the origin. The conclusion is that the transformation result depicted in figure 8.29c is used, and the extra relation is just "remembered" by the designer.

Figure 8.30 shows the result of the transformation. In this case just current source $I_{c2}$ is transformed. The others already have one terminal connected to ground.

Parallel-connected current sources can be merged into one source, resulting in the circuit shown in figure 8.31. For the supply current sources holds:

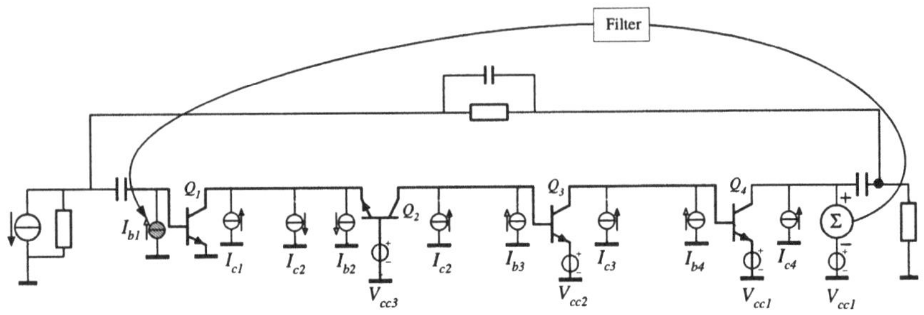


*Figure 8.30.* The results of the Current Source Shift.

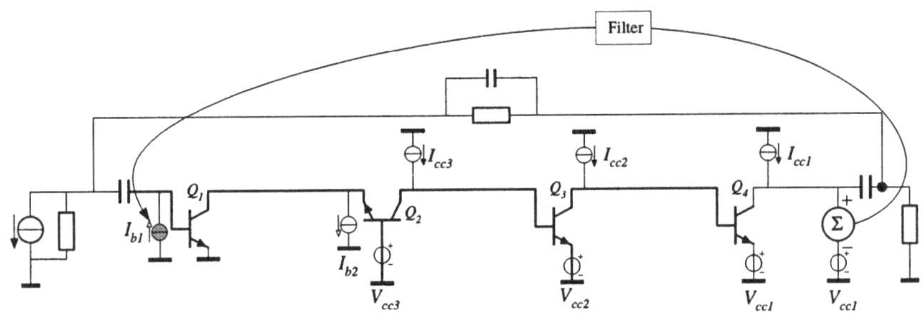


*Figure 8.31.* The circuit of figure 8.30 with merged current sources.

$$I_{cc1} = I_{c4} \tag{8.8}$$
$$I_{cc2} = I_{c3} + I_{b4} \tag{8.9}$$
$$I_{cc3} = I_{c2} + I_{b3} \tag{8.10}$$
$$I_{c1} - I_{c2} = 0 \tag{8.11}$$

It can be seen that the sources $I_{c1}$ and $I_{c2}$ cancel and just a small source $I_{b2}$ remains at that node. Frequently this source can be skipped, causing the bias current of $Q_1$ to increase a little bit. Usually this does not cause much biasing problems for $Q_1$.

### 8.9.3 Changing the device-type.

To minimize the required supply voltage, the type of one or more of the devices in the circuit can be changed from N-type to P-type or vice-versa. Figure 8.32 shows a cascoded FET. Both transistors are biased at the same drain current $I_d$. Using the Voltage Source Shift, two supply voltages are found. Both supply voltages are composed of two bias sources, each originating of a different transistor. This indicates that when the type of one of the transistors is changed from N-type to P-type, the sign of one of the sources would change. This could result in a lower supply voltage.

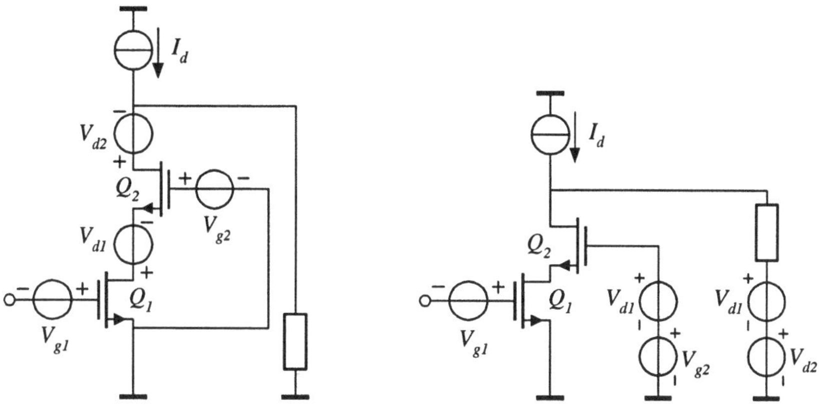


*Figure 8.32.* A biased cascode with two N-type devices, before and after the Voltage Source Shift.

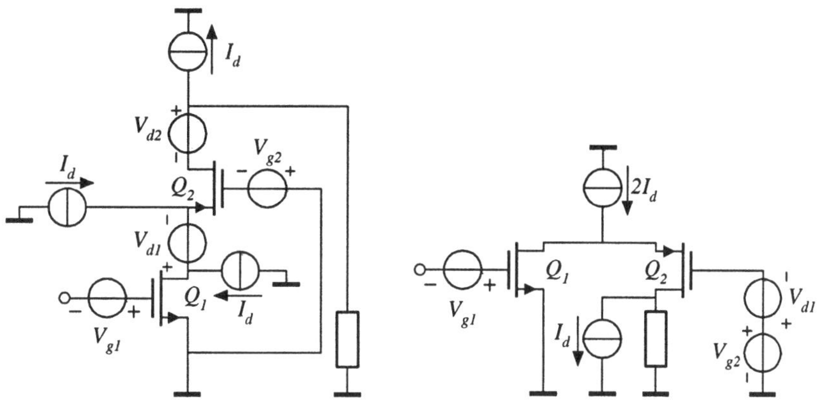


*Figure 8.33.* A biased cascode with one N-type and one P-type device.

Figure 8.33 shows the situation when $Q_2$ is made P-type. It can be seen that the sign of the sources $V_{g2}$ and $V_{d2}$ is changed. It can also be seen that two extra current sources appear with a value $I_d$, since also the sign of the bias current sources of $Q_2$ has changed. Now these sources do not cancel anymore. Using the transformations again results in the circuit at the right hand side. When $V_{d1}$ and $V_{d2}$ are chosen equal, they cancel. And also in the other branch the net supply voltage is lower. An extra current source with value $I_d$ is introduced and the value of the other current source is doubled.

From this it can be concluded that by swapping the device types, the supply voltage can be reduced at the expense of an increased current consumption and vice versa. *Note that the power consumption is not changed.* The product of the bias currents and voltages of the transistors remains constant. So, this method cannot be used for power reduction, but it can be efficiently used for supply voltage reduction without hampering the small-signal performance.

# 8.10 Biasing step 5: Implementation of the bias loops

## 8.10.1 Bias-loop filters

The purpose of the bias-loop filter is to separate the bias signals from the information carrying signal. Three different techniques exist that are frequently used to realize this:

- filtering in the frequency domain
- filtering in the common-mode/differential-mode domain
- filtering in the time domain

Each of these techniques will be discussed in the next sections.

## 8.10.2 Filtering in the frequency domain

When the bias and information signals each get their own part of the frequency spectrum, these two signals can not disturb each other. The bias voltages and currents are DC. To properly separate bias from signal, a frequency range, B, is reserved for biasing. The information carrying signals only are at frequencies beyond this band B (figure 8.34). In principle, the bias-signal bandwidth B is zero, however it is not possible to implement a filter this sharp. DC can never be a part of the information band.

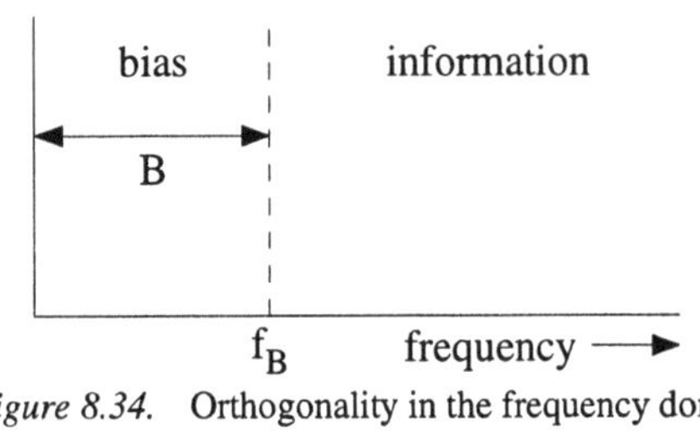


*Figure 8.34.* Orthogonality in the frequency domain.

## 8.10.3 Filtering in the common-mode/differential-mode domain

Filtering in the common-mode/differential-mode domain means making use of signal symmetry between different signal paths. Algebraic operations on the corresponding signals may yield either information of the bias signal only, or information of the information-carrying signal only. An often used method is the use of common-mode and differential-mode signals.

This technique is elucidated with an example of biasing a differential pair, figure 8.35. Due to the differential-mode signal source, $v_{dm}$, a current equal

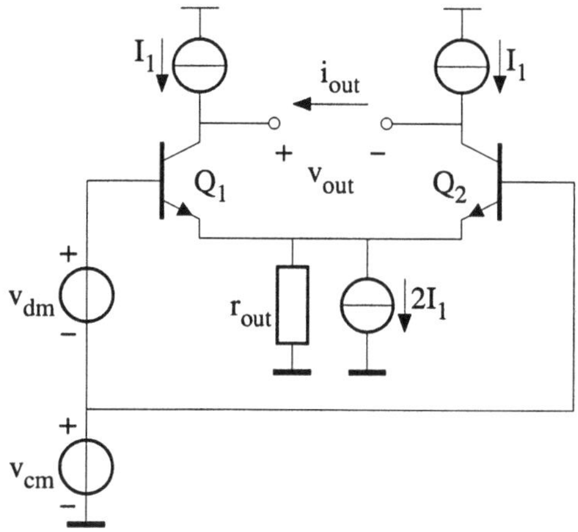


*Figure 8.35.* The biasing of a differential pair with the distinction of common-mode and differential-mode signals.

to $i_{dm} = 0.5g_m v_{dm}$ flows through the collector leads of both transistors, but in anti phase. In contrast, the common-mode source, $v_{cm}$, causes a current to flow in both collector leads equal to approximately $i_{cm} = 0.5v_{cm}/r_{out}$ with $r_{out}$ the output resistance of the tail-current source. The total collector currents thus equal:

$$\begin{aligned} I_{c1} &= +0.5g_m v_{dm} + 0.5v_{cm}/r_{out}, \\ I_{c2} &= -0.5g_m v_{dm} + 0.5v_{cm}/r_{out}. \end{aligned} \tag{8.12}$$

Each collector current has a common-mode and a differential-mode part. By combining both collector currents, the common-mode and the differential-mode signals can be extracted. Taking the *difference* of the two collector currents, the differential-mode output current is found:

$$i_{out_{dm}} = I_{c1} - I_{c2} = i_{dm} + i_{cm} + i_{dm} - i_{cm} = 2i_{dm} = g_m v_{dm}. \tag{8.13}$$

The *sum* of the two collector currents yields the common-mode output current:

$$i_{out_{cm}} = I_{c1} + I_{c2} = i_{dm} + i_{cm} - i_{dm} + i_{cm} = 2i_{cm} = v_{cm}/r_{out}. \tag{8.14}$$

For biasing purposes, mostly the bias signals are mapped on the common-mode signals, as the bias signals are often *in-phase* signals. Thus, to obtain the common-mode part an addition has to be done. Several very simple networks can do this addition. These will be discussed further in section 8.12.2 on page 283.

In balanced configurations nodes with a common-mode signal are inherently available and can be used for measuring the common-mode state of a circuit. In the differential pair of figure 8.35, the common-emitter node is such a common-mode node.

### 8.10.4 Filtering in the time domain

Time domain filtering is obtained when different, non-overlapping time slots are defined in which either the amplifier is used for signal processing or for biasing. Auto-zero technique is an example of this type of biasing. An amplifier using auto-zero technique is depicted in figure 8.36. In the first phase of the

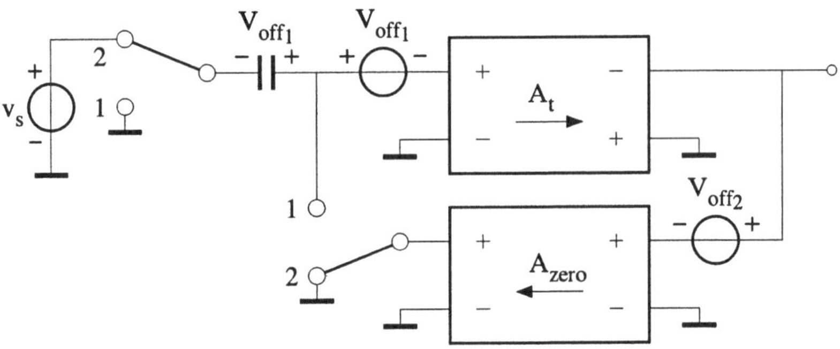


*Figure 8.36.* Biasing of an amplifier, $A_t$, with auto-zero technique.

auto-zeroing, the switches are in position 1: the signal source is disconnected and the output voltage of the amplifier $A_t$ is made zero by amplifier $A_{zero}$. In that case the input voltage of the amplifier equals the input offset voltage $V_{off_1}$. As the capacitor is connected now in parallel with the input of amplifier, $A_t$, the capacitor is loaded to a voltage equal to $V_{off_1}$. In the second clock phase the switches are in position 2: the signal source is connected and the output of the amplifier $A_{zero}$ is disconnected. Now the capacitor is in series with the input of amplifier $A_t$. The voltage on the capacitor is equal but in anti phase with the input offset voltage of the amplifier and the effective input offset voltage, the signal source experiences, is ideally zero.

However, when in phase 1 the output of amplifier is short circuited, the input offset voltage of amplifier $A_{zero}$, $V_{off_2}$, causes the output voltage of amplifier $A_t$ to be unequal to zero. An extra input offset voltage of $V_{off_2}/A_t$ is introduced at the input of amplifier $A_t$ and the capacitor is loaded to $V_{off_1} + V_{off_2}/A_t$. In phase 2 the extra input offset voltage caused by $V_{off_2}$ is not present anymore at the input of $A_t$. However, the capacitor was loaded to $V_{off_1} + V_{off_2}/A_t$ and thus the remaining input offset voltage is $V_{off_2}/A_t$. But still a large reduction is obtained.

The auto-zero technique can only be used when the offset voltage changes slowly with respect to the sampling rate of the input capacitor. Otherwise, this technique may worsen the offset voltage of the amplifier.

With respect to the other three biasing techniques, this technique is least influenced by mismatch of bias sources and the *only* technique where the separation is not made in the frequency domain so it is the only technique that can be used to implement a true DC amplifier. A further discussion of this technique and other time-domain techniques like chopper amplifiers is, however, beyond the scope of this book.

### 8.10.5 Frequency behavior of the bias loop

In chapter 6 and 7 the frequency behavior of the small-signal loop was discussed. It was found that to end up with the required frequency behavior, frequency compensation techniques have to be used.

During biasing, loops are introduced too. These loops can become unstable and consequently, the signal processing is terribly affected, see figure 8.37. Due to the unstable bias loop, the "small-signal amplifier" is periodically on

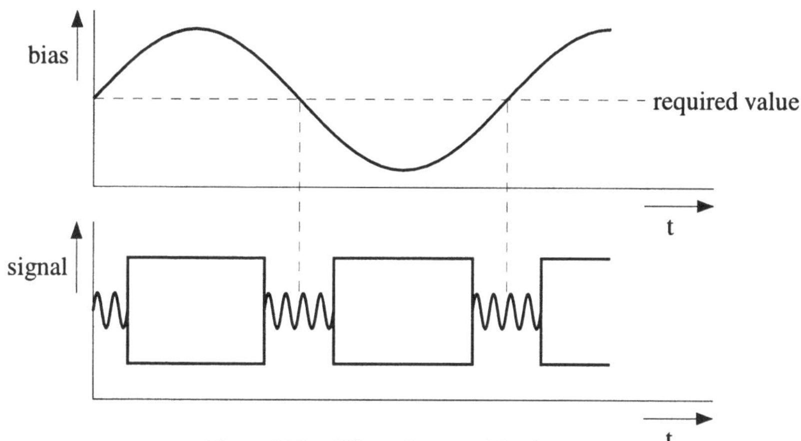


*Figure 8.37.* Effect of an unstable bias loop.

and off. When the bias is near the required value, the amplifier is able to do the small-signal processing. The rest of the time the amplifier is clipping and no signal processing can be done.

Mostly, the frequency behavior of the bias loop needs frequency compensation. The techniques introduced in chapter 6 and 7 apply to the bias loop too, but there is one major difference. For the small-signal loop the LP-product has to be as large as possible and frequency compensation techniques were not allowed to reduce the LP-product significantly. For the bias loop the bandwidth has to be below the information bandwidth such that the bias loop does not influence the small-signal behavior. The LP-product of the bias loop should predict a bandwidth that is *low* enough. This adds some new opportunities for frequency compensation.

For the frequency compensation of a bias loop the model of figure 8.38 can be used. The model represents a bias loop as an additional overall loop around the

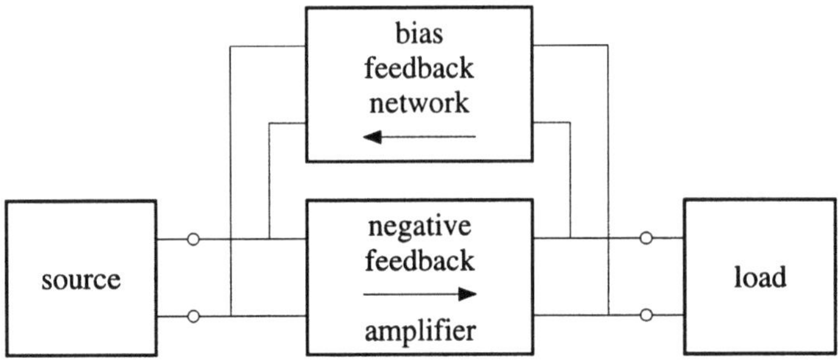


*Figure 8.38.* Model for the frequency compensation of a bias loop.

negative-feedback amplifier. The DC loop gain of the bias loop is determined by the DC transfer of the negative-feedback amplifier and the bias feedback network. Of course, the load and source impedance influence this DC loop gain when they are directly connected to the amplifier.

The poles and zeros of this loop are just the system poles (and zeros) of the negative-feedback amplifier and some relatively low-frequency poles (and zeros) introduced by the bias feedback network and, when used, by coupling capacitors. A pole-zero pattern of a bias loop may be as depicted in figure 8.39. The poles of the small-signal loop usually are non dominant. When this is not

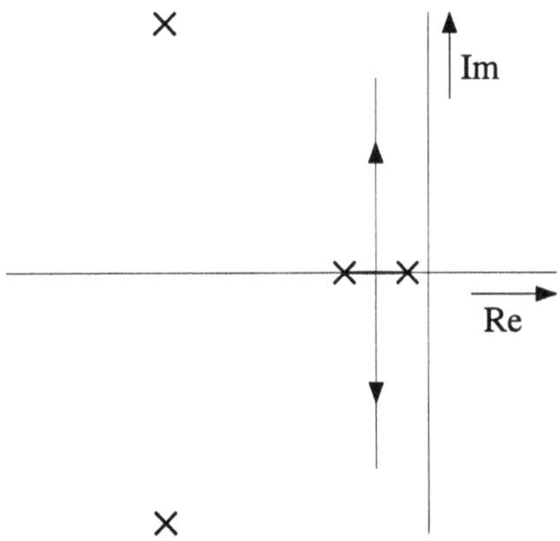


*Figure 8.39.* A pole-zero pattern and the root locus of a bias loop.

the case, they are influenced by the bias loop and a reduction of the LP-product of the bias loop is necessary.

The root locus of the bias loop is drawn in figure 8.39 too. The poles of the bias loop, that is not yet frequency compensated, can be high-Q poles or even poles in the right half plane. Two ways of frequency-compensation can be used:

- a subtle one, comparable to the method used in the small-signal design, using phantom zeros etc.
- a more "brute force" one, realizing a single dominant pole with a relatively high value capacitor. This method drastically reduces the LP-product of the bias loop.

The latter method generally requires larger capacitors than the more subtle methods and results in a high start-up time, but does not require much accuracy of the filter components. The first one requires more precision, but results in a much lower start-up time.

### 8.10.6 Example: A practical implementation of the loop filter

Figure 8.40 shows the example circuit of figure 8.30 in which the bias loop with the loop filter has been implemented in a simple way. The resistors $R_1$ and $R_2$ are chosen such that the base current of the first transistor generates the required voltage difference $V_{cc1} - V_{be1}$ across them. Capacitor $C_1$ takes care of the filtering and $R_{comp}$ is a small resistor that introduces the subtle frequency compensation (phantom zero) in the bias loop to keep it stable.

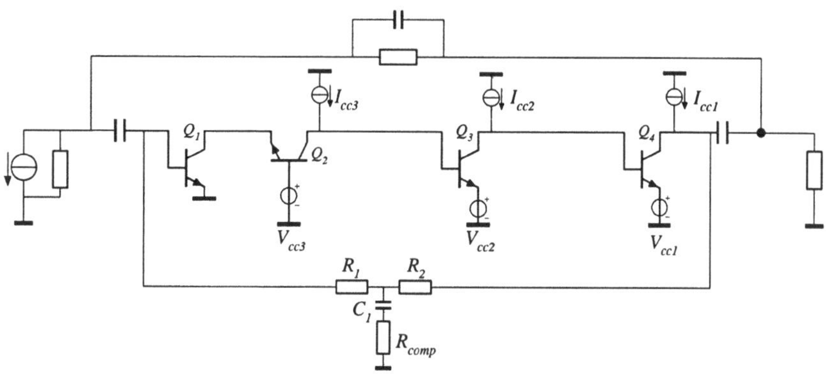


*Figure 8.40.* A biased circuit with a simple implementation of bias loop and filter.

This biasing method has two problems:

- it influences the small signal behavior because of loading effects.
- the accuracy depends on the accuracy of the base current of $Q_1$. This makes this method less reliable.

The resistors $R_1$ and $R_2$ load the circuit at the input and the output, so the filter influences the small-signal behavior. The values cannot be freely chosen because of the required voltage drop and the given value of the base current of $Q_1$. However, when by pure coincidence, the following equation holds:

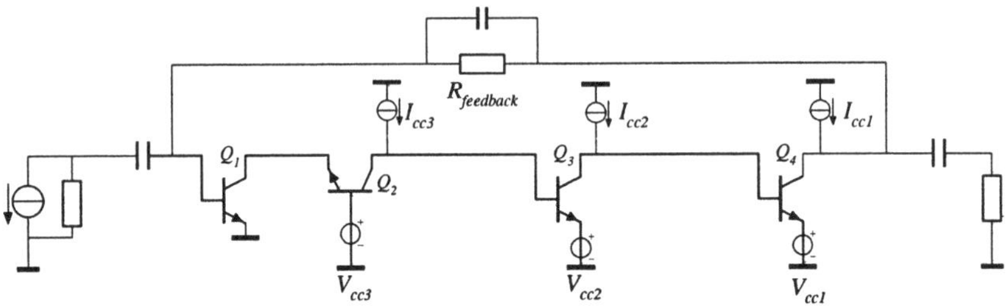


*Figure 8.41.* A biased with an even more simple implementation of bias loop and filter.

$$R_1 + R_2 \approx R_{feedback} \tag{8.15}$$

in which $R_{feedback}$ is the feedback resistor determining the small-signal transfer, the filter might be skipped and the circuit depicted in figure 8.41 results. Note that the feedback resistor is now connected to the other side of the coupling capacitor.

This method solves the loading problem, but does not solve the problem caused by the influence of the base current of $Q_1$. To solve this problem, the voltage measurement at the output must be done via a better voltage sensor. This is shown in figure 8.42. The base current of $Q_1$ does not play a role

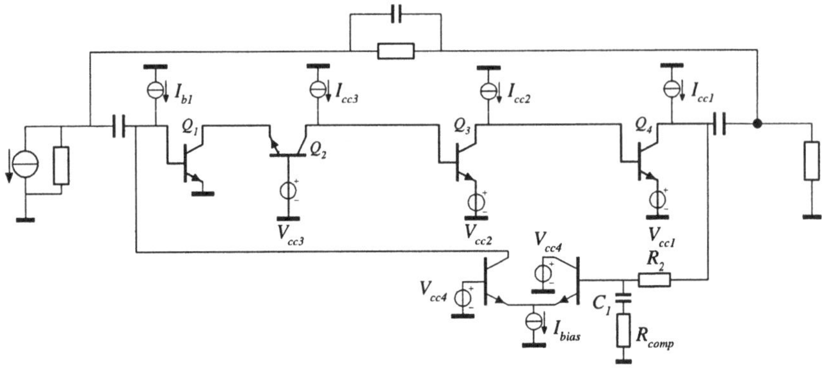


*Figure 8.42.* A biased with more robust implementation of the bias loop and filter.

anymore and the value of $R_2$ can be chosen more freely, so the loading effect can be minimized. Though the other options look very attractive because of their simplicity, probably the solution shown in figure 8.42 is preferred because of its reliability.

## 8.11 Biasing step 6: Implementation of the bias sources

Ideal voltage and current sources have been used for the biasing sofar. After finishing biasing step 4, via a simulator the signal behavior and the biasing can be verified, both should be correct. Subsequently, the ideal sources can be replaced by circuit implementations. These practical sources are likely to have some influence on the signal behavior of the amplifier via their finite output impedances, their non-linearity and their noise contributions. For bias sources at the input, the noise behavior is most likely effected and should be checked. A bias source at the output can easily influence the output capability of the amplifier and, of course, all sources can have a negative influence on the LP-product of the amplifier. Unfortunately, often a fine tuning of the frequency compensation is necessary after the implementation of the bias circuit. However, when the frequency compensation is systematically done, this does not have to cost a lot of time.

It is important to introduce the non-ideality of the bias sources gradually and one-by-one. For example, for current sources it is a good idea to first just place a capacitor in parallel to the ideal current source that models the output capacitance of the source. For voltage sources, in a similar way it is a good idea to, as a first step, put the expected impedance of the source in series with the ideal voltage source. When this first step shows satisfactory results, the practical circuits implementing the bias sources can be added.

In Chapter 9 more details are given on the practical implementation of bias sources.

## 8.12 Biasing differential amplifiers

In principle, this section could be very short. The biasing method described before will lead to a correctly biased circuit. It is not necessary to take into account that an amplifier is a differential amplifier, to obtain correct biasing. However, because of the symmetry in a differential circuit, some extra ways to do offset measurements are possible, and some more opportunities are found for skipping the control of controlled voltage sources. In this section, these extra options will be discussed. Also it will be discussed why even a differential amplifier cannot be a true DC amplifier, though many designers think it can be.

### 8.12.1 Offset measurements in differential amplifiers

Floating nodes have the property that they are only connected to the rest of the circuit via branches with a current-mode character. In differential amplifiers, many times a cut set of the circuit contains nodes that have well defined voltages between them, but all branches from the cut set to the rest of the circuit have a current mode character: a "floating node cluster".

Figure 8.43 shows an example of such a floating node cluster in a differential amplifier stage. The three nodes have a well defined voltage relation with

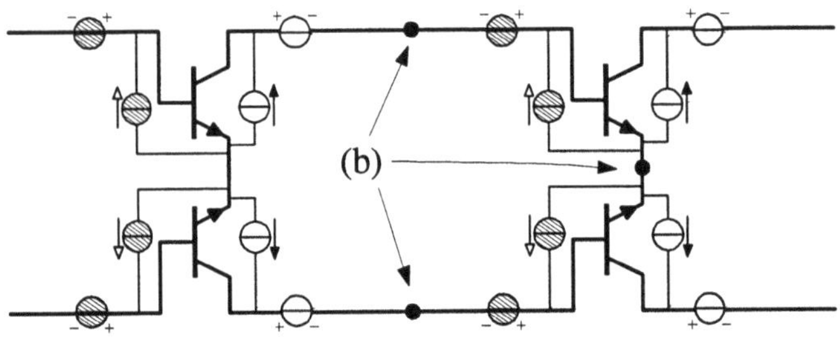


*Figure 8.43.* A floating node cluster (b).

respect to each other via the base-emitter junctions of the transistors, but do not have a well defined voltage relation to the rest of the circuit. Normally, a control loop that measures the voltage difference between one of the nodes of the cluster and an arbitrary other node in the circuit, not being a member of the same cluster, can control the collector currents with this information. In this way the voltage difference between the node cluster and the rest of the circuit can be well defined.

In a differential amplifier, usually complementary signal values are found on some of the nodes in a floating node cluster, whereas these nodes have equal bias values. Then, by summing the voltages at these nodes, the information carrying signals are filtered out and just the biasing information results, see figure 8.44. An adder sums the voltage of two nodes. No information carrying

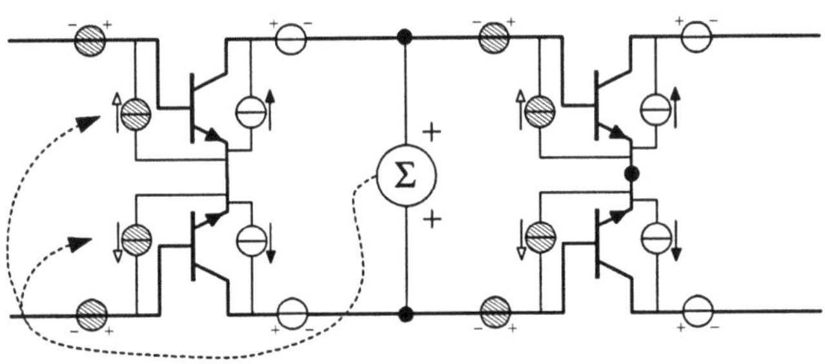


*Figure 8.44.* Measurement of the average voltage of a floating node cluster via an adder.

signal is found at the output of the adder, but just the bias voltage. So, the adder also acts as the loop filter that separates biasing from the information carrying signals. For the information carrying signals *this* bias loop is not effective; it does not change the small-signal transfer of the circuit.

The adder does not filter in the frequency domain, so one might think that this biasing method yields a true DC amplifier. However, there is a second loop that tends to be forgotten. There are *two* controlled current sources, controlled via *one* loop in figure 8.44. The second loop, is a loop that should control the differential DC offset current. This differential offset could be measured via a subtractor. Unfortunately, also the information carrying signals will be present at the output of this subtractor. Filtering in the *frequency domain* is the only method to separate the differential bias offset from the information carrying signals. Thanks to good matching properties of modern technology, the differential offset usually is so small that this second loop can be skipped. It is however not possible to distinguish a differential bias offset from a DC component of the information carrying signal. This is the reason why even a differential amplifier is not truly a DC amplifier.

In a similar way the control of the controlled voltage sources can be dealt with. Skipping the controls and setting the sources at their nominal value leads to a differential offset, that may be small as a result of the good matching properties of modern technology. However, also these bias offsets cannot be distinguished from DC components of the information carrying signal either.

### 8.12.2 Adders

Several very simple networks can be used to do the voltage addition of two floating nodes. The adders are connected in parallel with the signal output of a stage, so when they have a finite impedance they form an extra load in the small-signal circuit. The input impedance of an adder is therefore an important figure of merit. The other figure of merit is the quality of the addition, which means the amount of information carrying signal that "leaks" to its output. The most frequently used adders are depicted in figure 8.45. These adders are, in order of appearance starting at the top:

- the ideal adder;
- two normal resistors;
- two bipolar transistors;
- four diodes;
- four pinch resistors.

The last two adders are made of four elements in order to secure that always at both sides of the adder, at least one element (junction) is reverse biased, otherwise the input impedance of the adder would become too low. It should be noted that the used pinch resistor can be seen as a JFET of which the gate is connected to the source or drain.

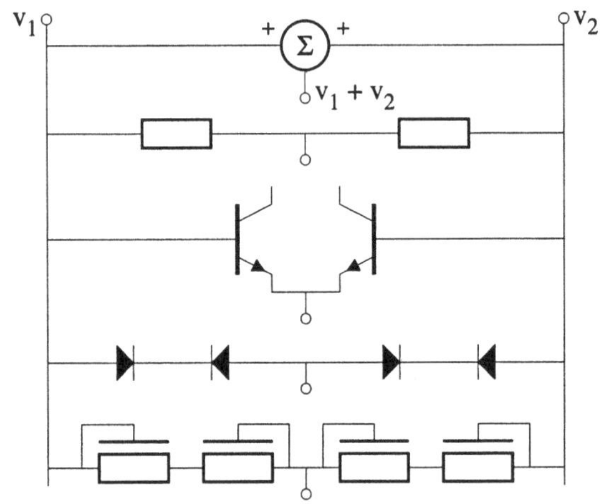


*Figure 8.45.* Simple networks which are able to perform an addition.

When the two sides of the adder are not symmetric, conversion from differential mode to common mode occurs, the two anti-phase differential-mode signals do not cancel completely anymore. Matching is important. Inspecting figure 8.44, it can be seen that the node to which the emitters of the right hand side differential pair are connected can be seen as the output of an adder that is coincidentally already available in the circuit. And indeed, very often this is a very useful adder. Only in the case when the concerning differential pair is driven by large signals, this may cause some asymmetric behavior due to the non-linearity of the transistors. Then the adder becomes "leaky" for the signal which causes interference of the bias loop with the signal behavior. It is up to the designer to decide whether this is tolerable or not. When the leakage is not tolerable, a separate adder has to be used. Then the extra loading is the side-effect that has to be evaluated.

### 8.12.3 Using a current mirror for biasing

A way to deal with floating nodes at the output of a differential pair, is the use of a current mirror. The configuration is depicted in figure 8.46. The two collector bias current sources are connected across the complete circuit. Now the collector of the left hand transistor is fixed at a well defined voltage via the mirror (when the emitters of the current mirror are at a well defined voltage of course), the collector at the right hand side is still floating, but in principle shows no offset current. The application of a current mirror can be a convenient way to bias a differential stage when a single sided signal output is desired.

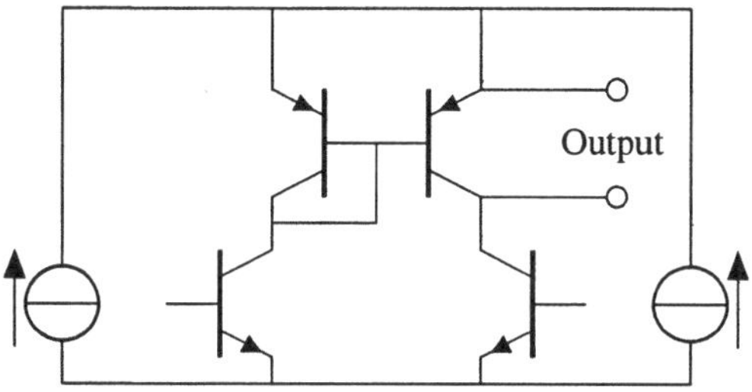


*Figure 8.46.* A current mirror used to bias a differential pair.

## 8.13 True DC amplifiers

DC amplifiers are actually amplifiers that can handle signals at very low frequencies. At these very low frequencies it may be difficult to implement the time constants for the bias filters. These may become very large. Then it is no longer possible to use frequency domain filtering. The two options that are left are filtering in the common-mode / differential-mode domain or in the time domain. Also advanced techniques can be used like chopping. For all these options it is necessary to know in advance, before the start of the small-signal design, that they are to be used, because it has an significant impact on the circuit topology, even for the small signal design. So, actually the very first design step that is taken is the decision on the way the bias filtering will be implemented.

## 8.14 Exercises

### Exercise 8.1

Given the transimpedance amplifier including the bias sources and the bias loop, as depicted in figure 8.47. The biasing loop is not performing as it should.

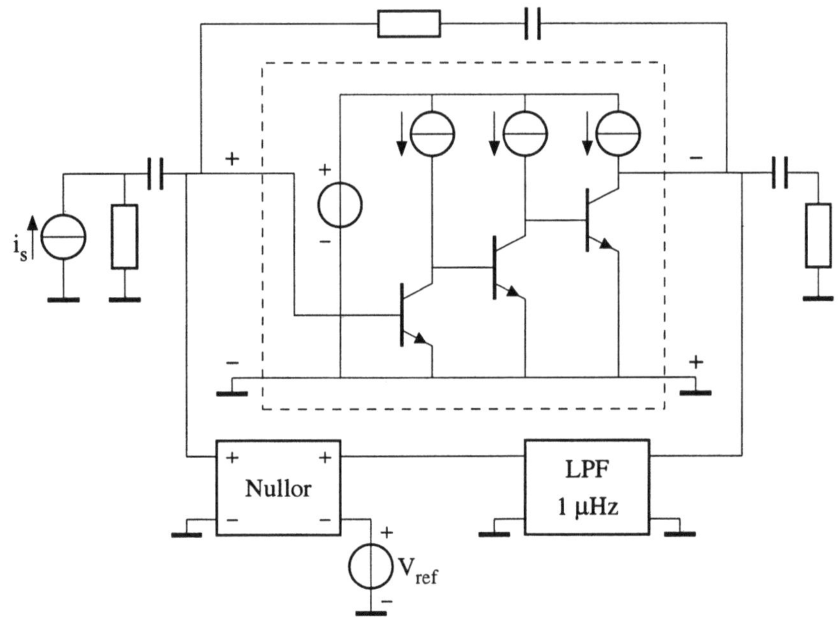


*Figure 8.47.* A transimpedance amplifier including bias sources and bias loop (LPF = Low-Pass Filter)

1. Explain why the loop is not functioning.
2. What has to be changed in order to get it functioning correctly?

## Exercise 8.2

Given the signal schematic of a transimpedance amplifier (figure 8.48).

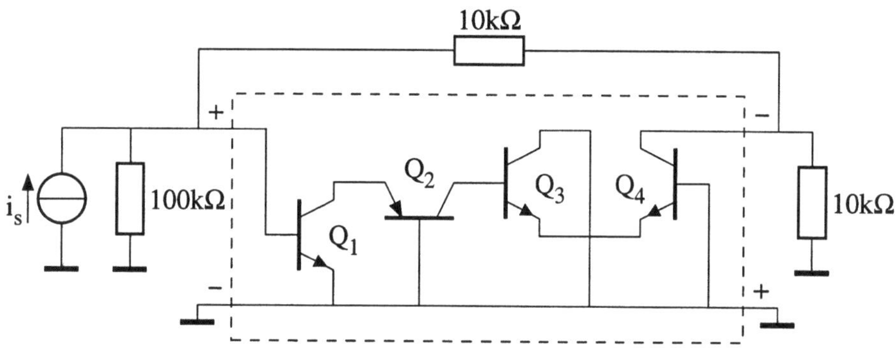


*Figure 8.48.* Signal schematic of a trans-impedance amplifier.

Which bias sources have to be added to get a correct biased circuit ?
Bias quantities:

- $I_{c1} = 100\ \mu\text{A}$
- $I_{c2} = 100\ \mu\text{A}$
- $I_{c3} = 1\ \text{mA}$
- $I_{c4} = 1\ \text{mA}$
- $\beta_{F,1-4} = 100$
- $V_{be,1-4} = 0.6\ \text{V}$
- $\hat{v}_{out} = 2.5\ \text{V}$

# 9

# DC SOURCES

## 9.1 Introduction

In the previous chapter, biasing has been done with ideal bias sources. In this chapter, some information will be presented on the practical implementation of these sources. Implementing the bias sources is the final design step that yields the amplifier that can be implemented completely with real-life components. So, this chapter discusses small circuits to implement:

- voltage sources
- current sources

The next section of this chapter starts with the description of the ideal voltage and current sources. Using this description, implementations of simple constant voltage sources are given. To obtain in-depth information on advanced sources like bandgap references etc., see reference [4, 8, 9]

The current sources are treated after the voltage sources, because in practice current sources are usually derived from a voltage source via a (trans)conductance. Implementations and non-ideal effects are discussed.

The bias sources supply the power the circuit needs to function. In a practical circuit a separation can be made into two classes of sources:

- Sources that truly supply the power. This power may be coming from chemical energy (batteries), coal, oil, gas or nuclear energy (via the wall outlet) etc. The number of sources like this is minimized. They are usually referred to as the "power supply" and tend to be voltage sources.[1]

[1]The easiest way to imagine current sources as true power supplies is probably to think of "magnetically charged" super-conducting inductors that are short-circuited in storage and have a circuit between their terminals when they are in use. This is not (yet?) a feasible option at normal temperatures.

- Sources that *transfer* power from the power supply to another part of the circuit where either the current or the voltage is kept at a fixed value. All of the current sources and most of the voltage sources in a circuit, especially the floating ones tend to be of this type.

A designer has to decide which one of the sources shall become the power supply, the true power provider. Especially in battery powered applications, this does not need to be the "standard choice". A designer should be aware of this freedom of choice. The ease with which the other sources can be derived from the power supply is an important factor to take into account too.[2]

## 9.2 Ideal voltage and current sources

The output signal of ideal sources is independent of the load, temperature and all other environmental influences and on top of that it is free of noise.

### 9.2.1 The ideal voltage source

In figure 9.1, the output signal versus the load current of the ideal voltage source is depicted in the V-I plane. As can be seen, the output voltage $V_{ref}$ is independent of the load current $I_{load}$. Since the output resistance of a volt-

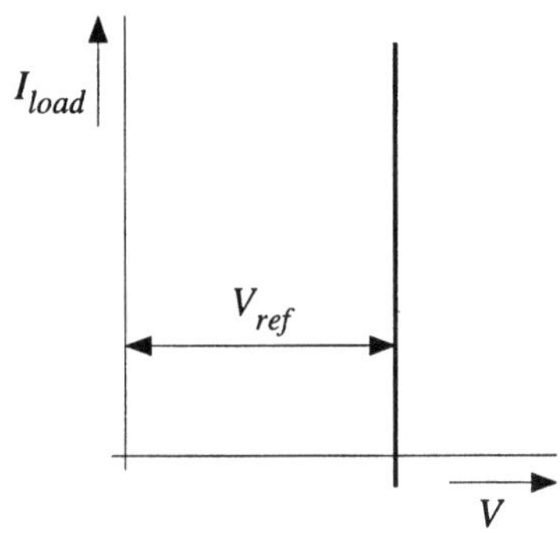


*Figure 9.1.* The output voltage $V_{ref}$ versus the load current $I_{load}$.

age source is defined as the ratio of output-voltage variation and load-current variation, the output resistance $r_{out}$ of the ideal voltage source equals zero:

$$r_{out} = \frac{dV_{ref}}{dI_{load}} = 0\,\Omega. \tag{9.1}$$

[2]An example of unorthodox placement of true power sources is the use of a small battery as a floating voltage source, a level shift. Especially when this level shift does not need to supply much power (e.g. a small lithium battery connected to a gate), this level shift may function well over the expected lifetime of the circuit. There even seems to be a possibility to integrate small batteries that can be used for this purpose [33]

All power supplied is concentrated at DC (0 Hz). The signal-to-noise ratio of the output voltage is infinite. The power-density spectrum $S_v$ of the noise voltage at the output equals:

$$S_v = 0 \text{ V}^2/\text{Hz}. \tag{9.2}$$

### 9.2.2 The ideal current source

In figure 9.2, the output current $I_{ref}$ of the ideal current source is depicted in the I-V plane as a function of the load voltage. The output current is independent

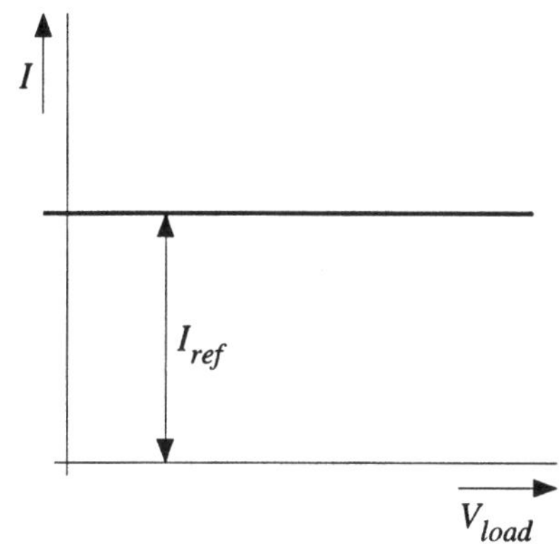


*Figure 9.2.* The output current $I_{ref}$ as a function of the load voltage $V_{load}$.

of this voltage. Therefore the output conductance $g_{out}$ of the ideal current source is zero:

$$g_{out} = \frac{dI_{out}}{dV_{load}} = 0 \text{ S}. \tag{9.3}$$

Just like the ideal voltage source, the power-density spectrum $S_i$ of the noise current at the output is zero:

$$S_i = 0 \text{ A}^2/\text{Hz}. \tag{9.4}$$

From this it can be seen that biasing with ideal sources does (should!) not influence the small-signal behavior of the designed amplifier at all.

## 9.3 Practical voltage sources

When the voltage source to be implemented is intended to be the power supply, there is not much design effort for the amplifier designer. It is just a matter of selecting the most appropriate one or of being forced to use one by, for example, "digital designers".

The work is in the voltage sources of the second type, that transfer power from the power supply to other parts of the circuit. Power-supply variations are not allowed to penetrate to the output of these voltage sources. Thus when the power supply only supplies the power in a "raw" form, the voltage source must transfer this power to another part of the circuit at a well-defined voltage, supplying any current that is demanded by the circuit. The figure of merit for this function is the Power Supply Rejection Ratio or PSRR for short. The PSRR is defined as:

$$PSRR \triangleq \frac{dV_{power}}{dV_{ref}} \tag{9.5}$$

with $V_{power}$ being the supply voltage. When this PSRR is not infinite, the source can be seen as an extra input to the amplifier. An input for trouble, unfortunately, not for the intended signal. So, when shifting the sources, (section 8.9) care has to be taken to shift the sources to the least sensitive places.

The output voltage of a realistic voltage source is depicted in figure 9.3. In

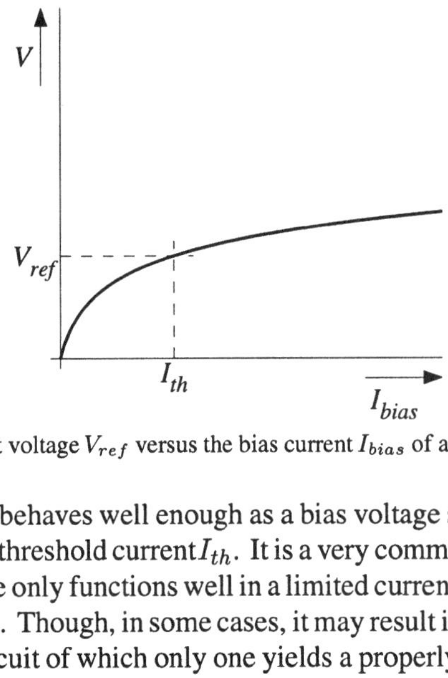


*Figure 9.3.* The output voltage $V_{ref}$ versus the bias current $I_{bias}$ of a realistic voltage source.

this case the source behaves well enough as a bias voltage source when the bias current is above the threshold current $I_{th}$. It is a very common phenomenon that a bias voltage source only functions well in a limited current range, so, generally this is not a problem. Though, in some cases, it may result in multiple bias-point solutions for the circuit of which only one yields a properly functioning circuit. Details on this can be found in [34].

Several implementations of voltage sources are discussed in the next section.

### 9.3.1 The resistive divider

The schematic of the voltage source implemented by a resistive divider is depicted in figure 9.4. The output voltage $V_{ref}$ is a fraction of the power-supply voltage $V_{CC}$:

$$V_{ref} = \frac{R_2}{R_1 + R_2} V_{CC}. \tag{9.6}$$

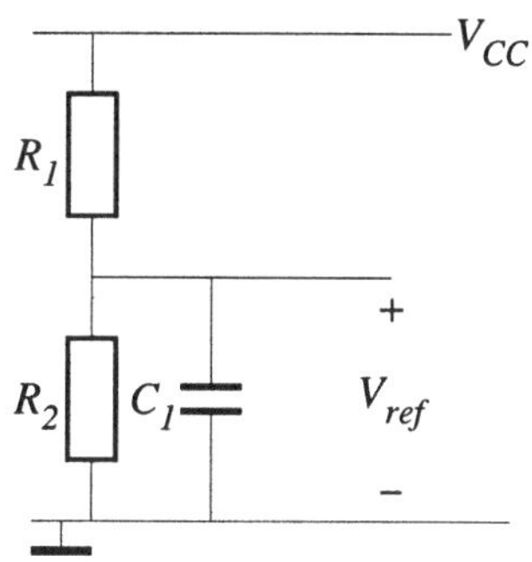


*Figure 9.4.* A voltage source implemented by a resistive divider.

The accuracy of this source is determined by the accuracy of the power-supply voltage and the matching of the two resistors.

The output resistance, $r_{out}$, of this source equals the parallel connection of the two resistors:

$$r_{out} = \frac{R_1 R_2}{R_1 + R_2}. \tag{9.7}$$

To obtain a low output resistance with this source, low resistances may have to be used. This may results in a too high current consumption. The relation becomes more clear when the total supply current, $I_{bias}$, is expressed as a function of the output resistor. The following relation is found:

$$I_{bias} = \frac{V_{CC}}{R_1 + R_2} = \frac{1}{r_{out}}\left(1 - \frac{V_{ref}}{V_{CC}}\right) V_{ref}. \tag{9.8}$$

Thus for a given output voltage and power-supply voltage, the bias current is inversely proportional to the output resistance.

For most amplifiers, DC is not a very interesting part of the frequency characteristic. So, different constraints may exist for the impedance level for biasing purposes and the impedance level in the signal band. The latter constraint may be more severe. Fortunately, in the signal band the impedance can be made lower without creating a higher power consumption with a capacitor. This is illustrated in figure 9.4 with capacitor $C_1$. Now $C_1$ determines the output impedance in the information band.

The output noise is determined by the thermal noise of the parallel connection of the two resistors. The power spectral density spectrum, $S_v$, of the noise voltage equals:

$$S_v = 4kT(R_1 \parallel R_2). \tag{9.9}$$

Again, for a low-noise behavior, low resistances and thus a high current consumption are required. With capacitor $C_1$ this power-density spectrum can be reduced in the information band.

The PSRR is determined by the resistors and is given by:

$$PSRR = 1 + \frac{R_1}{R_2}. \tag{9.10}$$

To obtain a high power supply rejection ratio, the ratio $R_1$ and $R_2$ should be large. However, for a given output voltage, this ratio is fixed. Again a capacitor, $C_1$ in figure 9.4, can solve this problem.
(A capacitor in parallel with $R_1$ would of course completely ruin the PSRR.)

### 9.3.2 The non-linear divider

To improve the performance of the source, a non-linear resistor can be used for $R_2$. The principle is depicted in figure 9.5. The non-linear function is the

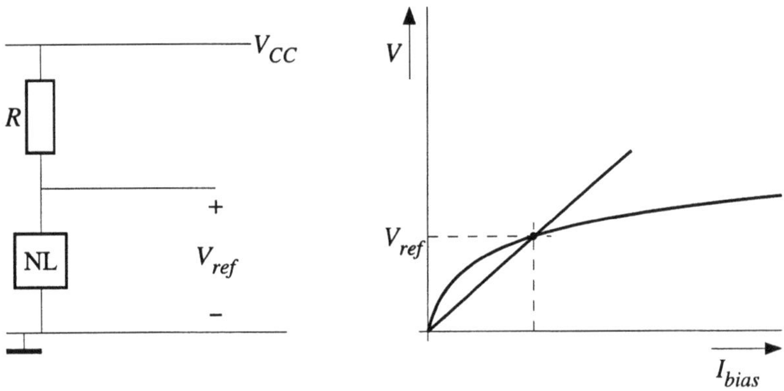


*Figure 9.5.* A voltage source using a non-linear resistor.

I-V characteristic of the non-linear device (NL), the linear one is of the resistor ($R$). The output voltage is given by the intersection point of the two functions.

With this non-linear device, the large-signal behavior (the generation of an output voltage) and the small-signal behavior (creation of a low output resistance) combine more efficient than in the linear case. The output resistance of the source is approximately the small-signal resistance of the non-linear device. This can be much lower than for the resistive divider.
The noise behavior of this source with respect to the linear divider with the same current consumption generally improves.

As seen in the previous section, the PSRR equation (9.10) increases when the value of $R_2$ decreases. For this source, $R_2$ is replaced by the small-signal resistance of the non-linear device. The PSRR will be considerably higher.

Various devices can be used for the non-linear part of the divider. Four examples are shown in figure 9.6. They are:

a. a diode-connected bipolar transistor;

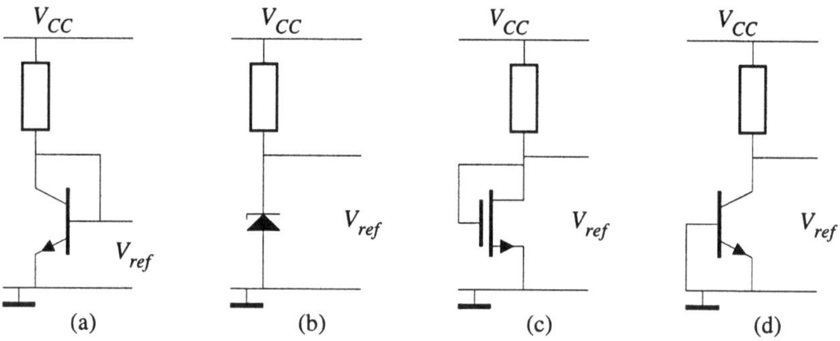


*Figure 9.6.* Four voltage sources using a non-linear device

b. a diode at reverse Breakdown;

c. a diode-connected normally-off FET;

d. a bipolar transistor used at punch-through.

#### 9.3.2.1 A diode-connected bipolar transistor

In figure 9.6a, the diode-connected bipolar transistor is used as the non-linear device. The output voltage equals the base-emitter voltage of the transistor that ranges, typically, from 0.5V to 0.9V, depending on the collector-current and the emitter size. For higher voltages, more diode-connected transistors in series can be used. The output resistance is approximately:

$$r_{out} \approx \frac{kT}{qI_{bias}} \tag{9.11}$$

with $I_{bias}$ being the collector bias current of the transistor. The higher the current, the lower the output resistance.

The noise performance of the base-emitter junction reference is found by transforming all the noise sources to the output. The noise performance is dominated by the thermal noise of the base resistance$r_b$ and the collector shot noise. The power-density spectrum of the noise voltage is approximately:

$$S_{V_{ref}} = 4kT\left(r_b + \frac{1}{2g_m}\right). \tag{9.12}$$

Reduction of the noise is possible by choosing a transistor with a lower base resistance (this can be done by taking several transistors in parallel) or by choosing a higher collector current. Which noise source is dominant depends on the specific circuit. In low-power circuits, mostly the collector shot noise is dominant and the noise of the base resistance is negligible.

For instance, a typical value for the base resistance of a minimal sized transistor is 500 Ω. When the transistor is biased at 1$\mu$A, the equivalent noise resistor at the output, representing the collector shot noise, equals 13 kΩ at room temperature. Thus the noise of the base resistance is negligible.

When low-noise voltage references are needed, the use of base-emitter junctions is the correct choice. This is easily seen when the noise of the base-emitter voltage reference is compared with the noise of a voltage source that is made with a resistor and a current source.

Example:
A reference of 600mV is made with a large transistor (for a high absolute accuracy). Assume 100$\mu$A is available for the biasing of the transistor and its base resistance is 150Ω. The total equivalent noise resistor equals 280Ω. This voltage can also be realized by a current flowing through a resistor. When the same bias current is used the required resistor equals:

$$R = \frac{U}{I} = \frac{0.6V}{100\mu A} = 6000\Omega. \tag{9.13}$$

Of course, the equivalent noise resistor is equal to this 6000Ω. The noise power of the voltage reference made with the base-emitter junction is more than a factor 20 lower than the one made with the resistor.

#### 9.3.2.2 A diode at reverse breakdown

A voltage source can also be realized with a reverse-biased diode at breakdown. Zener diodes are optimized for this use. The V-I characteristic of a Zener diode is plotted in figure 9.7. The forward behavior is comparable to that of a normal diode. In reverse mode, at a specific voltage, the current starts to increase rapidly. This specific voltage is called the reverse-breakdown voltage, $V_{br}$. Beyond this voltage the Zener diode behaves like a voltage source (c.f. figure 9.1).

The reverse breakdown is due to two distinct mechanisms, *avalanche multiplication*, which causes an avalanche breakdown and the *Zener effect*, which causes a Zener breakdown. Although in diodes any of the two mechanisms may be dominant, the diodes are always referred to as Zener diodes.

#### 9.3.2.3 Diode-connected normally-off FETs

When a normally-off FET is used as the non-linear component in the divider, as shown in figure 9.5, the output voltage is determined by the intersection point of $V_{GS}$ versus $I_R$ and $V_R$ versus $I_R$ (see figure 9.8).

The output resistance, $r_{out}$, of this source equals approximately:

$$r_{out} = \frac{1}{g_m} \tag{9.14}$$

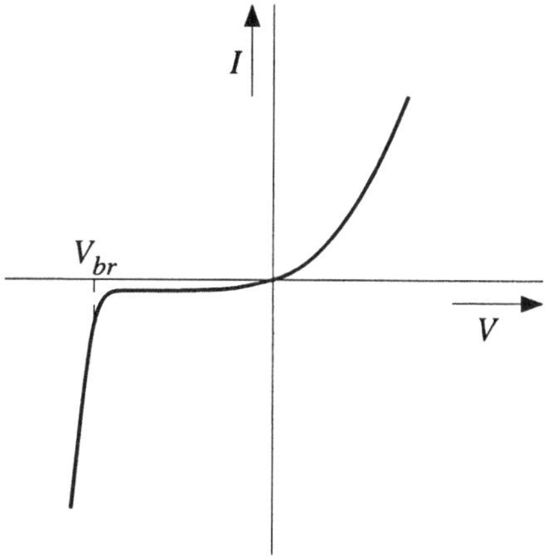


*Figure 9.7.* The V-I characteristic of a Zener diode.

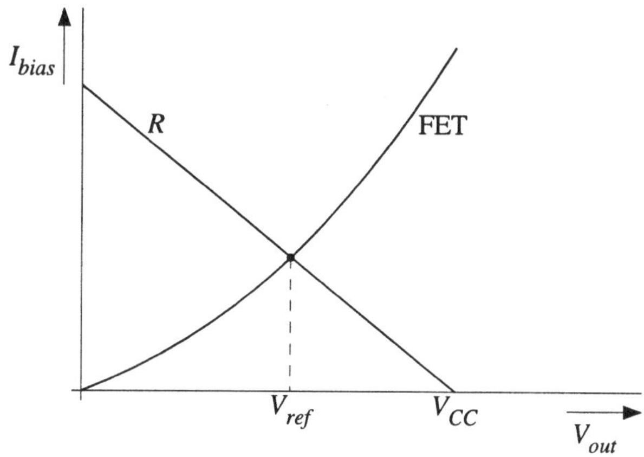


*Figure 9.8.* The output voltage of the voltage source using a normally-off FET.

with $g_m$ the transconductance of the FET. For lower output resistances, again more current is required. The noise of this source is due to the thermal noise of the resistor and the drain noise of the FET. The PSRR equals:

$$PSRR = \frac{1}{1 + g_m R}. \tag{9.15}$$

#### 9.3.2.4 A transistor used at punch-through

When a junction is biased in reverse mode, the depletion layer becomes wider for higher reverse voltages. In a transistor, the base-collector junction is mostly reverse biased. The depletion layer of the base-collector junction reduces the effective base width (modelled by the forward Early voltage). When the reverse voltage is increased such that the depletion layer of the base-collector junction touches the depletion layer of the base-emitter junction, the effective base width is reduced to zero. An electric field now exists across this depleted area and transports every carrier that enters the region to the other side. Similar to the current through a collector-base depletion layer, the field cannot influence the number of carriers transported. The current is determined by the supply of carriers at the depletion layer boundaries. Since the emitter is highly doped, the current can become very large and an external current-limiting resistor has to be connected in series with the collector lead (see figure 9.6). For an increasing current, the voltage across the resistor increases, consequently, the voltage across the transistor decreases. At the biasing point, the voltage is such that the base-emitter and base-collector depletion regions just touch each other.

The output voltage of this circuit is indirectly determined by the number of available carriers. A small increase in the reverse base-collector voltage results in a very large increase in the output current. Thus, the output resistance of this source is very low.

### 9.3.3 Floating voltage sources

Floating voltage sources (level shifts) can be easily made in a branch that already supports a DC bias current. Placing a resistor or a non-linear element in this branch can result in the required voltage drop. All design considerations described above also apply for these floating sources. Only the biasing resistor ($R_1$ in figure 9.4) is omitted.

### 9.3.4 Conclusions on voltage sources

The simplest implementation derives a voltage from the supply voltage by means of a resistive divider, or by placing a resistor in a branch that already supports a DC bias current.

Using a non-linear device produces a better performance. The forward-biased junction has shown to be a very good candidate for this. A low-noise behavior combined with a low output resistance is feasible.

## 9.4 Current sources

Current sources are all derived from a voltage reference with the aid of a (trans)conductance. This is depicted in figure 9.9. When the voltage source is assumed to be ideal, the power-density spectrum $S_i$ of the noise current at the output equals:

$$S_i = \frac{4kT}{R} \tag{9.16}$$

and can in the limit be zero when a source with an infinitely high voltage is used such that an infinitely high resistance can be used. In practice, the noise may be

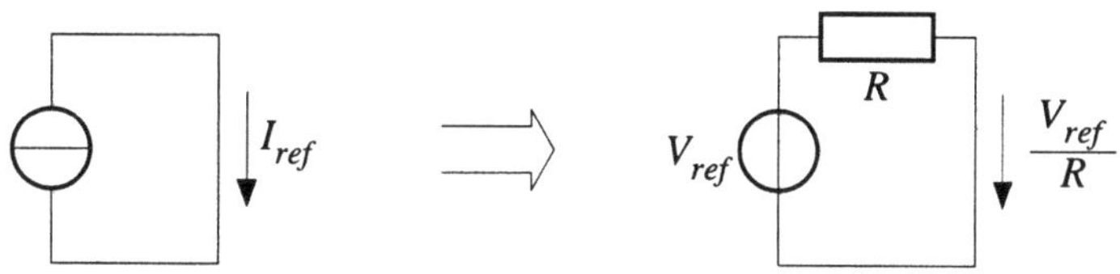


*Figure 9.9.* Derivation of a current source from a voltage source.

worse than given in equation (9.16). This is because a current flowing through a resistor can cause 1/f noise due to the granular structure of the resistive material.

When both a low output conductance and a high output current are needed, this type of current source gives problems because high reference voltages are required. Then it is better to realize a low output conductance with an active circuit.

### 9.4.1 An active current source using a transconductance

The basic configuration of an active current source is given in figure 9.10 in which a nullor is used. The current is again given by:

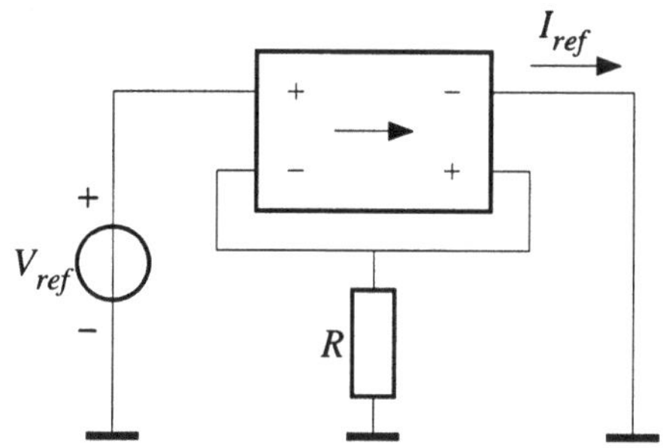


*Figure 9.10.* A current source realized with a transconductance amplifier and a reference voltage.

$$I_{ref} = \frac{V_{ref}}{R}. \tag{9.17}$$

Actually this is just a very "normal" transconductance amplifier, with a DC signal as input, so, all properties discussed in the previous chapters also apply for this circuit. With the nullor in place, the properties are ideal. So, the output conductance is zero and independent of $R$. The output noise is just the noise produced by the resistor.

The difference between the amplifiers discussed before and this one is the effort to keep the circuit that implements the nullor as simple as possible; to restrict it to just one transistor. When this does not yield the desired performance, all techniques discussed in the previous chapters may be used to improve it.

#### 9.4.1.1 Using a bipolar transistor as nullor

The simplest nullor implementation, a CE stage, results in the series stage of figure 9.11. The output current is given by (the base current is ignored):

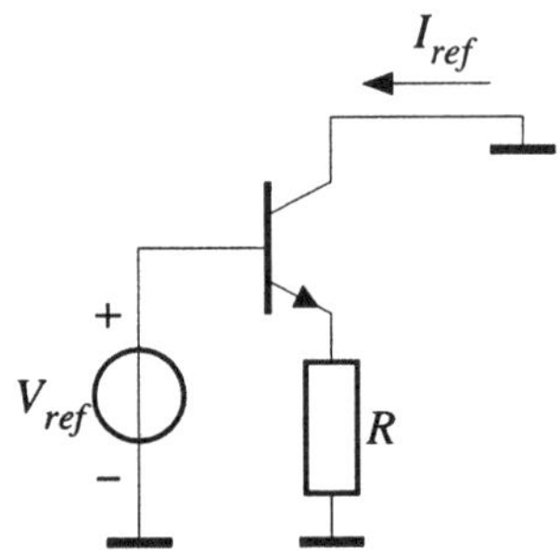


*Figure 9.11.* The nullor implemented with a single CE stage.

$$I_{out} = \frac{V_{ref} - V_{be}}{R}. \tag{9.18}$$

As the base-emitter voltage is temperature dependent, the output current of the source is temperature dependent too. The output resistance (being $1/g_{out}$) equals:

$$r_{out} = r_o + R + R\frac{\beta_f r_o - R}{r_b + r_\pi + R}. \tag{9.19}$$

This function is plotted in figure 9.12. The output resistance is normalized to $r_o$ and the feedback resistor $R$ to $r_\pi$. For values of $R$ much larger than $r_\pi$ plus $r_b$, the expression for the output resistances reduces to:

$$r_{out} = (\beta_f + 1)r_o. \tag{9.20}$$

Usually, this results in very high values for $R$. For 70% of the maximum output resistance, $R$ needs to be approximately 2 times larger than $r_\pi$. Then, the voltage across the feedback resistor equals:

$$V_R = I \cdot R = I \cdot 2r_\pi = I \cdot 2\beta\frac{kT}{qI} = 2\beta\frac{kT}{q} \tag{9.21}$$

For $\beta = 100$ and at room temperature, the voltage across the resistor is about 5V! This 5V is in strong contrast with modern supply voltages. Another limitation is given by the required resistance. In low-current applications, the output current can easily be in the order of $nA$s or $\mu$As, which is demanding for feedback resistors in the order of MΩs and GΩs.

Increasing the output resistance, without the need for very high resistances, can be realized in two ways:

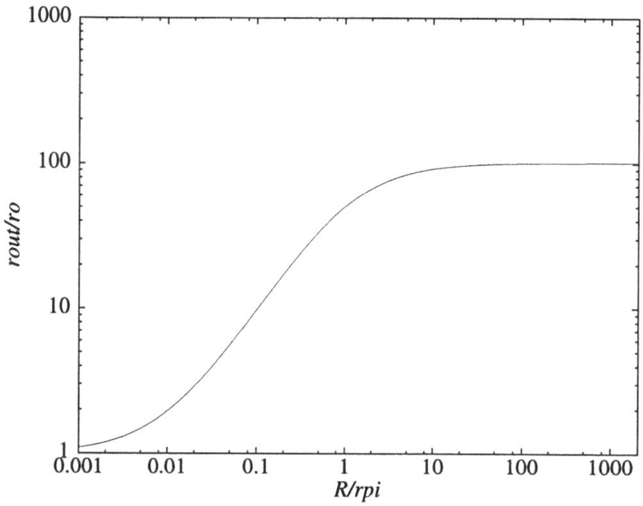


*Figure 9.12.* The normalized output resistance as a function of the normalized feedback resistor for $\beta$=100.

- increasing the loop gain ($\beta + 1$), a better approximation must be made for the nullor;
- increasing the output resistance of the active part ($r_o$).

For the first option, the next step is a two-stage nullor implementation. The strategy to follow is similar to that for normal amplifier design. The output resistance will increase by a factor $\beta$. The latter option is obtained when the CE stage is cascaded with a CB-stage, see figure 9.13. Without the resistor, R=0Ω, the output resistance of the transistor combination stage is approximately $\beta r_o$. For larger values of $R$, it can increase to approximately $\beta^2 r_o$. When a still higher output resistance is required, the loop gain has to be increased. An additional CB-stage does not increase the output resistance any further.

#### 9.4.1.2 Using FETs

The current source with one FET is depicted in figure 9.14. The output resistance equals:

$$r_{out} = R + (1 + g_m R) r_d \tag{9.22}$$

with $r_d$ the small-signal output resistance of the FET and $g_m$ its transconductance factor. In contrast with the output resistance of the bipolar source, which is limited to $(\beta_f + 1)r_o$, the output resistance of the FET source approaches infinity for a feedback resistor approaching infinity. But, again, high resistances are required. To increase the output resistance, without the need for high re-

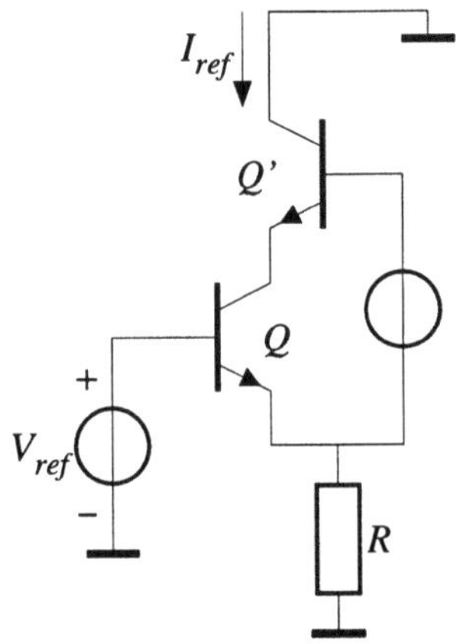


*Figure 9.13.* An active current source cascaded with a CB-stage.

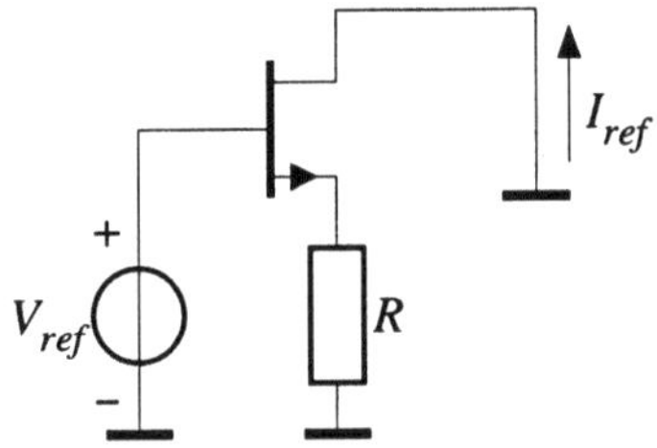


*Figure 9.14.* An active current source with a FET.

sistances, the same options as described for the bipolar version are open. An example of the second option is shown in figure 9.15. The output resistance of

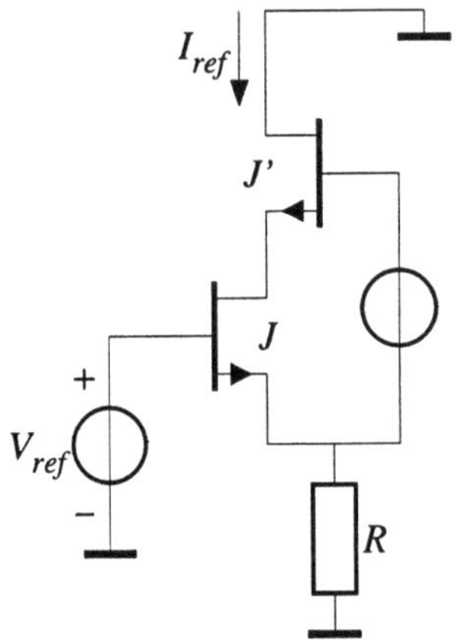


*Figure 9.15.* A FET current source cascaded with a CG-stage.

the CS-CG combination with the resistor set to R=0Ω is:

$$r_{d,cascode} = g_m r_d \cdot r_d = \mu r_d. \tag{9.23}$$

The output resistance is increased by a factor equal to the voltage-gain factor of the FET. For each additional CG-stage, the output resistance increases a factor $\mu$, in contrast to the bipolar implementation where a second CB-stage does not increase the output resistance any further.

The difference in the behavior of the current source made by the bipolar transistor and the FET is caused by the nature of the effect that causes the finite output resistance of these devices. More details on this can be found in [21].

#### 9.4.1.3 Noise behavior

The current source with all its noise sources is shown in figure 9.16. The

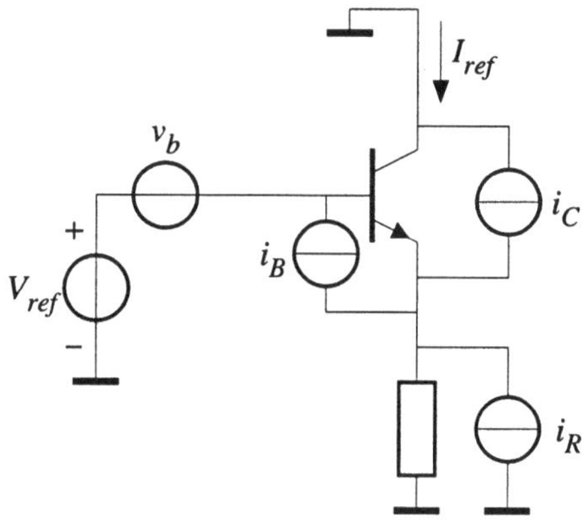


*Figure 9.16.* The noise sources in the active bipolar current source.

power-density spectrum of the noise sources is given by:

$$S_{v_b} = 4kTr_b \tag{9.24}$$
$$S_{i_B} = 2qI_B \tag{9.25}$$
$$S_{i_C} = 2qI_C \tag{9.26}$$
$$S_{i_R} = \frac{4kT}{R} \tag{9.27}$$

The power-density spectrum of the the equivalent noise current at the output, $S_i$, is given by:

$$S_i = \frac{2qI_C}{(1+\beta R/r_\pi)^2} + \frac{2qI_B}{(1+r_\pi/\beta R)^2} + \frac{4kT(R+r_b)}{(R+r_\pi/\beta)^2}. \tag{9.28}$$

The first term represents the effect of the collector shot noise. For very high resistor values, this term disappears completely. The second term is due to the shot noise of the base current. For very high values of $R$ this term reduces to $2qI_B$. The last term accounts for the noise of the feedback resistor and the base resistance. This term vanishes for high values of the resistor. The expression is plotted for a varying resistor value in figure 9.17. The noise-power density has been normalized to $2qI_C$ and the feedback resistor has been normalized to $r_\pi$.

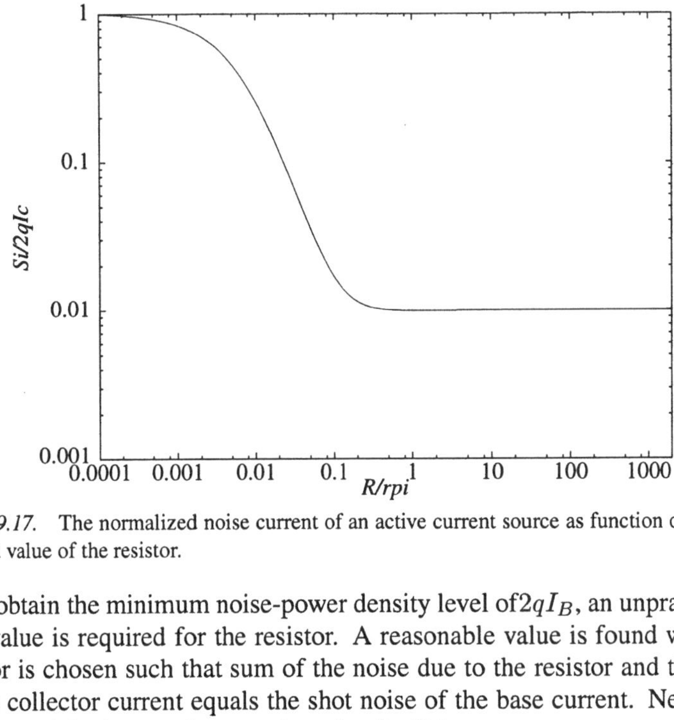


*Figure 9.17.* The normalized noise current of an active current source as function of the normalized value of the resistor.

To obtain the minimum noise-power density level of $2qI_B$, an unpractically high value is required for the resistor. A reasonable value is found when the resistor is chosen such that sum of the noise due to the resistor and the noise due to collector current equals the shot noise of the base current. Neglecting the noise of the base resistance, the value for $R$ is:

$$R = \frac{2\beta_f}{g_m}. \tag{9.29}$$

With $\beta_f = 100$ and $kT/q \approx 26\text{mV}$ the voltage across $R$ equals:

$$V_R = I_C R = 2\beta \frac{kT}{q} \approx 5\text{V} \tag{9.30}$$

This is the same value as found with the calculations done for the output resistance as a function of the feedback resistors, see page 300. Apparently, high-quality current sources are in all respects difficult to implement for modern supply voltages. More power will have to be spent in low-voltage circuits to make up for the degraded performance of the current sources. *It is always wise to check whether the use of a DC-DC converter, including its inefficiency,*

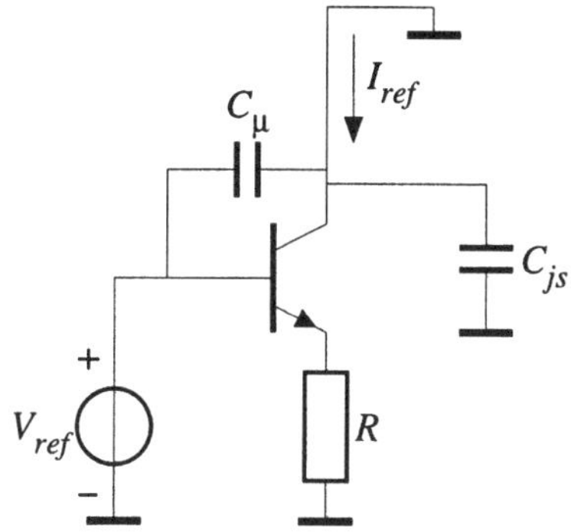


*Figure 9.18.* The parasitics of the active current source.

*to increase the supply voltage leads to a net decrease in the total power consumption.*

#### 9.4.1.4 HF behavior

The output impedance of the current source has to remain high in the complete information band. So, there are the same bandwidth considerations as there are for the amplifier itself even though the source is "just processing a DC signal". Figure 9.18 shows the source with its parasitic capacitances. The capacitor $C_{js}$ is only found for transistors in an integrated circuit and can be separately modelled in simulators like PSPICE. Unfortunately it is directly connected to ground. The output impedance is given by:

$$z_{out}(j\omega) = \frac{r_{out}}{1 + j\omega r_{out}(C_\mu + C_{js})} \tag{9.31}$$

and has a pole at

$$p = \frac{-1}{r_{out}(C_\mu + C_{js})}. \tag{9.32}$$

For frequencies beyond $1/[2\pi r_{out}(C_\mu + C_{js})]$ the output impedance is dominated by the parallel capacitance $(C_\mu + C_{js})$. The effectiveness of cascading the current source with a CB-stage, at relatively high frequencies, depends on the values of $C_\mu$ and $C_{js}$. Note that the CB-stage might introduce a capacitor $C_{js}$ directly connected to ground too. Four different situations are depicted in figure 9.19. Curve 1 shows the output impedance of a single CE stage. Curve 2 shows the output impedance of the source when the resistor in series with the emitter is used. The low-frequency impedance is increased. The high-frequency impedance is not increased. Both capacitors $(C_\mu + C_{js})$, are still in parallel with the high output impedance of the series stage (see figure 9.18). Curve 3 shows the output impedance of the source including the resistor, cascaded with

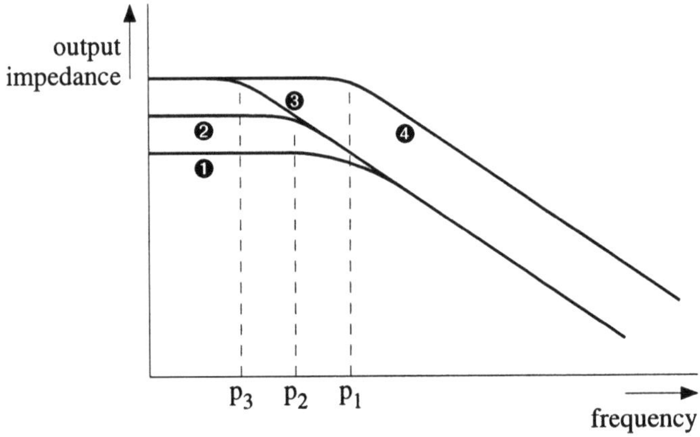


*Figure 9.19.* The effect of a cascading CB-stage on the high-frequency output impedance: 1) without resistor and CB-stage, 2) with a resistor but without CB-stage, 3) with both a resistor and a CB-stage, 4) similar to 3 but without substrate capacitors.

a CB-stage. The low-frequency output impedance is increased any further. The high-frequency output impedance is again not affected. The substrate capacitor of the CB-stage, which is directly connected to ground, is now in parallel with the output of the source (It is assumed that $C_{js}$ is much larger than $C_{\mu}$). Curve 4 shows the output impedance in a case similar to 3) but without substrate capacitors. The high-frequency impedance is drastically increased in the same way it is expected for any "normal" amplifier. The influence of the substrate capacitor can be reduced much in modern IC-processes. Of course, it is a special point of attention for every IC-designer.

When all the measures are taken, the output impedance may still be too low at (very) high frequencies. A very straightforward method used in HF design is putting a resistor (or even an inductor) in series with the current source. The output impedance is now, at high frequencies, dominated by this resistor instead of by the parasitic capacitance.

### 9.4.2 The peaking current source

The peaking current source is a special type of current source. It can be an efficient circuit to reduce the influence of the power-supply voltage on the output current. So, it might be considered in cases where a high PSRR is required. The circuit is shown in figure 9.20. The relation between $I_1$ and $I_{ref}$ is given by:

$$\ln\left(\frac{I_{ref}}{I_{S2}}\right) = \ln\left(\frac{I_1}{I_{S1}}\right) - \frac{I_1 R_1}{\frac{kT}{q}} \tag{9.33}$$

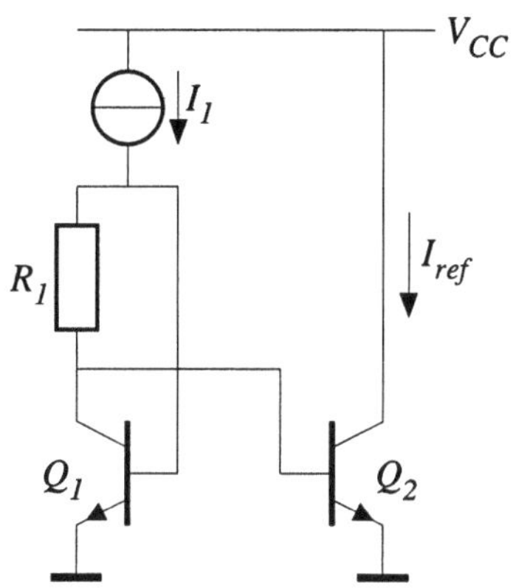


*Figure 9.20.* The peaking current source.

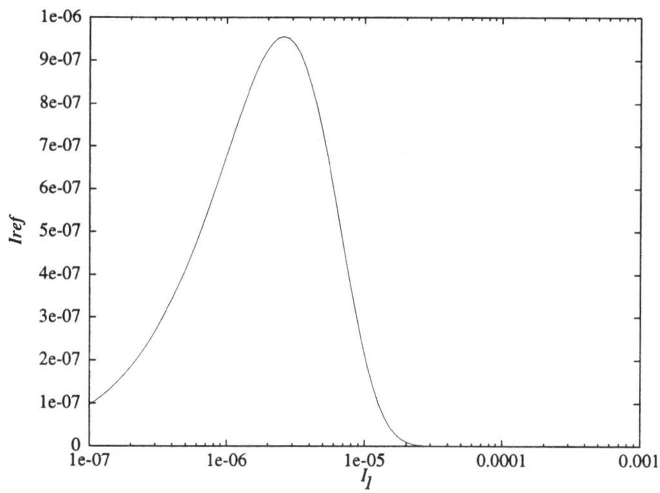


*Figure 9.21.* The relation between the input and output current ($I_1$ and $I_{ref}$) of the peaking current source.

or

$$I_{ref} = I_1 \left( \frac{I_{S2}}{I_{S1}} \right) \exp \left( -\frac{I_1 R_1}{\frac{kT}{q}} \right) \tag{9.34}$$

with $I_{S1}$ and $I_{S2}$ the saturation currents of $Q_1$ and $Q_2$, respectively. The function is depicted in figure 9.21 with $\frac{kT}{q} = 26$ mV, $I_{S1}/I_{S2} = 1$ and $R_1 = 10\,\text{k}\Omega$. The curve shows a peak, responsible for the name "peaking current source". At this extreme, a variation in $I_1$ is not transferred to $I_2$. The source

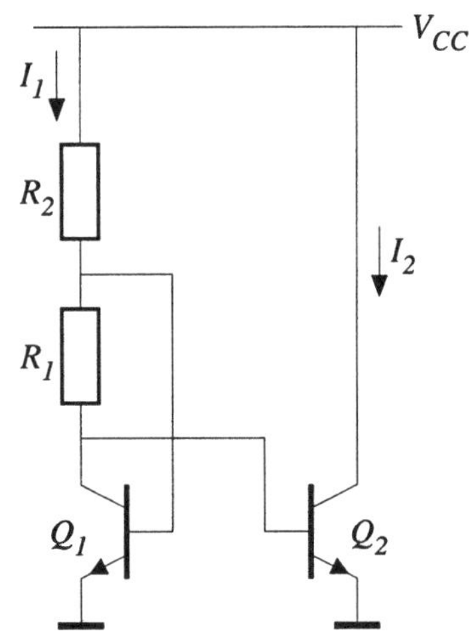


*Figure 9.22.* An example of a peaking current source.

is biased at this extreme when:

$$RI_1 = \frac{kT}{q}. \tag{9.35}$$

When the voltage across $R$ is equal to the thermal voltage, a change in the current $I_1$ is totally suppressed in the output current. At this extreme, the ratio between the input and output current is given by:

$$\frac{I_2}{I_1} = \frac{1}{\mathrm{e}} \left( \frac{I_{S1}}{I_{S2}} \right). \tag{9.36}$$

An example of a peaking current source of 100 $\mu$A is given in figure 9.22. The current in the left branch, $I_1$, must be 272 $\mu$A. At 300 K, $R_1$ must have a value of 95 $\Omega$, equation (9.35). The current source on top of the left branch (figure 9.20) is implemented by resistor $R_2$ and must have a value of:

$$R_2 = \frac{V_{CC} - V_{BE}}{272\ \mu\mathrm{A}} \approx 15.8\ \mathrm{k}\Omega \tag{9.37}$$

in the case of $V_{CC} = 5$ V and $V_{BE} = 0.7$ V.

Only resistor $R_1$ needs to be accurate. $R_2$ does not need to be accurate because small changes in $R_2$ can be seen as small changes in $I_1$ and these are not transferred to $I_2$.

As a consequence of the suppression of small changes in $I_1$, this type of current source exhibits a very high PSRR. More improvements can be found in [21].

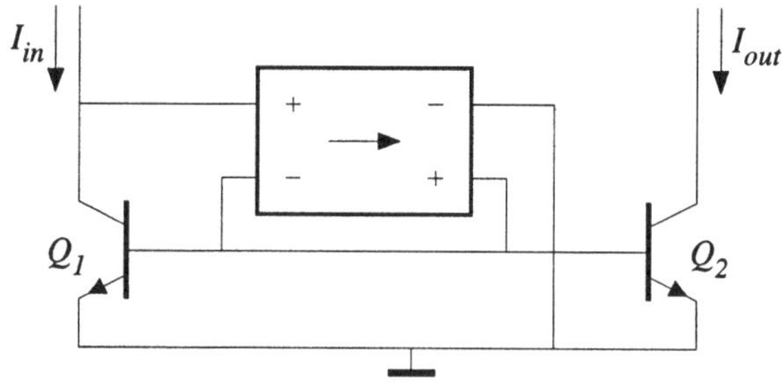


*Figure 9.23.* The basic current mirror.

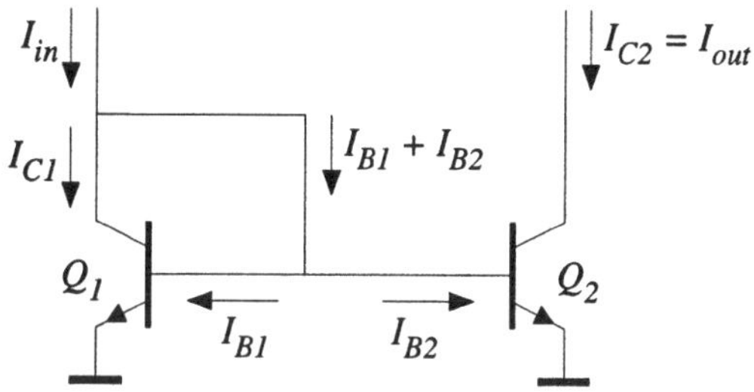


*Figure 9.24.* The simplest implementation of a current mirror.

### 9.4.3 The current mirror

The basic current mirror, based upon two bipolar transistors, is shown in figure 9.23. The nullor forces the input current to flow completely through the collector terminal of $Q_1$ and the output transistor mirrors that current.

The simplest implementation of the mirror is depicted in figure 9.24. In this current mirror, the nullor is implemented by just a wire. This results in an error in the transfer.

In figure 9.24, the currents flowing in the mirror are indicated. The output current is equal to:

$$I_{out} = I_{C1} = I_{in} - \frac{I_{C1}}{\beta_1} - \frac{I_{C2}}{\beta_2}. \tag{9.38}$$

Consequently, the transfer equals:

$$\frac{I_{out}}{I_{in}} = \frac{1}{1 + 2/\beta}. \tag{9.39}$$

The higher the current-gain factor is, the smaller the difference between the input and output current and the closer the transfer approaches 1.

The influence of the base currents can be minimized by using a better implementation for the nullor shown in figure 9.23, which is shown in figure 9.25. The nullor is now implemented with a CE stage. The input and output current

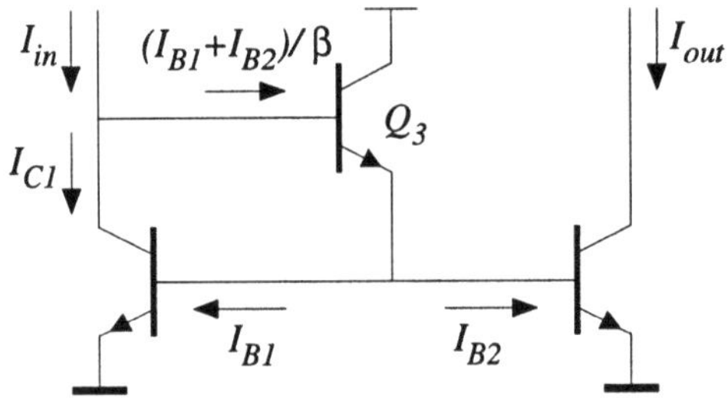


*Figure 9.25.* A current mirror with a reduced influence of the base currents.

are:

$$I_{in} = I_{C1} + \frac{I_{C1} + I_{C2}}{\beta^2} \tag{9.40}$$

$$I_{out} = I_{C1} \tag{9.41}$$

The influence of the base current is decreased by a factor $\beta$. This is exactly the increase in the loop gain in the mirror due to $Q_3$.

The output transistor basically works similar as the current sources described in the previous section. The reference voltage now originates from the input transistor. *This implies that resistors in series with the emitters improve the performance (output impedance, noise behavior) of a current mirror in the same way they do for a simple current source and should therefore be seriously considered.*

#### 9.4.3.1 High-frequency behavior of the current mirror.

To examine the high-frequency behavior of the current mirror, the current-gain factor of the transistors, is described by:

$$\beta_f = \beta_0 \cdot \frac{1}{1 + j\omega\tau_f\beta_0}. \tag{9.42}$$

The pole at $-1/(2\pi\beta_0\tau_f)$ represents the finite bandwidth of the transistors. Substitution of this expression for $\beta$ in the expression for the transfer of the current mirror, equation (9.39), results in:

$$\frac{I_{out}}{I_{in}} = \frac{1}{1 + 2/\beta_0} \cdot \frac{1}{1 + 2j\omega\tau_f}. \tag{9.43}$$

The high-frequency behavior is given by a pole at half the transit frequency of the transistor. Of course, a substrate capacitance and a collector bulk resistance if present have influence on the high-frequency behavior too.

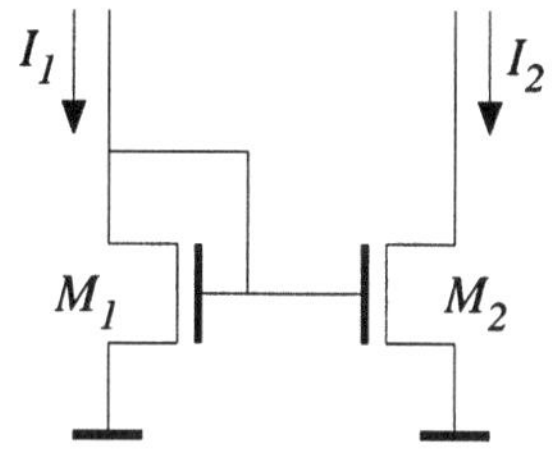


*Figure 9.26.* A MOS current mirror.

### 9.4.4 The MOS current mirror

In the previous sections, the current mirrors discussed were built with bipolar transistors, but they can be implemented with MOS transistors just as well. The MOS current mirror is shown in figure 9.26. The transistors must always be in the saturation region. The gate-source voltage is higher than the threshold voltage (normally-off device) and the gate-drain voltage is zero.

The behavior of this current mirror is analogous to its bipolar equivalent. The error due to the channel-length modulation ("Early effect") is generally larger, because the output resistance of MOS transistors is lower than that of a bipolar transistor, so cascading with a CG-stage is favorable. In contrast with the bipolar current mirror, the MOS mirror does not suffer from errors due to DC gate currents.

## 9.5 Conclusions on current sources

Several implementations of current sources have been discussed. The simplest current source uses a single resistor to derive the current from a voltage source. This source, however, requires a high supply voltage in the case of a high output impedance combined with a high output current. The active current source is able to realize a high output impedance and a high output current with a reasonable supply voltage at the expense of extra noise. When there is 5 V across the resistor in the current source, noise performance and output impedance are close to the optimal values for this type of source. This makes it problematic to implement high performance current sources in low-voltage applications. The use of a DC-DC converter could be considered.

The performance of the current mirror has been discussed too. Also in this case the use of emitter resistors is advised: the output impedance increases, the mirror factor is closer to unity and the output noise level and the sensitivity for matching errors in the transistors reduce.

# 10

# DESIGN EXAMPLE

In this chapter, a design is treated to show the application of the presented design methodology. The different steps in the design methodology are treated separately. The design is supposed to be realized on a PCB, so inductors and large capacitors can be used. For the active devices, it is assumed that the selection needs to be done from the ones listed in table 10.1. These devices

| Transistors | |
|---|---|
| Bipolar (N) | BC550B |
| | 2N3904 |
| Bipolar (P) | BC560B |
| | 2N3906 |
| JFET (N) | J108 |
| | J112 |
| JFET (P) | J174 |
| | J270 |

*Table 10.1.* The list of active devices from which needs to be selected.

can be considered as general purpose devices that are easily available and most simulators comprises those devices in their standard libraries.

## 10.1 Amplifier specification

The source for the amplifier to be designed is a piezo-electric sensor. It is to be used to measure pressure variations. When the pressure exposed to the piezo-electric element changes, the charge on the element changes, when keeping the voltage across the sensor constant. For this application the expected maximum effective value of the signal charge is 500 pC. The signal bandwidth is 50 Hz to 1.5 MHz. The output impedance of the sensor can be modelled with a capacitor of 10 nF. From the application follows that one-sided grounding is required for the sensor. Table 10.2 summarizes the specifications of the source.

| Source | |
|---|---|
| Max. effective signal | 500 pC |
| Signal Bandwidth | 10 Hz - 1.5 MHz |
| Impedance | 10 nF |
| Grounding | One sided |

*Table 10.2.* The specifications of the signal source.

The load for the amplifier can be modelled as a parallel connection of 10 kΩ and 100 pF. To obtain optimal signal quality for the load, it should be voltage driven. It requires a maximum effective signal voltage of 0.5 V. Also a single-sided grounding is required for the load. Finally, the load cannot stand a too large DC current. Therefore, a maximum DC current of 100$\mu$A is specified. Table 10.3 summarizes these specifications.

| Load | |
|---|---|
| Max. effective signal | 0.5 V |
| Impedance | 10 kΩ//100 pF |
| Grounding | One sided |
| Max. DC current | 100 $\mu$A |

*Table 10.3.* The specifications of the load.

To make the set of specification complete, the transfer of the required amplifier is considered. Its transfer should be such that given the maximum effective input signal, the maximum effective output signal is obtained. The signal-to-noise ratio of the output signal should be higher than 70 dB whereas the small-signal bandwidth should be at least 1.5 MHz with a maximum-flat magnitude response. The power bandwidth needs to be only 500 kHz. Table 10.4 summarizes these specifications.

## 10.2 Step 1: Determination input and output quantity

According to the specifications, the output signal of the piezo-electric sensor is in the charge domain. However, circuit design is in terms of voltages and current, i.e. either the sensor should be shorted by the amplifier input or it should be left as an open by the amplifier input. Thus, either a Norton or a Thevenin equivalent should be used for the sensor. Figure 10.1 depicts both

| Transfer | |
|---|---|
| SNR | > 70 dB |
| Bandwidth | > 1.5 MHz |
| Response type | MFM |
| Power bandwidth | >500 kHz |

*Table 10.4.* The specifications of the transfer.

options. To be able to choose between those two options, the relation between

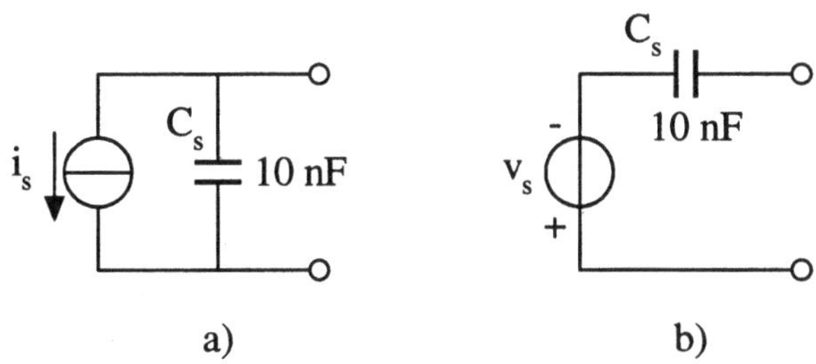


*Figure 10.1.* a) A current-source representation and b) a voltage-source representation of the piezo-electric sensor.

the electrical signal and the physical signal should be considered. In the case of the current-source representation, the relation between the current and the charge is given by:

$$i_s = \frac{dq_s}{dt} \tag{10.1}$$

in which $i_s$ is the signal current and $q_s$ is the signal charge. Clearly, the charge and current are unambiguously related to each other. For the case of the voltage source representation, this is slightly different. The signal voltage as a function of the signal charge is given by:

$$v_s = \frac{q_s}{C_s} \tag{10.2}$$

in which $v_s$ is the signal voltage and $C_s$ is the output capacitance of the sensor. Thus, the signal voltage depends on the signal chargeánd on the output capacitance of the sensor. As a result of the physical mechanism in the piezo-electric sensor, $C_s$ is dependent on the signal as well. Note that due to the pressure the sensor deforms slightly, resulting in a slightly different output capacitance.

As the relation between the signal current and the signal charge is the most accurate one, the current-source representation is chosen to be the appropriate one.

For the load, the discussion is straightforward. According to the specifications, a voltage drive yields the optimum signal quality.

## 10.3 Step 2: Synthesis of the feedback network

The amplifier to be designed needs to have a current input and a voltage output. Thus, the input of the feedback network needs to be connected in parallel with the output of the amplifier to sense the output voltage. Further, the output of the feedback network needs to be connected in parallel with the input for comparing the input-signal current with the feedback-signal current. The basic configuration for this type of amplifier is depicted in figure 10.2.

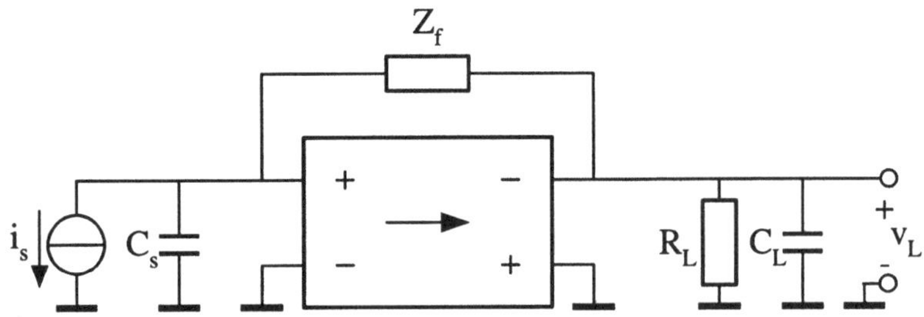


*Figure 10.2.* The basic configuration of a negative-feedback amplifier for a current input and a voltage output. $Z_f$ is the feedback impedance and $R_L$ and $C_L$ are the load resistance and capacitance, respectively.

The feedback impedance sets the value of the gain of the amplifier. For this configuration the asymptotic gain equals:

$$A_{t\infty} = \frac{v_L}{i_s} = Z_f \tag{10.3}$$

The required transfer follows from the specifications as:

$$A_{t\infty} = \frac{\hat{v}_{L-eff}}{\hat{i}_{s-eff}} = \frac{500\text{ pC}}{0.5\text{ V}} = 1\text{ nF} \tag{10.4}$$

So, the feedback impedance should be a capacitor of 1 nF in order to have a frequency independent transfer from the sensor-signal charge to the output voltage.

After this step the designed amplifier has got the required gain and is ideal with respect to noise, distortion and bandwidth. It should be noted that when checking the transfer of the current design with a simulator, the transfer from *current* to voltage equals:

$$A_t = \frac{1}{sC_s} \tag{10.5}$$

in which $s$ is the Laplace variable. This integration is because of considering the signal current as the input instead of the signal charge.

## 10.4 Step 3: Design of the first nullor stage: noise

Now the design of the feedback network is done, the next steps concentrate on the design of a nullor implementation, such that an implemented amplifier is obtained that meets all the specifications.

The first step in the design of a nullor implementation is to design the input stage for noise. The amplifier configuration is as depicted in figure 10.3. The

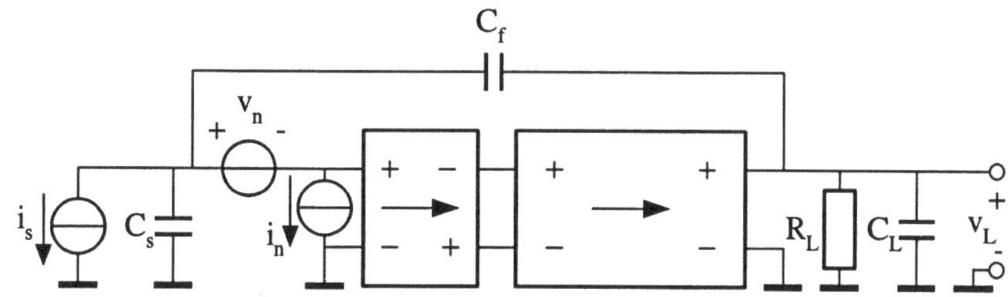


*Figure 10.3.* The amplifier configuration for designing the input stage with respect to noise.

input stage is depicted as a separate two port. The noise of the input stage is modelled by the two sources, $v_n$ and $i_n$. For this configuration an equivalent noise source needs to be determined. This can be either at the input or at the output. As for this example the information is in the charge domain, it is most convenient to determine the SNR at the output of the amplifier. Therefor, first an equivalent input noise current source is determined which is subsequently, transformed by the gain of the amplifier to the output.

For the input stage, the type of device is still left open. It is assumed that the noise of the input stage is modelled by two noise sources. As soon as an equivalent source is determined for the total amplifier these noise sources should be made specific depending on the type of input stage.

Obtaining an equivalent noise source at the output of the amplifier is done in the following six steps:

1. Starting point for determining the equivalent noise source is to identify the noise sources. The result is depicted in figure10.4. Besides the equivalent

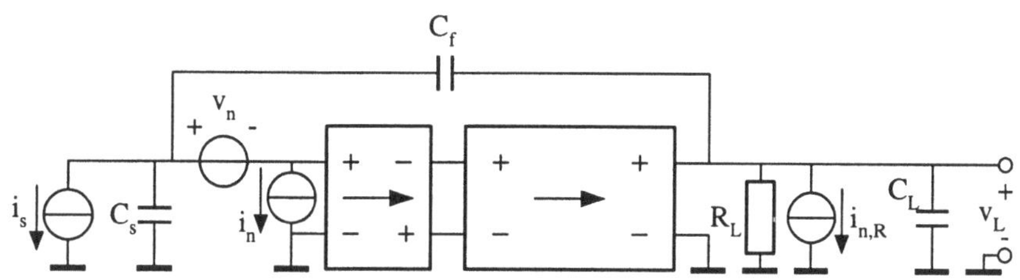


*Figure 10.4.* Identification of the noise sources.

noise sources of the input stage, only the load resistance generates noise. This noise is modelled with a current noise source,$i_{n,R}$.

2. The noise source of the load resistance is in parallel with the output of the amplifier, so the chain matrix of the amplifier can be used to transform this source to the amplifier input. As parameters $B$ and $D$ are zero for this amplifier, the resulting equivalent noise sources are zero. Thus, the noise of the load resistance can be ignored.

   The equivalent noise current of the input stage ($i_n$) is already in parallel with the signal source, thus does not need any further transformation for the moment.

   The equivalent noise voltage source of the input stage ($v_n$) is shifted through the input node, into the feedback network and in series with the signal source. Note that now two correlated noise sources are obtained. This correlation should be taken into account when determining the power spectral density of the total equivalent noise source.

   Figure 10.5 depicts the transformations.

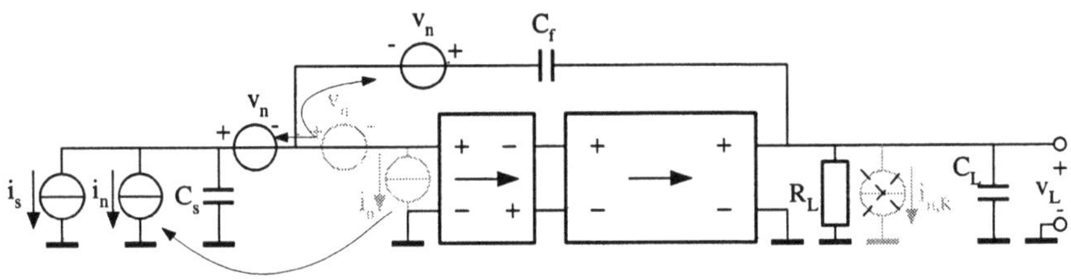


*Figure 10.5.* Transforming and moving $v_n$, $i_n$ and $i_{n,R}$.

3. The noise voltage source $v_n$ in series with the signal source is transformed into a current source, via the Norton-Thevenin transform, yielding a current source equal to:

$$i_n = sC_s v_n \tag{10.6}$$

   The same goes for the noise voltage source in series with the feedback network. It is also transformed into a noise current source, according to:

$$i_n = sC_f v_n \tag{10.7}$$

   These transformations are illustrated in figure 10.6.

4. Subsequently, the current source in parallel with the feedback network is split into a current source in parallel with the input and a current source in parallel with the output, see figure 10.7.

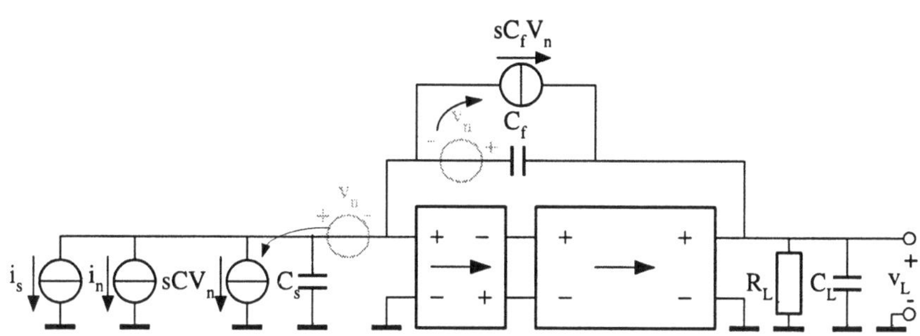


*Figure 10.6.* Transforming $v_n$ into a current source.

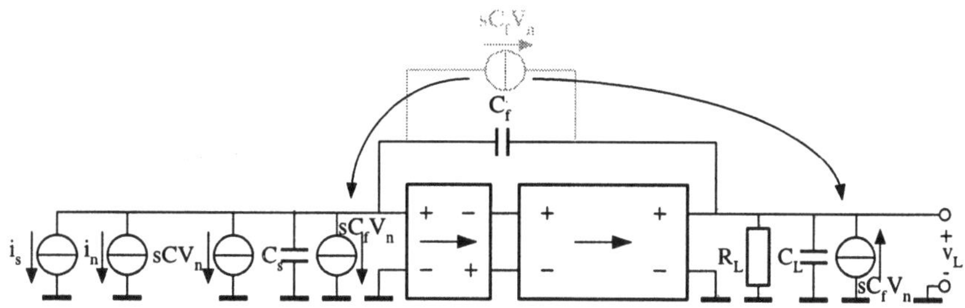


*Figure 10.7.* Splitting the current source which is in parallel with the feedback network.

5. The current source in parallel with the output can be ignored as the current source due to $R_L$ could be ignored. The sources left now are current sources in parallel with the signal source. Thus the equivalent input noise current source, $I_{n,eq}$, is given by (see figure 10.8):

$$i_{n,eq} = i_n + s(C_s + C_f)v_n \tag{10.8}$$

Intuitively, this source can be explained as follows. Source $i_n$ is directly

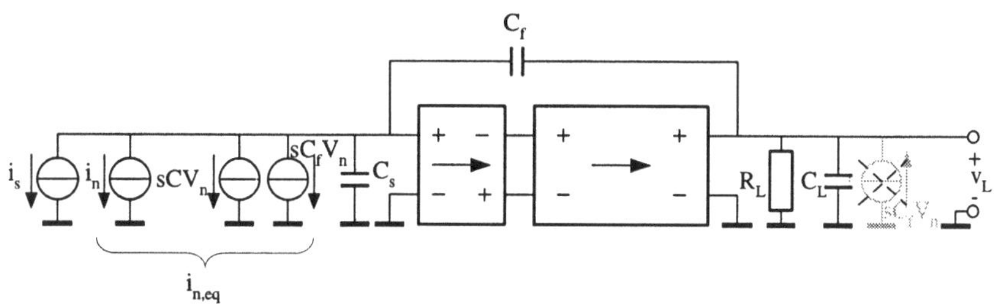


*Figure 10.8.* Obtaining the equivalent input noise current source.

in parallel with the signal source whereas the source $v_n$ is transformed via the sum of the source capacitance and the feedback capacitance to a current. This is in line with the short cut as described in section 4.5.5.

6. Finally, the equivalent source at the input can be transformed to an equivalent noise voltage source at the output of the amplifier, $v_{n,eq}$. This source is given by:

$$v_{n,eq} = A_{t\infty} \cdot i_{n,eq} = \frac{i_n}{sC_f} + \left(1 + \frac{C_s}{C_f}\right) v_n \qquad (10.9)$$

This equivalent noise source is depicted in figure 10.9.

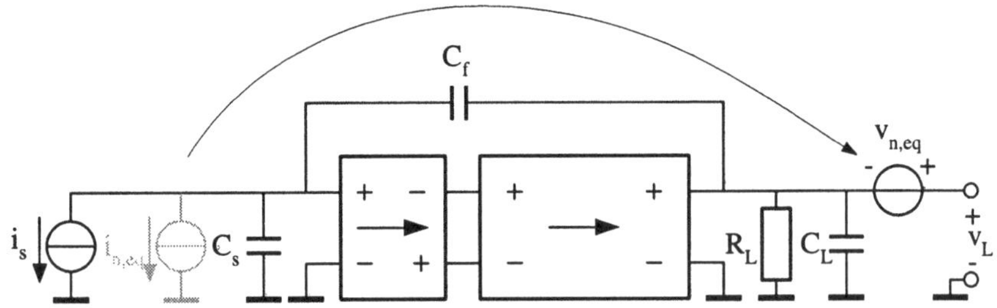


*Figure 10.9.* The equivalent noise voltage source at the output of the amplifier.

The expression of the equivalent noise source is in terms of the equivalent noise sources of the input stage. Depending on the type of transistor, a specific expression is obtained that can be optimized. This optimization yields basic different results for bipolar and JFET, as will be seen subsequently.

**Bipolar input stage.** For a bipolar transistor the relevant noise sources are the base shot noise, $i_b$, the collector shot noise, $i_c$, and the thermal noise of the base resistance, $v_{n,b}$, see figure 10.10. Mostly, the 1/f noise can be ignored. If not, it can be taken into account as a factor in the expression for the collector shot noise. Here it is assumed that the 1/f noise can be ignored.

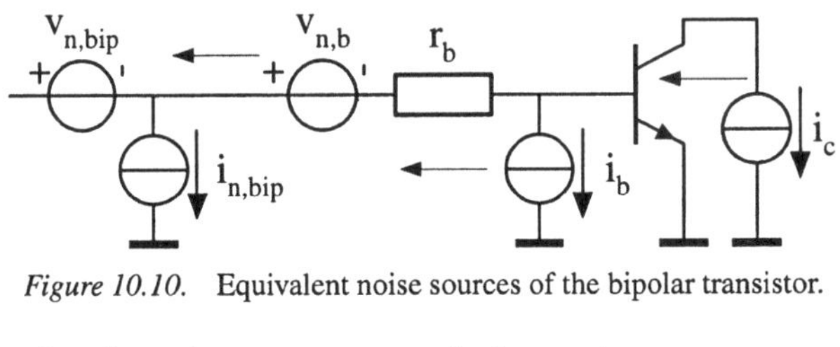


*Figure 10.10.* Equivalent noise sources of the bipolar transistor.

Transforming the noise sources to equivalent noise sources, $v_{n,bip}$ and $i_{n,bip}$ at the input of the transistor (see figure 10.10), yields the following two sources:

$$v_{n,bip} = v_{n,b} + i_c\left(\frac{1}{g_m} + \frac{r_b}{\beta_f}\right) \approx v_{n,b} + \frac{i_c}{g_m} \qquad (10.10)$$

$$i_{n,bip} = i_b + \frac{i_c}{\beta_f} \approx i_b \qquad (10.11)$$

The approximations can be made if $\beta_f \gg 1$ and $r_\pi \gg r_b$, which often hold. Using the approximated expressions for the noise sources of the bipolar transistor in equation (10.9), yields:

$$v_{n,eq} = \frac{i_b}{sC_f} + \left(1 + \frac{C_s}{C_f}\right) \cdot \left(v_{n,b} + \frac{i_c}{g_m}\right) \tag{10.12}$$

The power spectral density is given by:

$$S_{vn,eq} = \frac{2qI_b}{(2\pi f C_f)^2} + \left(1 + \frac{C_s}{C_f}\right)^2 \cdot \left(4kTr_b + \frac{2qI_c}{g_m^2}\right) \tag{10.13}$$

Subsequently, the total power in the signal band can be found by integration of this expression over the band, yielding:

$$P_{vn,eq} \approx \frac{qI_c}{2\pi^2 C_f^2 f_l \beta_f} + \left(1 + \frac{C_s}{C_f}\right)^2 \cdot \left(4kTr_b + \frac{2qV_T^2}{I_c}\right) \cdot f_h \tag{10.14}$$

in which $f_h \gg f_l$ and $1/f_l \gg 1/f_h$ are used. Further, the expression was rewritten in terms of $I_c$.

This expression is plotted for the four bipolar transistors in figure 10.11. The used parameters are listen in table 10.5.

| Transistor | $\beta_f$ | $r_b$ [$\Omega$] | $I_{c,opt}$ [$\mu$A] | $P_{n,min}$ [V$^2$] | SNR [dB] |
|---|---|---|---|---|---|
| BC550B | 375 | 300 | 140 | 1.5e-9 | 82 |
| 2N3904 | 180 | 150 | 90 | 1.3e-9 | 83 |
| BC560B | 290 | 130 | 120 | 1.1e-9 | 84 |
| 2N3906 | 200 | 150 | 100 | 1.2e-9 | 83 |

*Table 10.5.* The transistor parameters, the optimal collector current and maximum SNR for the four bipolar transistors.

Clearly, all the transistors can meet the SNR specification. The shape of the curves is typical for bipolar transistors. At the lower collector current the equivalent noise voltage of the bipolar transistor is dominant as it is inversely proportional to the collector current. Note that for this example this part of the curve does depend on the transistor only via the collector current. Consequently, for the four transistors these parts of the curves are equal. For the higher currents, the equivalent noise current of the bipolar transistor is dominant as it is proportional to the collector current. Comparing the curves for the currents, shows indeed different levels, proportional to the current-gain factor of the transistor. Somewhere in between these two ranges, a minimum can be found. The curve around this minimum is relatively flat. Analytically, this minimum

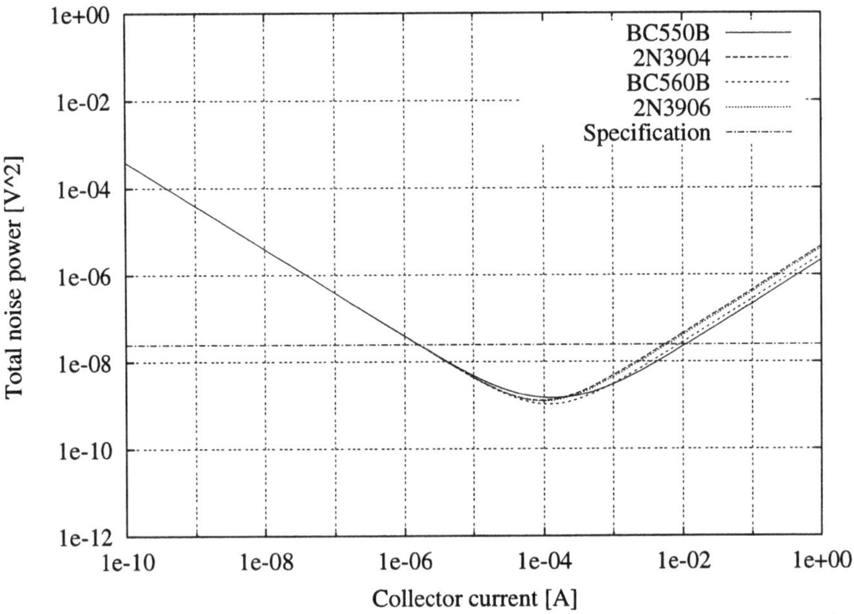


*Figure 10.11.* The total noise power, equation (10.14), for the four bipolar transistor listed in table 10.1. The required transistor parameters are listed in table 10.5. The SNR specification of 70 dB corresponds to a noise level of $P_{n,spec} = 2.5e - 8\ \mathrm{V}^2$.

can be found from equation (10.14). The optimal collector current at which this optimum is obtained is given by:

$$I_{c,opt} = 2\pi\sqrt{f_h f_l}\,(C_s + C_f) \cdot \sqrt{\beta_f} V_T \tag{10.15}$$

and the minimum noise power (at this optimal collector current is given by):

$$P_{n,min} = 4kTr_b\left(1 + \frac{C_s}{C_f}\right)^2 \cdot f_h + \frac{2kT}{\pi}\sqrt{\frac{f_h}{f_l\beta_f}}\left(\frac{C_s + C_f}{C_f^2}\right) \tag{10.16}$$

which yields for the BC550B, for instance:

$$I_{c,opt} = 0.14\ \mathrm{mA} \tag{10.17}$$

$$P_{n,min} \approx 1.5 \cdot 10^{-9}\ \mathrm{V}^2 \tag{10.18}$$

The corresponding signal-to-noise ratio equals:

$$SNR = 20\log_{10}\left(\frac{v_{max-eff-out}}{v_{n,min}}\right) = 20\log_{10}\left(\frac{0.5}{39\ \mu\mathrm{V}}\right) = 82\ \mathrm{dB} \tag{10.19}$$

From expression (10.16) follows that for obtaining the lowest noise level, low base resistances and high current-gain factors are preferable. Further, a reduction in the base resistance gives a proportional reduction in the noise level, whereas for the current-gain factor this proportionality is inverse with the square root. The optimal collector current is proportional to the square root of the collector current.

The results for the four bipolar transistor are listed in table 10.5. Conclusion from the table is that the transistor with the lowest base resistance has the lowest noise level and the transistor with the highest current-gain factor has the highest optimal collector current.

**JFET input stage.** For the JFET here the relevant noise source is considered to be the thermal noise of the channel, $i_d$, only. Also for the JFET, when 1/f noise can not be ignored, it is easily taking into account as a factor in the expression for the channel noise. Further, the gate resistance is also assumed to be negligibly small.

The equivalent noise sources at the input of the JFET are given by (see figure 10.12):

$$v_{n,jfet} = \frac{i_d}{g_m} \tag{10.20}$$

$$i_{n,jfet} = \frac{s(C_{gs}+C_{dg})}{g_m} i_d \tag{10.21}$$

Note that both sources have an inversely proportional dependency on the$g_m$ of the JFET. More detailed expressions can be found in section 4.4.6

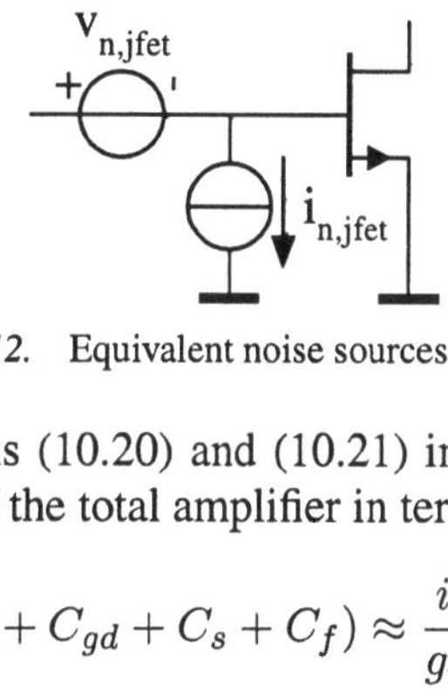


*Figure 10.12.* Equivalent noise sources of the JFET.

Substituting expressions (10.20) and (10.21) in equation (10.9) yields the equivalent noise source of the total amplifier in terms of the JFET parameters:

$$v_{n,eq} = \frac{i_d}{g_m C_f}(C_{gs}+C_{gd}+C_s+C_f) \approx \frac{i_d}{g_m}\left(1+\frac{C_s}{C_f}\right) \tag{10.22}$$

in which $C_{gs}$ and $C_{gd}$ are the gate-source capacitance and the gate-drain capacitance, respectively. The approximation holds as for the JFETs to be considered, $C_s \gg C_{gs} + C_{gd}$, see table 10.6. The power spectral density is given by:

$$S_{vn,eq} = \frac{4kTc}{g_m}\left(1+\frac{C_s}{C_f}\right)^2 \tag{10.23}$$

| JFET | $C_{gs}$[pF] | $C_{gd}$[pF] | $V_{th}$[V] | $I_{dss}$[mA] | $P_{n,min}$[$V^2$] | SNR[dB] |
|---|---|---|---|---|---|---|
| J108 | 48 | 48 | -4.5 | 155 | 2.9e-10 | 99 |
| J112 | 6 | 6 | -2.0 | 23 | 8.7e-11 | 95 |
| J174 | 9 | 7 | -6.5 | 34 | 1.9e-10 | 91 |
| J270 | 15 | 11 | -1.0 | 4 | 2.5e-10 | 90 |

*Table 10.6.* The transistor parameters, the optimal drain current and maximum SNR for the JFETs.

in which $c$ is a constant an equals about $2/3$. Clearly, the larger $g_m$ the lower the noise. Integration of the power spectral density yields the total noise power:

$$P_{vn,eq} = \frac{4kTc}{g_m}\left(1 + \frac{C_s}{C_f}\right)^2 f_h \tag{10.24}$$

in which $f_h \gg f_l$ is used. For the four FETs this expression is plotted in figure 10.13. Note that in the plot the maximum drain current is not taken into account. Further, like for the bipolar transistors, all the JFETs can meet the SNR specification. For $g_m$ the following expression was used:

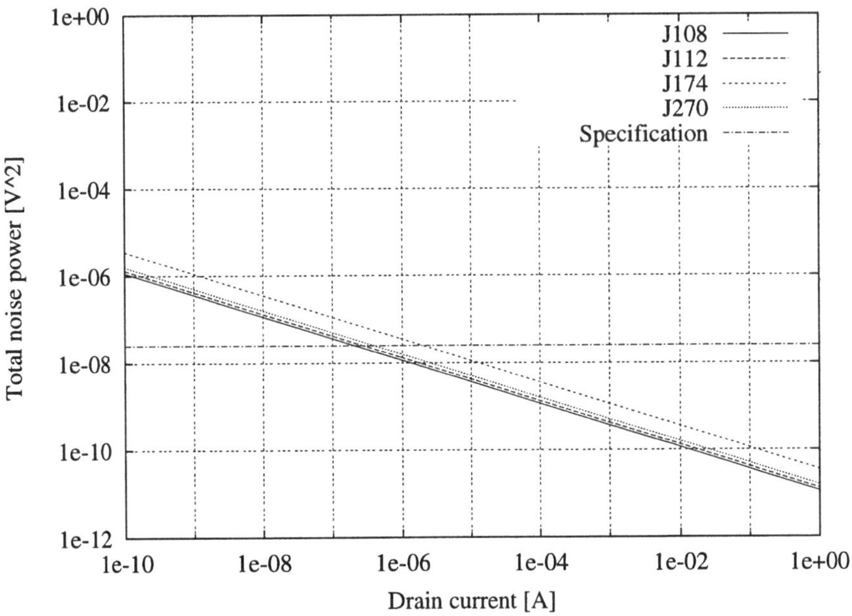


*Figure 10.13.* The total noise power, equation (10.24), for the four JFETs listed in table 10.1. The required transistor parameters are listed in table 10.6. The SNR specification of 70 dB corresponds to a noise level of $P_{n,spec} = 2.5e - 8$ V$^2$.

$$g_m = -\frac{2}{V_{th}}\sqrt{I_{dss}I_d} \tag{10.25}$$

in which $V_{th}$ is the threshold voltage and $I_{dss}$ the maximum drain current (current at $V_{gs}$=0 V).

As was already indicated, the larger the drain current ($g_m$), the lower the noise level. In two ways the noise behavior of the FETs differ from each other. First, because of dependency of the $g_m$ on the threshold voltages, FETs with a different threshold voltage, have got a different $g_m$ while they have the same drain currents. Second, the higher the maximum drain current is, the further the decrease in noise can be pushed.

The minimum noise levels are described by:

$$g_{m,opt} = \frac{-2I_{dss}}{V_{th}} \tag{10.26}$$

$$P_{n,min} = \frac{-2kTcV_{th}}{I_{dss}}\left(1+\frac{C_s}{C_f}\right)^2 f_h \tag{10.27}$$

Thus, the FETs with the highest ratio of $|V_{th}|$ and $I_{dss}$ have the lowest noise level. Table 10.6 summarizes the noise performance of the four JFETs.

**Choice input-stage transistor.** From the calculations follow that each of the 8 transistors can meet the specifications. Question is now what transistor to choose. Basically, one could say lets choose JFET J108 that can reach 99 dB SNR at a drain current of 155 mA. However, the specifications are 70 dB, so why consuming more current than required. In the plots for the bipolar transistors and JFETs, figures 10.11 and 10.13, the minimum required current for each of the transistors to meet the SNR specification is on the order of 1$\mu$A, for the the JFET slightly lower than for the bipolar.

As each of the transistors can meet the SNR specification, the choice of device, consequently, is based on other arguments than noise.

An argument could be the current consumption (the cost). However, no maximum current consumption is specified and as indicated before, the 8 devices require more or less the same current for meeting the SNR specification.

Subsequently, as SNR and cost do not give an argument for choosing a device, the other two quality aspects are considered: distortion and bandwidth. For low distortion a high loop gain is important. An important parameter for the loop gain is the $f_T$ of a device, i.e. up to what frequency is it capable of supplying gain. Table 10.7 summarizes the $f_T$ of the devices at 1 mA of collector/drain current. From the table follows that the bipolar transistor, 2N3904, has the highest $f_T$. Therefore, this transistor is chosen. Its bias current is chosen to be the optimal value: $I_c$=90 $\mu$A. It should be noted, however, that this choice does not need to be definite. When later for some other reason the bias current needs to be changed, this is possible within a range of about: 1.5$\mu$A to 3 mA.

| Transistor | $f_T$ [MHz] |
|---|---|
| BC550B | 100 |
| 2N3904 | 280 |
| BC560B | 64 |
| 2N3906 | 200 |
| J108 | 24 |
| J112 | 120 |
| J174 | 36 |
| J270 | 38 |

*Table 10.7.* The transit frequencies, $f_T$, of the eight transistors.

For the collector-emitter biasing voltage of the input stage, no constraint is found with respect to noise. Thus, it can be chosen such that it is minimum, i.e. equal to the base-emitter voltage (signal voltage swing is negligible) to prevent saturation: $V_{CE} = 0.7$ V.

The topology of the amplifier after this step is depicted in figure 10.14.

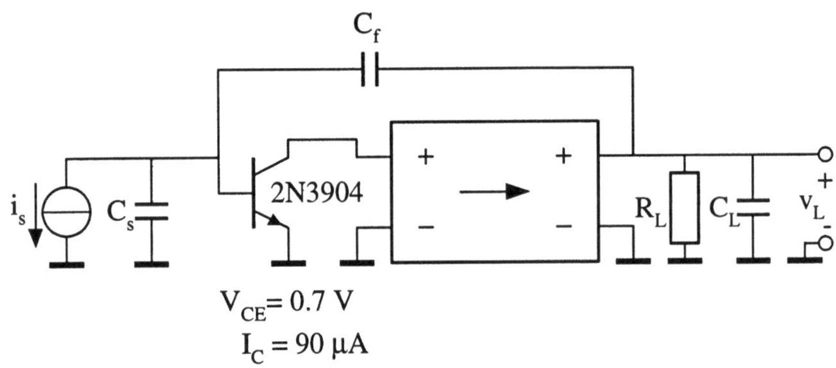


*Figure 10.14.* The amplifier topology after the design of the input stage (note no biasing sources are depicted).

## 10.5 Step 4: Design of the last nullor stage: clipping distortion

Next step in the design procedure is the design of the output stage for clipping distortion. As discussed in chapter 5, the design of the output stage is basically finding the required quiescent point such that the output stage never clips for the specified signal range.

The circuit topology for this design step is depicted in figure 10.15.

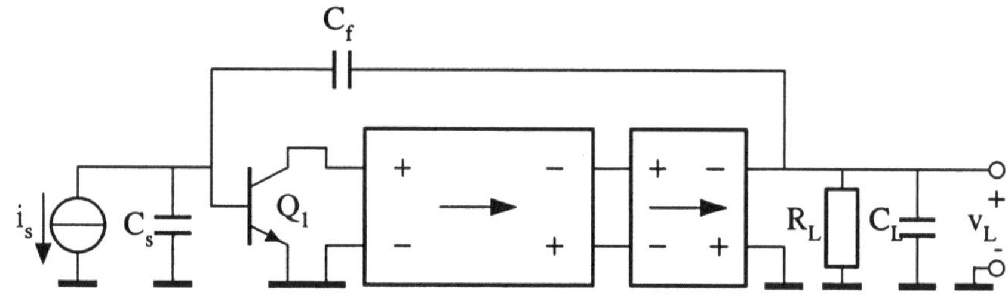


*Figure 10.15.* The amplifier topology for designing the output stage.

The output stage is loaded by the feedback capacitor, $C_f$, and the load impedance, $R_L$ in parallel with $C_L$. The input of the amplifier is at ground level, so for calculating the maximum load conditions for the output stage, the feedback network can be considered to be in parallel with the load impedance. Thus, the total impedance loading the output stage is given by:

$$Z_{load,output} = C_f // C_L // R_L \qquad (10.28)$$

For the output signal of the amplifier is specified a maximum effective value of 0.5 V. This means a peak value of 0.7 V. This is the first information for designing the output stage. Depending on the choice of transistor, i.e. JFET or bipolar, a specific margin needs to be taken into account.

For determining the maximum output current the worst case condition needs to be considered. As for higher frequencies the total loading impedance reduces, the maximum current that needs to be supplied by the output stage is found at the highest frequency at which the maximum signal needs to be supplied, i.e. the power bandwidth. According to the specification the power bandwidth is 500 kHz. The total load impedance as function of the complex frequency $s$ is given by:

$$Z(s)_{load,output} = \frac{R_L}{1 + sR_L(C_L + C_f)} \qquad (10.29)$$

Thus, for relatively low-frequency this impedance equals $R_L$, whereas beyond the frequency given by:

$$f_{pole,load,output} = \frac{1}{2\pi R_L(C_L + C_f)} = 14.5 \text{ kHz} \qquad (10.30)$$

the impedance is dominated by the capacitors, even mainly by the feedback capacitor. Thus, the lowest impedance that needs to be driven is a result of the parallel connection of the two capacitors:

$$|Z_{load,output}|_{f=500 \text{ kHz}} \approx 290 \ \Omega \qquad (10.31)$$

Consequently, a maximum output current that can occur equals:

$$I_{output,max} \approx 2.4 \text{ mA} \qquad (10.32)$$

Note that the largest portion of this current is necessary for driving the feedback network.

Now the maximum output signals are obtained, the output transistor needs to be chosen. It should be able to cope with these maximum signals, i.e. $v_{out,max} = 0.7$ V and $i_{out,max} = 2.4$ mA. Maximum voltage ratings are not specified by the SPICE parameters. However, these ratings are easily found in transistor data books and it can be concluded that each of the eight transistors can easily drive that voltage. With respect to the maximum output current of the transistors, the respective parameters for the bipolar transistor and JFET are different. For the JFET the maximum drain current is specified by $I_{DSS}$, see table 10.6, whereas the maximum collector current for the bipolar transistor is specified by its high-level injection current, $IKF$. The maximum current ratings for the eight transistors are summarized in table 10.8. Clearly, the J270

| Transistor | $IKF$ [mA] | $I_{DSS}$ [mA] |
|---|---|---|
| BC550B | 87 | |
| 2N3904 | 67 | |
| BC560B | 103 | |
| 2N3906 | 80 | |
| J108 | | 155 |
| J112 | | 23 |
| J174 | | 34 |
| J270 | | 4 |

*Table 10.8.* The maximum current ratings for the eight transistors.

is not able to supply the output current (note that the peak-peak signal current is 4.8 mA). The other transistors can easily supply the current. For even higher currents, also the bulk resistances, $R_E$ and $R_C$, needs to be considered (the voltage drop across those resistors might become too high leading to quasi saturation). For the eight transistor these bulk resistances are below 10Ω and can cause no harm.

For choosing one of the seven transistors, again the other quality aspects need to be considered. Of course, the noise of the output stage does not effect the amplifier performance. However, the bandwidth capability of the transistor is important for the next design step, i.e. bandwidth. Therefor, also here the transistor is chosen based on its $f_T$. According to table 10.7 transistor 2N3904 has the highest $f_T$ at 1 mA. It is likely to have also the highest $f_T$ at the required bias current for the output stage. Therefor, also for the output stage transistor 2N3904 is chosen.

Transistor 2N3904 is an NPN device and thus the minimum collector-emitter bias voltage should be:

$$V_{CE,min} = V_{out,max} + V_{be} = 0.7\text{ V} + 0.7\text{ V} = 1.4\text{ V} \qquad (10.33)$$

This expression is found by constraining $V_{BC} < 0$.

The maximum bias current is chosen 50% beyond the minimum. Thus,

$$I_C = 1.5 \times I_{out,max} = 3.6\text{ mA} \qquad (10.34)$$

The amplifier topology after this design step is depicted in figure 10.16.

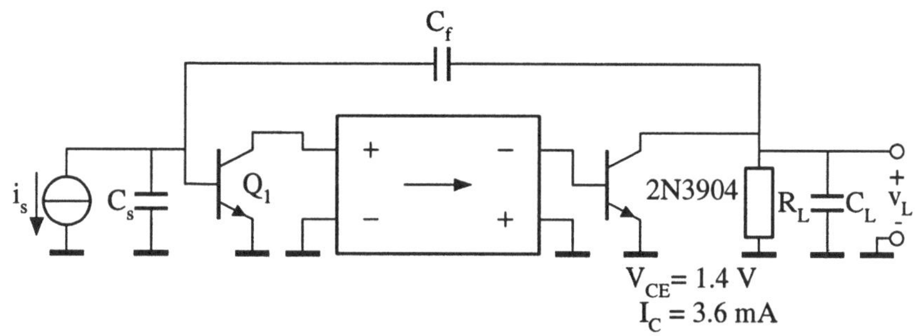


*Figure 10.16.* The amplifier topology after the design of the output stage.

## 10.6 Step 5a: Design of the intermediate stage: LP-product

The amplifier is designed now for noise and clipping distortion. Next step is to design the dynamic behavior. This step comprises two main items. In step 5a, the maximum bandwidth capabilities are investigated by means of the LP-product. Subsequently, step 5b involves the actual frequency compensation. Step 5a can be seen as the feasibility study for step 5b.

To be able to determine the LP-product of the amplifier designed so far, the small-signal parameters need to be known. For finding the small-signal parameters of a transistor biased at a certain current, the simple circuit of figure 10.17 can be used. The transistor in the figure is diode connected and the current source pulls a current equal to $I_E$ out of the emitter. As in most practical cases the current-gain factor of the transistor is much larger than one, the collector current equals approximately the emitter current. With a Spice-like simulator, an operating-point analysis yields the small-signal parameters for the transistor.

The small-signal parameters for the input and output stage were determined in this way. The resulting parameters are listed in table 10.9.

For determining the LP-product of the amplifier, it is assumed that the intermediate stage is just a pair of wires connecting the input and output stage. For each of the amplifying stages the simplest model is chosen, i.e. $r_\pi$, $c_\pi$ and

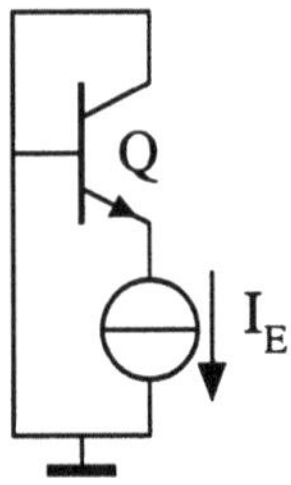


*Figure 10.17.* The auxiliary circuit for determining the small-signal parameters for a transistor biased at a emitter current (≈ collector current) $I_E$.

| Small-signal parameter | Input stage 2N3904 @ $I_C$=90 $\mu$A | Output stage 2N3904 @ $I_C$=3.6 mA |
|---|---|---|
| $r_\pi$ [kΩ] | 52 | 1.2 |
| $c_\pi$ [pF] | 7 | 46 |
| $g_m$ [mA/V] | 3.5 | 132 |
| $\beta_f$ | 180 | 152 |
| $r_o$ [kΩ] | 1500 | 38 |
| $c_\mu$ [pF] | 3 | 3.7 |
| $r_b$ [Ω] | 150 | 150 |

*Table 10.9.* The small-signal parameters for the input and output stage for the respective bias conditions.

$g_m$. The corresponding small-signal diagram is depicted in figure 10.18[1]. For obtaining a negative loop gain, the input stage is assumed to be implemented by means of a differential pair, having almost same small-signal parameters as the original CE stage (double $r_\pi$ and half $c_\pi$ and $g_m$). Of course, the differential pair could also be located at the output. However, as this circuit diagram is only used for determining the LP-product, the specific choice for the location of the differential pair is not important yet. When it appears that the LP-product is high enough, a decision must be made where to place the differential pair.

As the feedback element is a capacitor, the DC loop gain is zero. Clearly, the negative-feedback loop of this amplifier has got a zero in the origin. As explained in section 6.7, for design purposes, this zero is counteracted by resistor

[1]Basically, the first circuit that could be investigated on its LP-product merits is the amplifier with a single-stage nullor implementation with the transistor biased at $I_C = 3.6$ mA. This implementation would meet the noise and the clipping distortion specifications. However, in most practical cases more than one amplifying stage is required for sufficient LP-product. In the special case that the LP-product of the two stage implementation is much too high, one could revert to a single stage implementation. The collector bias current needs to be equal to the highest collector bias current of the two stages. Of course, it needs to be checked whether the noise and clipping distortion specifications are still met.

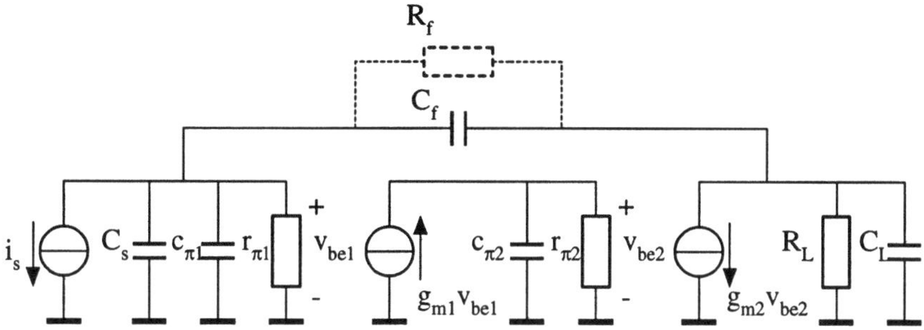


*Figure 10.18.* The small-signal diagram of the amplifier configuration as depicted in figure 10.16. The small-signal parameters are listed in table 10.9. It is assumed that the input stage is implemented with a differential pair for obtaining a negative loop gain. Resistor$R_f$, which is only for design purposes, is required for cancelling the zero in the origin in the feedback loop.

$R_f$. The value of $R_f$ should be:

$$R_f \approx r_{\pi 1}\frac{C_s}{C_f} \approx 1\ \text{M}\Omega \tag{10.35}$$

The loop gain as a function of the complex frequency,$s$, is found to be equal to (the loop was broken by assuming controlled source,$g_{m1}$, to be the independent source):

$$L(s) = \frac{-\beta_{f1}\beta_{f2}R_L}{[1+sr_{\pi 2}c_{\pi 2}][R_L+R_f+r_{\pi 1}+sR_L(R_fC_f+R_fC_L+r_{\pi 1}C_L)]} \tag{10.36}$$

as $C_f \gg C_L$ and $R_f \gg r_{\pi 1}, R_L$, this expression simplifies to:

$$L(s) = \frac{-\beta_{f1}\beta_{f2}R_L}{R_f}\cdot\frac{1}{[1+sr_{\pi 2}c_{\pi 2}][1+sR_LC_f]} \tag{10.37}$$

From this expression the DC loop gain, $L(0)$, and two poles, $p_1$ and $p_2$ are found to be:

$$L(0) = -280 \tag{10.38}$$
$$p_{l,1} = -15.9\ \text{kHz} \tag{10.39}$$
$$p_{l,2} = -2.9\ \text{MHz} \tag{10.40}$$

and thus the LP-product for the second-order system equals:

$$LP_2 = 1.3\cdot 10^{13}\ \text{Hz}^2 \tag{10.41}$$

yielding a maximum bandwidth of:

$$B_2 = \sqrt{LP_2} = 3.6\ \text{MHz} \tag{10.42}$$

Of course, it needs to be checked whether both poles are dominant or not. For the case of the Butterworth position of the system poles holds:

$$p_{s,1} + p_{s,2} = -2 \cdot \frac{1}{2}\sqrt{2} \cdot 3.6\ \text{MHz} \approx -5.1\ \text{MHz} \tag{10.43}$$

The sum of the loop poles equals:

$$p_{l,1} + p_{l,2} \approx -2.9\ \text{MHz} \tag{10.44}$$

As

$$p_{l,1} + p_{l,2} > p_{s,1} + p_{s,2} \tag{10.45}$$

both poles are dominant.

One might wonder whether a single-stage amplifier would suffice or not, as the bandwidth of the second-order system is considerably larger than the specified bandwidth. In that case a single stage biased at 3.6 mA needs to be chosen, i.e. the current output stage. Then, the bias current is sufficient to prevent clipping distortion and according to figure 10.11 it also meets the noise specifications. Calculating the loop gain as a function of frequency yields:

$$L(s) = \frac{-\beta_{f2} R_L}{R_L + R_f} \cdot \frac{1}{1 + s(R_f // R_L)C_f} \tag{10.46}$$

Which results in ($R_f = 12$ kΩ):

$$L(0) = -72 \tag{10.47}$$
$$p_{l,1} = -29\ \text{kHz} \tag{10.48}$$

and thus

$$B_1 = 2.1\ \text{MHz} \tag{10.49}$$

As, this is relatively close to the specified bandwidth, it is a risk to go for this option. Especially, now the model is still rather ideal. Therefor, we proceed in this example with the two stage implementation.

Now the LP-product is sufficient, a decision should be made about which stage becomes the differential stage. It can be placed at the input as well at the output. For reasons of power efficiency it is chosen to be placed at the input. As the LP-product is high enough and an additional 3 dB noise is no problem, the two transistors are both biased at 90 $\mu$A instead of the double current to make its total behavior equal to the CE stage. The amplifier after this design step is depicted in figure 10.19.

## 10.7 Step 5b: Frequency compensation

The second part of the design of the amplifiers dynamic behavior, is to perform a frequency compensation such that the poles of the closed loop are

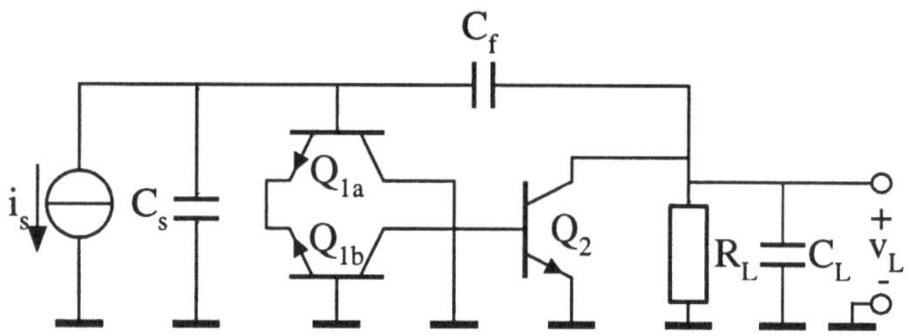


*Figure 10.19.* The amplifier topology after realizing sufficient LP-product.

in Butterworth. Preferred compensation techniques is, as explained in chapter 7, the phantom-zero technique. Given the DC loop gain and the loop poles as calculated in the previous section, the required phantom zero is located at (see equation (7.15)):

$$z_{ph} = \text{-5.9 MHz} \tag{10.50}$$

Inspecting the circuit topology of figure 10.19 shows that only at the input a phantom zero can be realized. In the feedback network a phantom zero is not possible as the feedback element is a capacitor. It is not possible by parallel connection of a compensation element to increase, beyond 5.9 MHz, the loop gain. At the output, the load capacitor introduces an attenuation for relatively high frequencies of only $\approx (C_L + C_f)/C_L = 1.1$. So, when realizing a phantom zero at the output, a pole at approximately -6.5 MHz is obtained also. The pole will cancel the phantom zero such that it is not effective. At the input however, the source capacitance realizes, for relatively high frequencies, an attenuation of $(C_s + 0.5c_{\pi 1})/(0.5c_{\pi 1}) \approx 2800$. Thus, an effective phantom zero can be realized at the input.

The phantom zero at the input is implemented by means of a resistor in series with the source (see figure 10.20). The resistor value is determined by:

$$n = \frac{-1}{2\pi C_s R_{ph}} \tag{10.51}$$

which yields a value of about 2.7 Ω. Using this resistor the closed loop poles are found at (resistor $R_f$ is removed from the circuit):

$$p_{s1} = \text{-0.6 Hz} \tag{10.52}$$

$$p_{s2,3} = (\text{-2.4} \pm 2.4j)\ \text{MHz} \tag{10.53}$$

$$\text{poles} < \text{-0.6 GHz} \tag{10.54}$$

$$\text{zeros} > |\pm 10\ \text{MHz}| \tag{10.55}$$

The first pole is the transfer pole for integrating the input charge to a voltage. The higher the DC gain of the nullor implementation, the closer it is to the origin. The two complex poles are indeed in Butterworth and the bandwidth of the amplifier is 3.4 MHz which is slightly lower than the predicted bandwidth. This is mainly because of the approximations made by deriving expression (10.37).

The topology with the frequency compensation is depicted in figure 10.20.

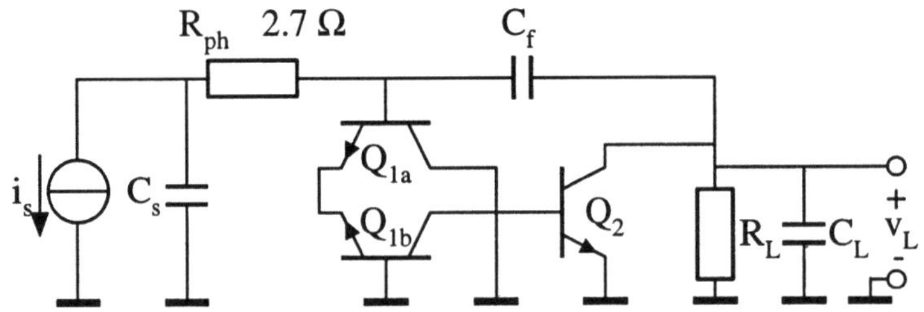


*Figure 10.20.* The amplifier topology including a phantom-zero frequency compensation by resistor $R_{ph}$.

Now the frequency compensation is done using the simple models, we need to check whether these simple models are valid in the current situation or not. This is done by investigating the effect of the parasitic elements,$C_\mu$, $r_o$ and $r_b$, on the system-pole positions. This is done by inserting these parasitic elements one at a time. When it appears that a parasitic element has a serious effect on the system-pole positions, countermeasures need to be taken. In most of the cases, inserting a current-buffer (subsequently implemented by a CG or CB stage), might solve the problem.

For the three transistors used in the design so far, the values of the parasitic elements are listed in table 10.10. Each of these elements were inserted into

| Element | $Q_{1a,b}$ | $Q_2$ |
|---|---|---|
| $c_\mu$ [pF] | 3 | 3.7 |
| $r_o$ [kΩ] | 1500 | 38 |
| $r_b$ [Ω] | 150 | 150 |

*Table 10.10.* The parasitic elements for the three transistors.

the circuit and the system poles and zeros were determined by means of a simulator. An over view of the results is given in table 10.11. From the table follows that the effect of $c_{\mu 2}$ needs some further investigation. The effect of the other elements is negligible. Inspecting the topology its is easy to understand why:

| Parasitic element | Integrator pole [Hz] | Band-edge poles [MHz] | Other poles [Mhz] | \|Zeros\| [Mhz] |
|---|---|---|---|---|
| None | -0.6 | $-2.4 \pm 2.4j$ | < -600 | > 10 |
| $c_{\mu 1a}$ | -0.6 | $-2.4 \pm 2.4j$ | < -600 | > 10 |
| $c_{\mu 1b}$ | -0.6 | $-2.3 \pm 2.4j$ | < -600 | > 10 |
| $c_{\mu 2}$ | -0.6 | $-2.9 \pm 1.3j$ | < -600 | > 10 |
| $r_{o1a,b}$ | -0.6 | $-2.4 \pm 2.4j$ | < -600 | > 10 |
| $r_{o2}$ | -0.7 | $-2.4 \pm 2.4j$ | < -600 | > 10 |
| $r_{b1a,b}$ | -0.6 | $-2.4 \pm 2.4j$ | < -150 | > 10 |
| $r_{b2}$ | -0.6 | $-2.4 \pm 2.4j$ | < -600 | > 10 |

*Table 10.11.* The system poles when one of the parasitic elements are inserted into the circuit. $R_f$ was removed from the circuit.

- $c_{\mu 1a}$ can be ignored as it is in parallel with $C_s$ which is much larger than $c_{\mu 1a}$.
- $c_{\mu 1b}$ can be ignored as it is in parallel with $c_{\pi 2}$ which is about an order of magnitude larger. A small change in the system pole is visible.
- The series connection of $r_{o1a}$ and $r_{o1b}$ is in parallel with $r_{\pi 2}$ which is much smaller then $r_{o1a} + r_{o1b}$.
- $r_{o2}$ is in parallel with $R_L$. $r_{o2}$ is about 4 times larger than $R_L$. So a noticeable effect on the low-frequency (extrapolated DC) loop gain might be expected. The consequence is that the integrator pole is somewhat further from the origin than originally. The effect on the system poles is negligible. The effect of $r_{o2}$ can be compared with resistive broad banding which does not lower the LP-product, i.e. the factor by which the corresponding poles reduces (becomes more negative) is equal to the factor by which the low frequency (extrapolated DC) loop gain decrease.
- Due to $r_{b1a,b}$ an additional pole is realized at:

  $$p_{rb} = \frac{-1}{2\pi(r_{b1} + r_{b2})(c_{\pi 1a}/2)} \approx \text{-150 MHz} \tag{10.56}$$

  This poles was indeed found. However, the pole is non dominant. It should be noted that for higher bias currents of the input stage the effect of $r_b$ becomes more pregnant. This is because for larger collector current this capacitor increases.
- $r_{b2}$ is in series with the output resistance of the differential pair which is much higher than $r_{b2}$.

Now we need to investigate whether countermeasures are available for the noticed effect due to $c_{\mu 2}$ or not. Again, the effect can largely be understood by inspecting the topology:

- $c_{\mu 2}$ might effect the loop poles via the Miller effect. This is easily checked by inserting an ideal current follower behind the output stage. When doing so, it appeared that the poles and zeros remained more or less at the same position and thus the ideal current follower had no effect. This is because the load of the output stage is a relatively large capacitor which shortens already for relatively low frequencies the output of the transistor and thus removing the Miller effect. Note that the magnitude of the two complex poles is about 3.2 MHz. So, only a fraction of the bandwidth will be lost.

  The right-half plane zero, due to $C_{\mu 2}$, is at +5.7 GHz; its effect can be ignored too. So, the noticed effect can be contributed to the fact that$c_{\mu 2}$ can be considered to be parallel connected to$c_{\pi 2}$ including a little Miller effect. The system poles can be placed again in Butterworth by slightly changing the phantom zero.

Adding all the parasitic elements and tuning the resistor implementing the phantom zero, yielded:

$$R_{ph} = 1\,\Omega \tag{10.57}$$

for poles in the Butterworth position. The resulting poles and zeros are listed in table 10.12. The circuit of the amplifier after the completely finished frequency compensation step is depicted in figure 10.21.

| Parasitic element | Integrator pole [Hz] | Band-edge poles [MHz] | Other poles [Mhz] | \|Zeros\| [Mhz] |
|---|---|---|---|---|
| None | -0.6 | -2.4 ± 2.4j | < -600 | > 10 |
| All | -0.7 | -2.2 ± 2.2j | < -150 | > 9 |

*Table 10.12.* The system poles with a changed phantom zero resistor in the case of none parasitic elements and all parasitic elements included.

## 10.8 Step 6: Biasing

Last step to complete the design is the biasing. Table 10.13 summarizes the bias points for the transistors. The values of $V_{BE}$ and $I_B$ where found with the help of a simulator. Note that the $V_{CE}$ of transistors $Q_{1a,b}$ where set to >0.67 V. This is because the constraint is $V_{CE} > V_{BE}$ and up until now, $V_{BE}$ was assumed to be 0.7 V.

**Biasing step 1: Identification and first implementation of the bias loops.** To realize the transistors required biasing point, 2 bias-voltage sources and two

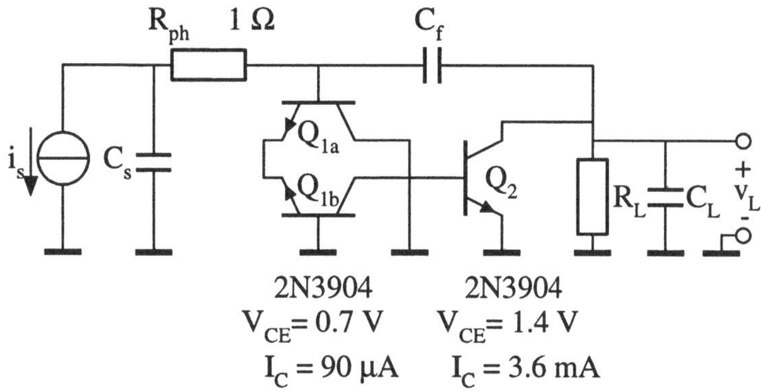


*Figure 10.21.* The amplifier topology after finishing the frequency compensation.

| Bias | $Q_{1a}$ | $Q_{1b}$ | $Q_2$ |
|---|---|---|---|
| $I_C$ [mA] | 0.09 | 0.5 | 3.6 |
| $V_{CE}$ [V] | ≥0.63 | ≥0.67 | >1.4 |
| $I_B$ [$\mu$A] | 0.5 | 2.8 | 23 |
| $V_{BE}$ [V] | 0.63 | 0.67 | 0.73 |

*Table 10.13.* The bias points of the three transistors.

bias-current sources are added to each of the transistor. As discussed in chapter 8, the base-emitter voltage source and the base current source are dependent whereas the collector-emitter voltage source and the collector current source are independent. The signal source has got an infinite impedance at DC, so, the control loop for $Q_{1a}$ and $Q_{1b}$ is both via the input bias current. The input bias voltage is controlled by sensing the input voltage of the amplifier. Likewise, the output transistor is biased by sensing the output voltage and controlling the base bias-current. The input bias voltage of the output stage is controlled via the base-emitter voltage source. All these bias sources and control loops are indicated in figure 10.22.

**Biasing step 2: The bias-current loops.** The obtained circuit, basically, biases correctly. The following steps aim on reducing the number of loops and sources.

The control of the input bias current of the two input transistors, can be realized by sensing the output voltage of the output transistor. In that case, the local control of the output stage can be omitted as it is comprised within the newly created loop. Of course, doing so, two errors are introduced. Firstly, by making the base-current source of the output stage independent, an uncertainty in the actual base current is compensated by the input stage, changing its bi-

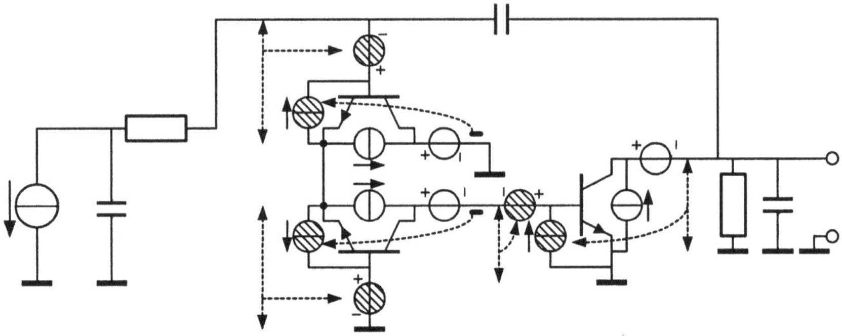

*Figure 10.22.* Identification of the bias sources and a first implementation of the bias loops. For sake of clarity, the labels at the bias sources are omitted.

asing current slightly. As the base current of the output stage is much smaller, and the expected uncertainty even more smaller, than the collector current of the input stage, this error can be assumed to be negligible. Secondly, as the control of both input base-currents rely on the same sensor, a mismatch in the two input transistor yields an input offset current. Matching of transistors can be good enough to make this offset current small. Figure 10.23 depicts the simplifications.

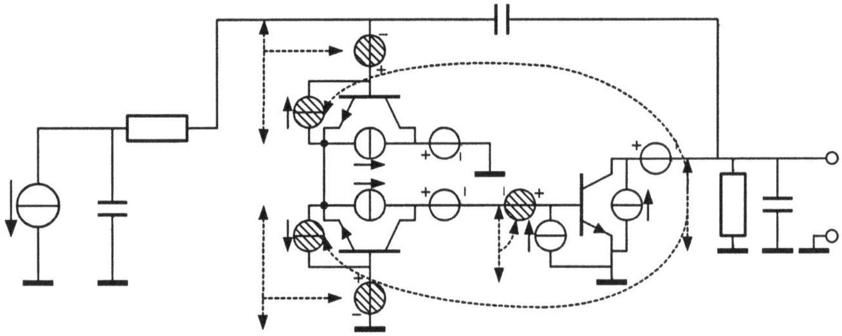

*Figure 10.23.* Reducing the number of bias-current loops.

**Biasing step 3: The bias-voltage loops.** For the bias-voltage loops, analogous simplifications can be made. The control of the base-emitter bias-voltage source can be omitted. The uncertainty in the actual base-emitter voltage can be easily compensated by the output voltage of the differential pair. It can easily compensate an error of 100 mV. The control of the two input voltage sources, can also be combined into one sensor. Again, use is made of the matching properties of transistors to simplify the bias loops. The resulting circuit diagram after the changes in the bias-voltage loops is depicted in figure 10.24.

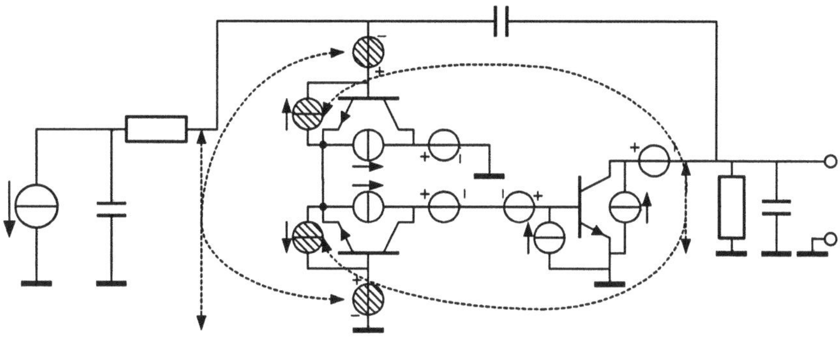

*Figure 10.24.* Reducing the number of bias-voltage loops.

**Biasing step 4: Reducing the number of bias sources.** The current sources that are not grounded are split into two current sources each having one terminal grounded. In this way, several current sources become parallel connected and can be combined into one source. For instance, at the common-emitter node of the differential pair, four current sources are connected: two for the collector currents and two for the base currents. These four sources can be combined into one source with magnitude of$2I_C$. In this way a slight error is made in the collector current. However, as the current-gain factor of the transistor is larger than hundred, the error is less than one percent. On top of this reduction, the tail current sources can be assumed to be independent.

Further inspecting the topology shows that two current sources are shorted by a voltage source. These current sources can be omitted. It should be noted however, that when later on the two shorting voltage sources are implemented, that they should be able to supply that current.

Having a first look to the voltage sources shows that two voltage sources are in series and can therefor be combined. These changes in the topology are indicated in figure 10.25. Further simplifications can be done as follows. The combined voltage source between the input and output has got a value:

$$V \geq 0.63 - 0.67 = -0.04\,\mathrm{V} \tag{10.58}$$

Thus, it can be chosen zero. The base-current source of the output stage is in parallel with a collector-current source of the input. Considering that the base-current is small compared with the collector current, the base-current source can be ignored. Finally, the bias-voltage source at the output can be shifted into the feedback network and in series with the load. The source shifting into the feedback network can be ignored as it is in series with a capacitor. This holds as long as the feedback capacitor is allowed to have a DC voltage, equal to the voltage of the ignored bias source, across its terminals. Here it is assumed that it is allowed. These final reductions are displayed in figure 10.26.

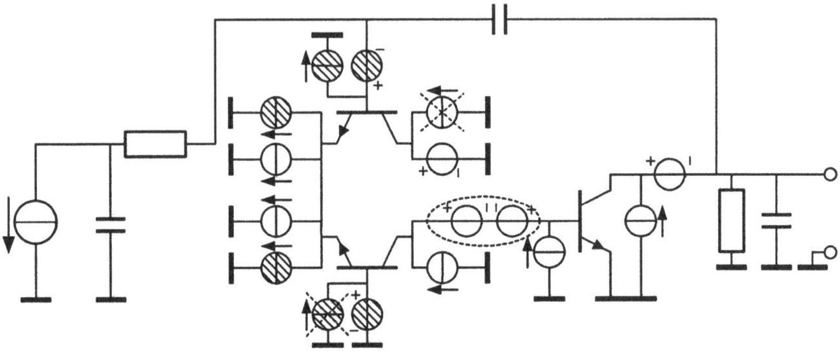

*Figure 10.25.* A first step in reducing the number of bias sources.

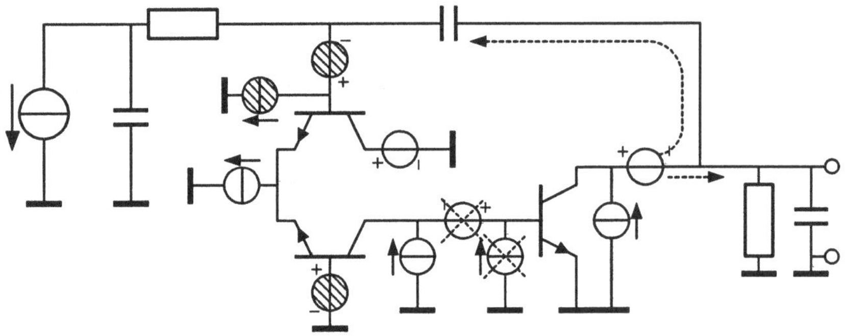

*Figure 10.26.* Some final movements for reducing the number of bias sources.

**Biasing step 5: Implementing the bias loops.** Inspecting the circuit diagram of figure 10.26, shows that two bias loops need to be implemented. One bias loop for controlling the input base current via sensing the load voltage. The second loop was for controlling the common-mode voltage of the two input bias-voltage sources via sensing the signal-source voltage.

The later loop is implemented relatively easy. The input voltage source of the upper transistor of the differential pair can be shifted into the feedback capacitor and into the signal source. In this way, the signal source and the feedback capacitor compensate for this input voltage source. Again, this is possible as long as both, the signal source and the feedback capacitor, may have some DC voltage across their terminals. An option could be to leave the nominal voltage at its place and only shift the controlled part into the signal source and the feedback network. In this example it is assumed that shifting the complete source is no problem. For the lower source, shifting only the controlled part is a good option. By doing so, it shifts in series with the input of the output stage where it can be ignored and it shifts in series with the signal source and the feedback network. Effectively, it means that the control of the voltage source of the lower differential-pair transistor can be ignored. The final circuit with the remaining loop to be implemented is depicted in figure 10.27. The bias loop requires a comparison of the load voltage with zero Volt. As a

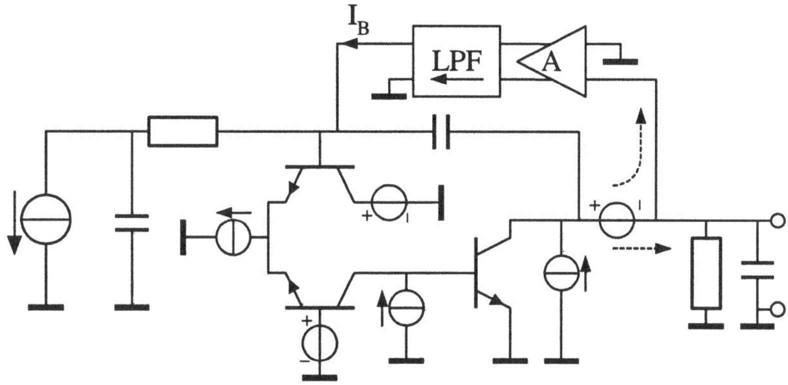


*Figure 10.27.* The bias loop to be implemented.

function of the error, the input base current is controlled. The bias loop also comprises a low-pass filter as it should only act in a band below the signal band. the output voltage source can be shifted through the output node, again in series with the load and in series with the input of the gain block of the bias loop. The voltage source in series with the load can be implemented by means of a capacitor, whereas the voltage sources in series with the input of the gain block can be shifted to the other terminal of the gain block such that it becomes a grounded reference source.

A straightforward implementation would be to replace the gain block by something like a differential pair and the low-pass filter by some T-network. However, as the gain in the loop is already considerably high, i.e. two amplifying stages in the amplifier, this differential pair can be omitted. Therefor, an implementation might be as depicted in figure 10.28.

For dimensioning the T-network the resistor values and the capacitor value should be determined. The resistor values are determined by the required collector-emitter voltage at the output. This voltage should be larger than 1.4 V. The DC output voltage of the T-network equals approximately 0.63 V. Thus the voltage drop across the T-network should be larger than 0.77 V. The direction of the required base current is corresponding with the required voltage drop across the T-network and thus:

$$R_{T1} + R_{T2} \geq \frac{0.77\ \mathrm{V}}{0.5\ \mu\mathrm{A}} = 1.5\ \mathrm{M}\Omega \qquad (10.59)$$

Further, capacitor $C_T$ shorts the common node of $R_{T1}$ and $R_{T2}$ to ground for frequencies beyond the bandwidth of the bias loop. Consequently,$R_{T1}$ is then

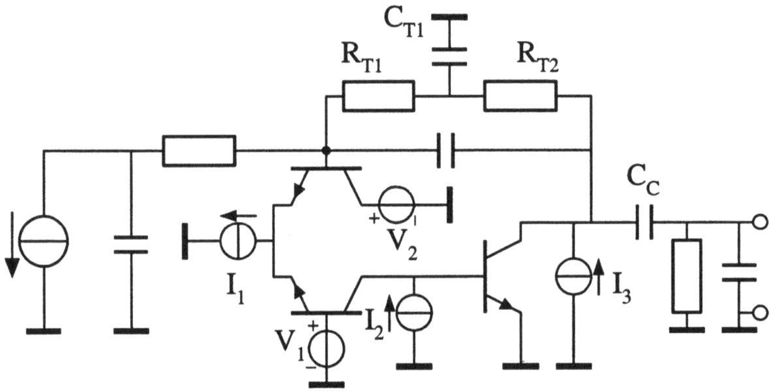


*Figure 10.28.* An implementation of the bias loop.

parallel connected to the input of the differential pair and $R_{T2}$ is connected in parallel with the output. Therefor, also the following constraints should hold:

$$R_{T1} \gg r_{\pi 1a} + r_{\pi 1b} = 104\,\text{k}\Omega \tag{10.60}$$

$$R_{T2} \gg R_L = 10\,\text{k}\Omega \tag{10.61}$$

Choosing the following values fulfills all the posed constraints:

$$R_{T1} = 1.2\,\text{M}\Omega \tag{10.62}$$

$$R_{T2} = 330\,\text{k}\Omega \tag{10.63}$$

resulting in

$$V_{CE2} = 0.63\,\text{V} + 1.53\,\text{M}\Omega \cdot 0.5\,\mu\text{A} = 1.4\,\text{V} \tag{10.64}$$

For dimensioning $C_T$ and $C_C$ we have to investigate the dynamic behavior of the bias loop. The requirement is that the bandwidth of the bias loop is below 10 Hz. The circuit diagram that can be used for this investigation does not need to comprise all the small-signal elements considered so far. As the bias loop is only active for relatively low frequencies, small-signal elements that are noticeable for relatively high frequencies only, can be omitted. Such elements are, for instance, $c_\pi$, $r_b$, et cetera. The thus obtained small-signal diagram is depicted in figure 10.29. This bias loop comprises three poles and three zeros. Approximate expressions for the three poles are:

$$p_{lb1} \approx \frac{-1}{2\pi R_{T1} C_T} \tag{10.65}$$

$$p_{lb2} \approx \frac{-1}{2\pi R_{T2} C_C} \tag{10.66}$$

$$p_{lb3} \approx \frac{-1}{2\pi (r_{\pi 1a} + r_{\pi 1b}) C_s} \approx -150\,\text{Hz} \tag{10.67}$$

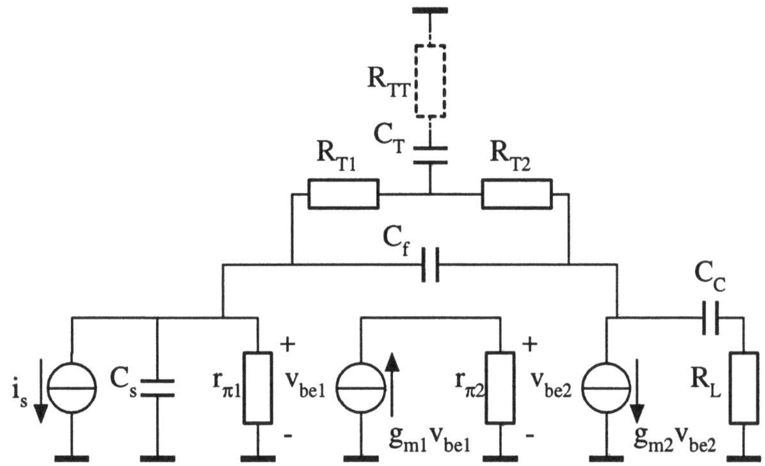


*Figure 10.29.* The small-signal diagram of the amplifier for determining the dynamic behavior of the bias loop. For the moment do not pay attention to resistor $R_{TT}$.

For one of the two zeros a simple expression can be obtained. The zero is a result of the load resistance in series with the coupling capacitor. It is located at:

$$z_{lb1} = \frac{-1}{2\pi R_L C_C} \tag{10.68}$$

The other two zeros are a result of the combination of the T-network of the bias feedback and capacitor $C_f$. The T-network is, so called, bridged by capacitor $C_f$. Such a network easily introduces two high-Q complex zeros in the left half plane. This is easily explained by the following reasoning. With the T-network a low-pass filter is realized for the bias loop. However, as a result of feedback capacitor, $C_f$, this low-pass filter is bridged and consequently, for relatively high frequencies the behavior of a high-pass filter is obtained. This requires two zeros, which appear to be highly complex! Because of the interaction the expression for the two zeros is not simple. For our purpose, however, we do not need this expression as will be seen.

Looking at the three poles, we can note the following. When doing the frequency compensation, the loop pole in the origin was shifted on top of the pole closest to the origin. This was done by means of $R_f$. Pole $p_{lb3}$ is thus the pole that should move to the origin and realize the integrator pole. The other two poles are a result of the bias loop we implemented. Therefor, for a correct functioning amplifier, $p_{lb1}$ and $p_{lb2}$ are closer to the origin than pole $p_{lb3}$. Looking at the three zeros the following can be remarked. The zero as a result of the combination of $C_C$ and $R_L$ is a phantom zero. Thus it creates also a pole in the input-output transfer at that the same frequency. Therefor, this zero should be below 10 Hz. As indicated the two complex zeros will often have a high Q. Considering the DC loop gain of the amplifier, it is not difficult to find the expression:

$$L(0) = -\beta_{f1}\beta_{f2} \approx -27000 \tag{10.69}$$

which is considerably high. As a result of these observations we can conclude that a typical root locus of the bias loop can be as depicted in figure 10.30. Clearly, as the number of zeros is equal to the number of poles and the loop

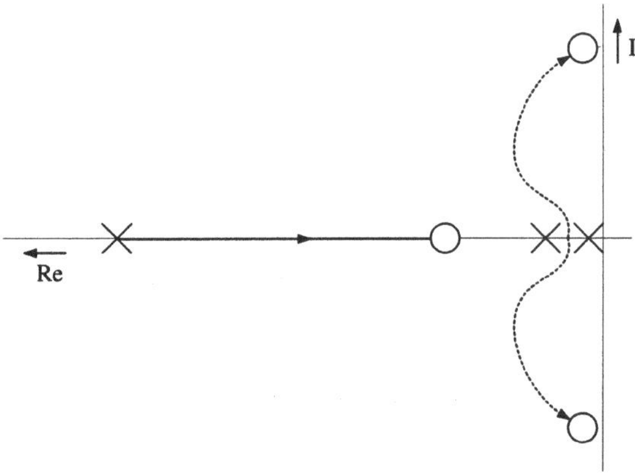


*Figure 10.30.* A typical root locus of a bias loop with a capacitive-bridged T-network.

gain is relatively high, the root locus ends at the zeros[2]. Thus, frequency compensation is done for this bias loop by placing the zeros at the location where we want to have the closed loop bias poles!

As can be seen from figure 10.30, $p_{lb3}$ still is responsible for the closed loop integrator pole. It moves to the zero determined by $R_L$ and $C_C$, $z_{lb1}$. For this example we going to place the integrator pole at 1 Hz. This requires capacitor $C_C$ to be:

$$C_C = \frac{1}{2\pi \cdot 1\ \text{Hz} \cdot 10\ \text{k}\Omega} \approx 16\ \mu\text{F} \tag{10.70}$$

For determining the value of $C_T$ an simulator was used that is able to determine poles and zeros from a small-signal diagram, numerically. For capacitor $C_T = 0.7\ \mu$F, the zeros are located at:

$$z_{l1,2} = (-0.7 \pm j9.5)\ \text{Hz} \tag{10.71}$$

Indeed, the Q of these zeros is high. The bandwidth of the bias loop will be approximately 10 Hz, however, the loop is close to oscillation. Main issue now

[2]This is, of course, in general true: the starting points of the root locus are at the poles and the ending points of the root locus are at the zeros. In many practical situations, however, one or more zeros are located at infinity and the DC loop gain is too low to realize that the closed loop poles are more or less at the location of the loop zeros

is thus to locate the two complex zeros at the position where we want to have the closed-loop bias poles. The complex zeros arise, as indicated, by capacitor$C_f$ bridging the low-pass filter. In order to be able to design the location of the zeros we need to understand somewhat more about the origin of their being complex. The interacting occurs because by means of capacitor$C_f$ the effective behavior of the bridged T-network changes from a single-pole behavior to a single-zero behavior, which requires two zeros. As these two zeros originate from the fact that one element is added, they are complex. Thus, when the behavior of the T-network is resistive at the point of take over, a real zero would be obtained. The T-network can be made resistive in the following way. For relatively low frequencies the single-pole behavior is, of course, realized by$C_T$. Thus, adding a resistor, $R_{TT}$, in series with $C_T$, yields a zero. This zero is determined by:

$$z_{l1} = \frac{-1}{2\pi R_{TT} C_T} \tag{10.72}$$

Subsequently, capacitor $C_F$ introduces the second zero when bridging the resistive T-network. When these two zeros are created at considerably different frequencies they will be both real. When, however, the effect is more or less at the same frequency, two complex zeros having equal real and imaginary parts are obtained. For having the zeros in Butterworth, the real part should be about -7 Hz. Thus the resistor should have a value:

$$R_{TT} = \frac{1}{2\pi 7\text{ Hz}12\ \mu\text{F}} \approx 33\ \text{k}\Omega. \tag{10.73}$$

By using these values the following three system poles are obtained:

$$p_{integrator} = -1\ \text{Hz} \tag{10.74}$$

$$p_{1,2} = (-6.5 \pm 6.1j)\ \text{Hz} \tag{10.75}$$

On top of that two low-frequency zeros are obtained at:

$$z_{l1} = -10\ \text{nHz} \tag{10.76}$$

$$z_{l2} = -0.8\ \text{Hz} \tag{10.77}$$

Clearly, these two zeros are required in order to have in the signal band a first-order role off, i.e. integrator behavior.

Table 10.14 summarized the required sources and components for the biasing.

**Biasing step 6:Implementing the bias sources.** Final step in the amplifier design is to implement the bias sources. According to figure 10.28 we need to implement two voltage sources and three current sources. For the current sources at the input, the noise contribution is important. For the current source

| | |
|---|---|
| $V_1$ | 0.63 V |
| $V_2$ | > 0.63 V |
| $I_1$ | 0.18 mA |
| $I_2$ | 0.09 mA |
| $I_3$ | 3.6 mA |
| $C_C$ | 16 $\mu$F |
| $C_T$ | 0.7 $\mu$F |
| $R_{TT}$ | 33 kΩ |
| $R_{T1}$ | 1.2 MΩ |
| $R_{T2}$ | 330 kΩ |

*Table 10.14.* Sources and components required for the bias loop.

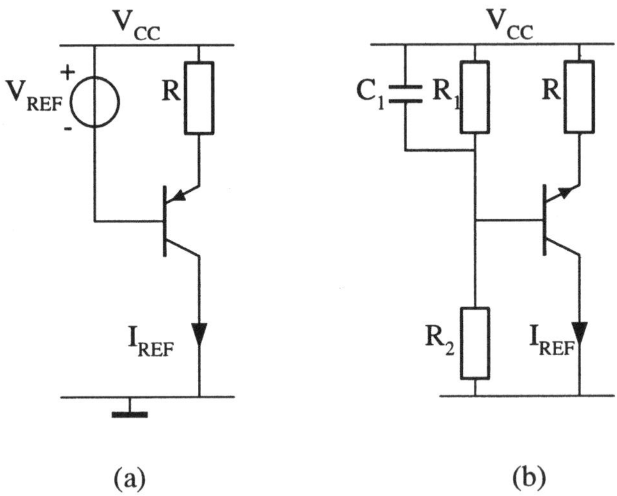


*Figure 10.31.* (a) The basic structure of the current source to be used (b) A typical implementation.

at the output, its output conductance is the key parameter. For each of these sources the basic structure as depicted in figure 10.31a is used. A straightforward implementation of this current source is depicted in figure 10.31b. The reference-voltage source is made by means of the supply voltage, a resistive divider and a capacitor. Note that the capacitor must be connected to the supply line. This is required for a high power-supply rejection. In this case the output impedance of the current source remains high for disturbances on the supply line. Would the capacitor terminal, that is now connected to the supply line, be connected to ground line, the feedback loop of the current source would be broken for frequencies higher than the pole due to this capacitor. As a result the current source behaves as a common-base stage for disturbances on the supply line. Clearly, a much worse power-supply rejection is obtained in the latter case.

The current source of figure 10.31b requires a voltage of about $V_R$ plus $V_{CE-sat}$. $V_R$ is determined by the level of output noise or output conductance. Minimum noise and output conductance is obtained for a voltage drop on the order of $5V$ (depending on $\beta_f$). However, in many cases a much lower voltage can be chosen which yields sufficient low noise level and output conductance. In this example we choose a voltage drop of 1 V.

For the minimum collector-emitter voltage again 0.7 V is chosen, which makes the minimum required voltage drop for the current source 1.7 V. Subsequently, by taking into account the voltage swing present on the node to which the current source is connected, a minimum value for the supply voltage can be determined. Clearly, the output current source requires the highest supply voltage. It should be larger than;

$$V_{node-bias} + V_{node-signal-peak} + V_{source} > 3.8 \text{ V} \tag{10.78}$$

A positive supply source $V_{CC} = 4$ V is chosen. For the negative supply voltage, in an analogous way, a value of maximal -2.5 V is found. However, for reasons of symmetry $V_{EE} = -4$ V is chosen. It should be noted that in this design no attention is paid to design for a single supply line. Would it be required to have a single supply line, by shifting bias sources and by changing polarity of one or more devices, all the node voltage can be kept beyond zero.

Resistors $R_1$ and $R_2$ are chosen such that the current through them is about ten times as large as the corresponding base current. Capacitor $C$ is subsequently determined by the time constant $2\pi(R_1//R_2)C$ which must be below 10 Hz.

The element values chosen for the three current sources are listed in table 10.15.

| Current Source | 90 $\mu$A | 3.6 mA | 180 $\mu$A |
|---|---|---|---|
| $Q$ | 2N3906 | 2N3906 | 2N3904 |
| $R$ | 10 k$\Omega$ | 220 $\Omega$ | 5.6 k$\Omega$ |
| $R_1$ | 180 k$\Omega$ | 4.7 k$\Omega$ | 100 k$\Omega$ |
| $R_2$ | 270 k$\Omega$ | 6.8 k$\Omega$ | 120 k$\Omega$ |
| $C$ | 0.22 $\mu$F | 10 $\mu$F | 0.47 $\mu$F |

*Table 10.15.* The component values for the three current sources.

Voltage source $V_2$ is easily implemented. As it should be $> 0.63V$, it can be replaced by the positive supply voltage of $4V$. Voltage source $V_1$ is implemented by a resistive divider in the way the reference voltage in the current sources are implemented. It should be noted, however, that it is not required to make its value exactly 0.63 V. An error in this voltage would yield an additional voltage across the signal source, the feedback capacitance and the collector-emitter branch of the transistor. For each of these elements it is no problem when the voltage changes several hundreds of mV.

The complete circuit diagram is depicted in figure 10.32.

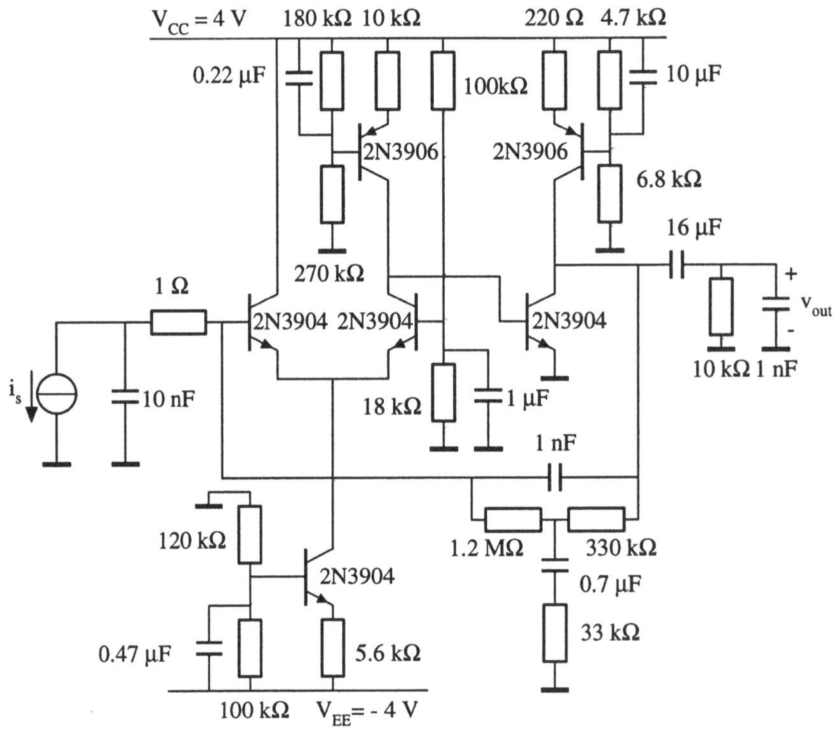


*Figure 10.32.* The complete circuit diagram of the charge amplifier.

## 10.9 Step 7: Verification

The circuit diagram of figure 10.32 was evaluated in a Spice-like simulator. The three main specifications: transfer, noise and maximum signal swing were considered. In the simulator the Thevenin representation of the signal source was used, as depicted in figure 10.33. By using this source, an additional differentiation is obtained because of the source capacitance. Consequently, the transfer of the amplifier should be flat over the band of interest, which makes verification more easy. The ideal gain from source voltage to output voltage equals:

$$A_{t\infty} = -\frac{10\ \text{nF}}{1\ \text{nF}} = -10 \tag{10.79}$$

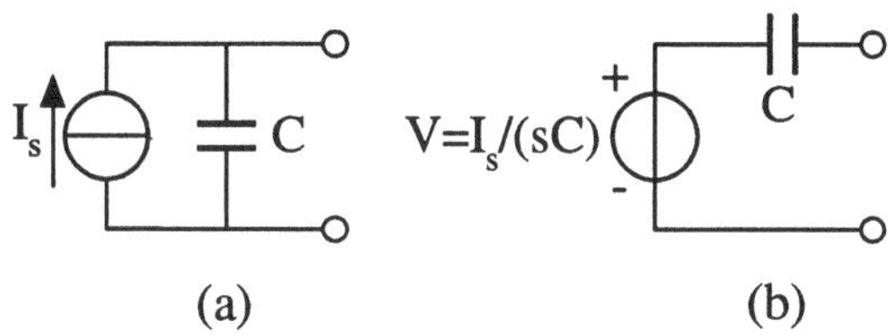


*Figure 10.33.* (a) The charge source and (b) its Thevenin representation.

The simulated gain and phase of the amplifier are depicted in figure 10.34. Indeed, the gain is almost 10, only a fraction of a percent difference. This is, of course, a result of the high, but not infinite, loop gain. Within the band the phase is -180 degrees, which corresponds with the inversion of the amplifier. The lower side of the band is, as designed at 10 Hz. The upper side of the band is at B = 2.4 MHz, which is somewhat lower than obtained B = 3.1 MHz after the complete frequency compensation. This lower bandwidth is mainly because of the output impedances, more specific the output capacitances, of the current sources that load the amplifier loop. Consequently, the LP-product is reduced. The value of the resistor, for implementing the phantom zero, showed to be correct.

The simulated noise level at the output of the amplifier resulted in a signal-to-noise ratio of 79 dB. A few dB less than designed SNR. Reasons are again a slight contribution of the bias sources and a slight difference in the designed and simulated small-signal parameters.

Finally, a full-power, $\hat{v}_{out} = 0.7\,\mathrm{V}$, transient at the power bandwidth, f=500 kHz showed a correctly functioning amplifier.

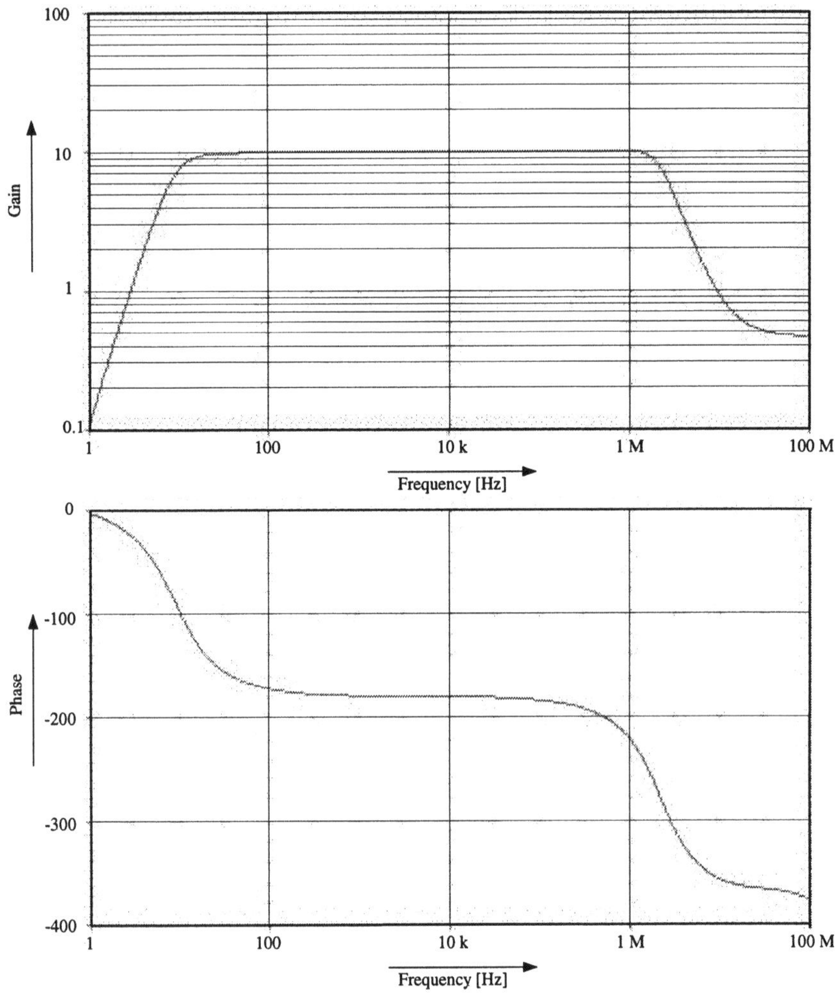


*Figure 10.34.* The simulated gain, upper graph, and phase, lower graph, of the amplifier when using a voltage signal source.

# References

[1] E.H. Nordholt. *Design of High-Performance Negative-Feedback Amplifiers* PhD thesis, Delft University of Technology, The Netherlands, 1980.

[2] J. Stoffels. *Automation in High-Performance Negative-Feedback Amplifier Design* PhD thesis, Delft University of Technology, The Netherlands, 1988.

[3] J. Davidse. *Analog Electronic Circuit Design.* Prentice Hall, New York, 1991.

[4] Arie van Staveren, Chris J.M. Verhoeven and A.H.M. van Roermund. *Structured Electronic Design: High-performance Harmonic Oscillators and Bandgap References*Kluwer Academic Publishers, Boston, ISBN 0-7923-7283-2, 2001

[5] J. Stewart. *Calculus: early trancendentals.* Brooks/cole Publishing Company, Pacific Grove, USA, 1999.

[6] E.H. Nordholt. Strategies and routines in analog design. In*Proceedings of the workshop Advances in Analag Circuit Design*, pages 369–396, April 1992.

[7] J. Stoffels, C., and Reeuwijk. Ampes: a program for the synthesis of high-performance amplifiers. In *Proceedings of the European Conference on Design Automation*, pages 474–479, March 1992.

[8] A. van Staveren, C.J.M. Verhoeven, and A.H.M. van Roermund. The influence of the reverse Early effect on the performance of bandgap references. *IEEE Transactions on Circuits and System I*, 43(5):418–421, May 1996.

[9] A. van Staveren, C.J.M. Verhoeven, and A.H.M. van Roermund. The design of low-noise bandgap references. *IEEE Transactions on Circuits and System I*, 43(4):290–300, April 1996.

[10] C.A.M. Boon. *Design of High-Performance Negative-Feedback Oscillators* PhD thesis, Delft University of Technology, The Netherlands, 1989.

[11] C.J.M. Verhoeven. *First-Order Oscillators.* PhD thesis, Delft University of Technology, The Netherlands, 1990.

[12] C.J.M. Verhoeven, J.R. Westra, and A. van Staveren. Low-noise oscillators. In J.H. Huijsing et al., editor, *Analog Circuit Design*. Kluwer Academic Publishers, Dordrecht, The Netherlands, 1996.

[13] J.R. Westra, R.T. Godijn, C.J.M. Verhoeven, and A.H.M. van Roermund. Coupled relaxation oscillators with highly stable and accurate quadrature outputs. In *Proceedings of the 1st IEEE-CAS Region 8 Workshop on Analog and Mixed IC Design*, pages 32–35, Pavia, Italy, 13-14 September 1996.

[14] J.R. Westra, C.J.M. Verhoeven, A. van Staveren, and A.H.M. van Roermund. Design principles for low-noise relaxation oscillators. In *Proceedings of Electronics 1996*, Sozopol, Bulgaria, September 1996.

[15] J.R. Westra, C.J.M. Verhoeven, A. van Staveren, and A.H.M. van Roermund. Design principles for low-noise harmonic second-order oscillators. In *Proceedings of Electronics 1996*, Sozopol, Bulgaria, September 1996.

[16] J.O. Voorman. *The Gyrator as a Monolithic Circuit in Electronic Systems*. PhD thesis, University of Nijmegen, Nijmegen, The Netherlands, 1977.

[17] B. Nauta. *Analog CMOS Filter for very-high Frequencies*. PhD thesis, Twente Technical University, Enschede, The Netherlands, 1991.

[18] G. Groenewold. *Optimal Dynamic Range Integrated Continuous-Time Filters*. PhD thesis, Delft University of Technology, The Netherlands, 1992.

[19] G.L.E. Monna. *Design of Low-Voltage Integrated Filter-Mixer Systems*. PhD thesis, Delft University of Technology, Delft, The Netherlands, 1996.

[20] C.E. Shannon. A mathematical theory of communication. *The Bell System Technical Journal*, 27(3):379–432 and 623–656, July 1948.

[21] W.A. Serdijn, C.J.M. Verhoeven, and A.H.M. van Roermund, editors. *Analog IC Techniques for Low-Voltage Low-Power Elecronics*. Delft University Press, The Netherlands, 1995, ISBN 90-407-1202-6.

[22] H.T. Friis Noise figures of radio receivers. *Proceedings of the IRE*, vol.32:419–422, 1944.

[23] A. van der Ziel. Thermal noise in field effect transistors. *Proc.IEEE*, pp1801–1812, Aug. 1962.

[24] Y. P. Tsividis. *Operation and modeling of the MOS transistor*. McGraw-Hill, New York, 1987.

[25] M.L. Lantz. *Systematic Design of Linear Feedback Amplifiers*. PhD thesis, Lund University, Lund, Sweden, 2002, ISBN 91-628-5395-3.

[26] M.L. Lantz and S. Mattisson. Nonlinearity of multi-stage feedback amplifiers. *In Proceedings of NDES 2002*, Çeşme, Turkey, pp. 49-52, June 2002.

[27] M.L. Lantz and S. Mattisson. Local feedback and nonlinearity of multi-stage feedback amplifiers. *In Proceedings of NDES 2002*, Çeşme, Turkey, pp. 29-32, June 2002.

[28] W. J. Rugh. *Nonlinear System Theory. The Volterra/Wiener Approach.* The John Hopkins University Press, Baltimore, US, 1981, ISBN O-8018-2549-0. Also freely available via web: http://www.ece.jhu.edu/ rugh/214/me/rugh.html

[29] P. Wambacq and W.M.C. Sansen. *Distortion Analysis of Analog Integrated Circuits.* Kluwer Academic Publishers, Boston, 1998, ISBN 0-7923-8186-6.

[30] E.K. de Lange, A. van Staveren, O. De Feo, F.L. Neerhoff, M. Hasler, and J.R. Long. Predicting nonlinear distortion in common-emitter stages for amplifier design using Volterra series. *In Proceedings of NDES 2002*, Çeşme, Turkey, pp. 41-44, June 2002.

[31] E.K. de Lange, O. De Feo, and A. van Staveren. Modelling differential pairs for low-distortion amplifier design. *In Proceedings of ISCAS 2003*, Bangkok, Thailand, pp. I-261-I264, May 2003.

[32] H.W. Bode. *Network Analysis and Feedback Amplifier Design.* Van Nostrand, New York, 1945.

[33] W.C.West et.al. *Micro-power Sources Enabling Robotic Outpost Based Deep Space Exploration.* Forum on Innovative Approaches to Outer Planetary Exploration 2001–2020 February 21–22, 2001 Lunar and Planetary Institute, Houston, Texas

[34] L.A.Sarmiento Reyes *A partition method for the determination of multiple DC operating points.* PhD thesis, Delft University of Technology, The Netherlands, 1994, ISBN 90-6275-974-2.

# Index

MIX
Papier aus verantwortungsvollen Quellen
Paper from responsible sources
FSC® C105338

If you have any concerns about our products,
you can contact us on
**ProductSafety@springernature.com**

In case Publisher is established outside the EU,
the EU authorized representative is:
**Springer Nature Customer Service Center GmbH**
**Europaplatz 3, 69115 Heidelberg, Germany**

Printed by Libri Plureos GmbH
in Hamburg, Germany